TDP-43 and Neurodegeneration

TDP-43 and Neurodegeneration

From Bench to Beside

Edited by

Vijay Kumar
Amity Institute of Neuropsychology and Neurosciences,
Amity University, Noida, Uttar Pradesh, India

Manoj Kumar Jaiswal
Department of Psychiatry,
Icahn School of Medicine at Mount Sinai,
New York, NY, United States

ELSEVIER

ACADEMIC PRESS
An imprint of Elsevier

Academic Press is an imprint of Elsevier
125 London Wall, London EC2Y 5AS, United Kingdom
525 B Street, Suite 1650, San Diego, CA 92101, United States
50 Hampshire Street, 5th Floor, Cambridge, MA 02139, United States
The Boulevard, Langford Lane, Kidlington, Oxford OX5 1GB, United Kingdom

Notices
Knowledge and best practice in this field are constantly changing. As new research and experience broaden our understanding, changes in research methods, professional practices, or medical treatment may become necessary.

Practitioners and researchers must always rely on their own experience and knowledge in evaluating and using any information, methods, compounds, or experiments described herein. In using such information or methods they should be mindful of their own safety and the safety of others, including parties for whom they have a professional responsibility.

To the fullest extent of the law, neither the Publisher nor the authors, contributors, or editors, assume any liability for any injury and/or damage to persons or property as a matter of products liability, negligence or otherwise, or from any use or operation of any methods, products, instructions, or ideas contained in the material herein.

Library of Congress Cataloging-in-Publication Data
A catalog record for this book is available from the Library of Congress

British Library Cataloguing-in-Publication Data
A catalogue record for this book is available from the British Library

ISBN: 978-0-12-820066-7

For information on all Academic Press publications visit our website at https://www.elsevier.com/books-and-journals

Publisher: Nikki Levy
Acquisitions Editor: Melanie Tucker
Editorial Project Manager: Billie Jean Fernandez
Production Project Manager: Niranjan Bhaskaran
Cover Designer: Mark Rogers

Typeset by TNQ Technologies

Contents

5. Prion-like behavior of TDP-43 aggregates and its implication to disease

Ankit Srivastava, Nikita Admane, Shiv Pratap Singh Yadav and Himanshi Kukrety

6. Search for functions of intrinsically disordered prion-like domains for FET proteins involved in amyotrophic lateral sclerosis and frontotemporal dementia

Oxana V. Galzitskaya

9. New opportunities for treatment of neurodegenerative disease through the modulation of TDP-43

Pasha Apontes

List of contributors

Nikita Admane, Department of Biological Sciences, BITS Pilani - K K Birla Goa Campus, Goa, India

Md Sheeraz Anwar, Department of Computer Science, Jamia Millia Islamia, New Delhi, India

Pasha Apontes, Department of Psychiatry, Icahn School of Medicine at Mount Sinai, New York, NY, United States

Vidhya Bharathi, Department of Biotechnology, Indian Institute of Technology Hyderabad, Kandi, Sangareddy, Telangana, India

Oxana V. Galzitskaya, Institute of Protein Research of Russian Academy of Sciences, Head of Laboratory, Pushchino, Russia; Institute of Theoretical and Experimental Biophysics of Russian Academy of Sciences, Pushchino, Russia

Amandeep Girdhar, Department of Biotechnology, Indian Institute of Technology Hyderabad, Kandi, Sangareddy, Telangana, India

Mohd Maksuf Ul Haque, Department of Computer Science, Jamia Millia Islamia, New Delhi, India

Manoj Kumar Jaiswal, Department of Psychiatry, Icahn School of Medicine at Mount Sinai, New York, NY, United States

Himanshi Kukrety, Department of Biotechnology, Jawaharlal Nehru University, New Delhi, India

Vijay Kumar, Amity Institute of Neuropsychology & Neurosciences, Amity University, Noida, Uttar Pradesh, India

Md Zubbair Malik, School of Computational and Integrative Sciences, Jawaharlal Nehru University, New Delhi, India

Basant K. Patel, Department of Biotechnology, Indian Institute of Technology Hyderabad, Kandi, Sangareddy, Telangana, India

Samir Rahman, Friedman Brain Institute, Icahn School of Medicine at Mount Sinai, New York, NY, United States; Department of Genetics and Genomic Science and Institute for Multiscale Biology, Icahn School of Medicine at Mount Sinai, New York, NY, United States

Saurabh Kumar Sharma, School of Computer & Systems Sciences, Jawaharlal Nehru University, New Delhi, India

R.K. Brojen Singh, School of Computational and Integrative Sciences, Jawaharlal Nehru University, New Delhi, India

Shiv Pratap Singh Yadav, Division of Nephrology, Indiana University School of Medicine (IUSM), Indianapolis, IN, United States

Ankit Srivastava, LPVD, Rocky Mountain Laboratories, NIAID, NIH, Hamilton, MT, United States

Anil Kumar Tomar, Department of Biophysics, All India Institute of Medical Sciences, New Delhi, India

Nidhi Verma, Medical School, University of Texas Health Science Centre, Department of Microbiology and Molecular Genetics, Houston, TX, United States

Savita Yadav, Department of Biophysics, All India Institute of Medical Sciences, New Delhi, India

Preface

Amyotrophic lateral sclerosis (ALS) and frontotemporal dementia (FTD) are common neurodegenerative disorders that are relentlessly progressive, fatal, and without effective treatment, representing a grave challenge to healthy human aging and modern healthcare systems. The last decades have marked a turning point in our knowledge of ALS and FTD, two diseases forming a clinical continuum and sharing common pathogenic mechanisms and genetic etiologies. In particular, the identification of TAR DNA-binding protein 43 (TDP-43) in 2006 as the cardinal protein in the most common subtypes of ALS and FTD represented a breakthrough discovery in these domains and was only a prelude to a series of important discoveries that followed. Aggregates of TDP-43 are hallmark of ALS and FTD, which share overlapping clinical, genetic, and pathological features. In addition, TDP-43 pathology is found in Alzheimer's disease, Lewy body dementia, Parkinson's disease, and chronic traumatic encephalopathy.

We are pleased to put together an issue of *TDP-43 and Neurodegeneration: From Bench to Bedside* in this challenging time of the COVID-19 pandemic. We have made the effort of identifying what we call "emerging milestones of the next century," building a bridge between the work of the discoverer and that of today's pioneers and explorers. Each chapter of this book not only illustrates the present state of the art but also reveals the challenges ahead in the field of TDP-43 research. The chapters in this book summarize TDP-43's associated legacy discovery, structure, function, biology, misfolding, aggregation, pathogenesis, and therapeutics, bringing together the latest research and also developments along with new hypotheses emerging in TDP-43 field. It has been designed to be read by people from a wide range of backgrounds from the basic biological, clinical, and neurosciences field and will be of interest to neuroscientists, neurologists, clinicians, human health researchers, drug discovery researchers, as well as cell and molecular biologists involved in ALS, motor neuron disease, and other neurodegenerative disorders research. This book will certainly improve our understanding of various pathological mechanisms involved in TDP-43 proteinopathies and thus ultimately lead to discoveries of new therapeutic cures. Thus, this book is valuable for a broad audience.

We have attempted to convey to the reader a feeling of the excitement engendered within the TDP-43 research by the new discoveries and the emerging ideas. We hope that the content of this book will stimulate provocative discussions and inspire novel avenues of investigation to further broaden our knowledge of TDP-43 role in health and disease with the ultimate goal of translating these novel findings into therapeutic interventions for patients. We hope to have inspired as well as to inform and in this small way contribute to the overall endeavor. We would like to extend our gratitude to all the authors who contributed chapters in this volume. We are also thankful to senior acquisitions editor Melanie Tucker and Joslyn Paguio (Elsevier/Academic Press) and editorial project manager Billie Jean Fernandez (Elsevier) for their constant support throughout the project. Project manager Niranjan Bhaskaran (Elsevier) and senior copyright coordinator Swapna Praveen (Elsevier) are also acknowledged for their contributions.

Vijay Kumar, PhD
Noida, UP, India
Manoj Kumar Jaiswal, PhD
New York, NY, United States

Chapter 1

TDP-43 and neurodegenerative diseases: past, present, and future

Manoj Kumar Jaiswal
Department of Psychiatry, Icahn School of Medicine at Mount Sinai, New York, NY, United States

Introduction

The transactive response DNA-binding protein of 43 kDa (TDP-43) is a highly conserved, ubiquitously expressed, 414 amino acid DNA/RNA-binding protein. It exhibits enormous specificity in binding to the mutual DNA/RNA microsatellite region (GU/GT)n. TDP-43 normally has a tightly autoregulated expression level and, after being synthesized in the cytoplasmic compartment, it shuttles between the nucleus and cytoplasm, whereas in neuropathological states, it primarily exits the nucleus and form the aggregates in the cytoplasm.[1,2] Identification of TDP-43 was initially observed in a screening experiments for protein factors where it is equipped to associate with the long terminal repeat transactive response element of human immunodeficiency virus 1 (HIV1) and subsequently discovered to bind with both DNA and RNA.[3–5]

A nebulous description of TDP-43 dates back to 1892 when Czech neurologist Arnold Pick published a first report on a progressive dementia which varied both clinically and pathophysiologically from Alzheimer's disease (AD).[6] The dementia that Pick first described was characterized by atrophy (Fig. 1.1), including significant loss of frontal and temporal lobes due to atrophy,[7] known as frontotemporal dementia (FTD). A century following these first descriptions, by 2006, most cases of FTD were largely characterized by (1) unusual accumulations of tau (FTD-tau) protein and (2) ubiquitin-positive inclusion bodies (FTD-U) without α-synuclein accumulation.[8,9] Until recently it was unclear whether FTD-U characterized a single disease symptoms or a broad spectrum of cases due to the lack of abnormal tau immunoreactivity. This suspense was straighten out after the two landmark discoveries in 2006. Firstly, a study reported on a subset of FTD-U cases associated with mutations

TDP-43 and Neurodegeneration. https://doi.org/10.1016/B978-0-12-820066-7.00007-2

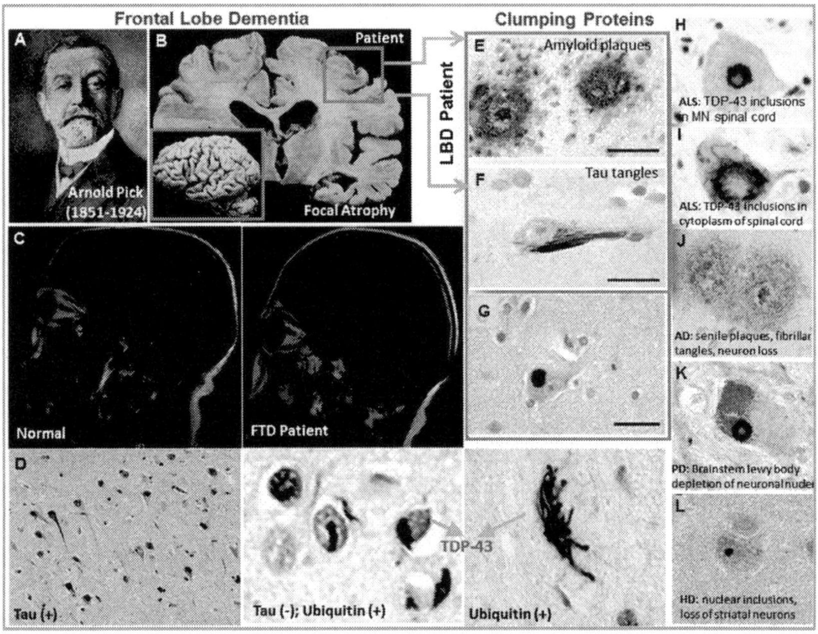

FIGURE 1.1 Different neurodegenerative disorders present disease-specific Misfolded proteins (MPs) and characteristic aggregation patterns with and without TDP-43 inclusions. (A) Portrait of Arnold Pick. (B) A representative coronal section of the human brain showing the left circumscribed temporal lobe atrophy in Lewy body dementia (LBD) patients. Brain lateral view displaying frontotemporal atrophy of the left hemisphere in LBD patients (inset). (C) A representative sagittal MRI scan of normal (control) and an age-matched FTD patients. Brain of a person with FTD shows a diffuse atrophy of the frontal lobe, temporal lobe, and parietal lobe, especially the temporal lobe and the hippocampus (right) compared to normal brain without any atrophy (left). (D) Photomicrographs showing the difference between FTD-ALS spectrum and ALS disorders. In FTD-ALS characteristics inclusions are either Tau (+) inclusions or Tau (−) and Ubiquitin (+) inclusions (left and middle, respectively), whereas in ALS only Ubiquitin (+)+ TDP-43 inclusions are appear in postmortem brain (right). (E) A photomicrograph of Aβ plaques in the cortex of an Alzheimer's disease (AD) patient. (F) A photomicrograph of Tau neurofibrillary tangle in a neuron of an AD patient. (G) A photomicrograph of α-synuclein inclusion in a neuron from a Parkinson's disease (PD) patient. Scale bars are 50 mm in (E) and 20 mm in (F–G). (H) A photomicrograph of TDP-43 inclusion in the nucleus of a motorneuron of the spinal cord from a patient with ALS. (I) A photomicrograph of TDP-43 cytoplasmic cellular inclusion in a motorneuron of the brain cortex from a patient with ALS showing disintegration of motorneuron. (J) A photomicrograph of disease-specific cellular markers for AD, showing senile plaques, fibrillary tangles, and neuronal loss. (K) A photomicrograph of disease-specific cellular markers for PD, showing Lewy bodies and depletion of dopaminergic neurons. (L) A photomicrograph of disease-specific cellular markers for Huntington's disease (HD), showing nuclear inclusions and loss of striatal neurons. *(A) Adapted with permission from Galimberti D, Scarpini E. Chapter 6 - frontotemporal lobar degeneration. In: Martin CR, Preedy VR, editors.* Diet and nutrition in dementia and cognitive decline *(San Diego): Academic Press; 2015. pp. 57-66, originally from Haymaker W, Schiller F. The founders of neurology.* Neurology *1953;3, 550550, first edition, 1953. Courtesy of Charles C Thomas Publisher, Ltd. Springfield, Illinois; (B) Adapted from Brito-Marques PRD, Mello RVD, Montenegro L. Classic Pick's disease type with ubiquitin-positive and tau-negative*

in the gene progranulin (GRN).[10,11] Secondly, a report identified TDP-43 as the main disease protein in both amyotrophic lateral sclerosis (ALS) and FTD with ubiquitin-positive inclusions.[12,13] Subsequently, presence of TDP-43 was discovered in the hippocampus region of the brains of patients with AD,[14–16] Lewy body dementia (LBD),[16,17] and argyrophilic grain disease.[18]

The bulk of TDP-43 protein is truncated in cytoplasmic inclusions, and its C-terminal domain is shown to be susceptible to aggregate formation. Previous studies have indicated that the mutations in the C-terminal region of the TDP-43 linked with both ALS and FTD are supposed to assist in phosphorylation and ubiquitination of the TDP-43 protein, leading to inclusion formation and, subsequently, triggering the neurodegenerative process. Discovery of TDP-43 as the key protein in both FTD and ALS is in the limelight not only because numerous parallels exist but also due to previously known but underappreciated associations among FTD and ALS. Various genetic mutations have been linked to both familial ALS and FTD. Three most important genes associated with FTD are (i) microtubule-associated protein tau (MAPT), (ii) GRN, and (iii) chromosome 9 open reading frame 72 (C9orf72). Two studies in 2011 found a mutation in the C9orf72 gene that occurred in both ALS and FTD.[19,20] This discovery was the first demonstration showing a strong genetic link between the ALS and FTD. Both GRN and C9orf72 mutations can have deposition of TDP-43. TDP-43 is a ubiquinated protein that has been found to accumulate in both ALS and FTD. Familial cases are far outnumbered by sporadic cases in both disorders (5%–10% of ALS and ~40% of FTD). In FTD, characteristics inclusions are either Tau (+) inclusions or Tau (−) and Ubiquitin (+) inclusions, whereas in ALS, only Ubiquitin (+) + TDP-43 (+) inclusions are present. Recent studies had shown that ~15% of patients with FTD meet criteria that also define ALS and, ultimately, developed MND,[9,21,22] whereas ~36% of FTD patients meet criteria for possible ALS, and ~50% of patients with disease, go on to developed some degrees of ALS and FTD cosegregated in some families.[22]

inclusions: case report. Arq Neuro Psiquiatr *2001;**59**:128–133; (C) Adapted and modified from Zhang W, Jiao B, Xiao T, Pan C, Liu X, Zhou L, Tang B, Shen L. Mutational analysis of PRNP in Alzheimer's disease and frontotemporal dementia in China.* Sci Rep *2016;**6**, 38435; (D) Adapted and modified with permission from Cairns NJ, Neumann M, Bigio EH, Holm IE, Troost D, Hatanpaa KJ, Foong C, White CL, 3rd, Schneider JA, Kretzschmar HA, et al. TDP-43 in familial and sporadic frontotemporal lobar degeneration with ubiquitin inclusions.* Am J Pathol *2007;**171**:227–40; Nolle A. van Haastert ES, Zwart R, Hoozemans JJ, Scheper W. Ubiquilin 2 is not associated with tau pathology.* PLoS One *2013;**8**:e76598; (E–H) were adapted and modified from Carbonell F, Iturria-Medina Y, Evans AC. Mathematical modeling of protein misfolding mechanisms in neurological diseases: a historical overview.* Front Neurol *2018;**9**:37; and (I and J), were adapted and modified from Feneberg E, Gray E, Ansorge O, Talbot K, Turner MR. Towards a TDP-43-based biomarker for ALS and FTLD. Mol* Neurobiol *2018;55:7789–801. and (K and L) were adapted and modified from Surguchov A, Emamzadeh FN, Surguchev AA. Amyloidosis and longevity: a lesson from plants.* Biology *2019;**8**:43, respectively.*

In this chapter, we précis past discoveries, current evidence vis-à-vis the physiological and pathophysiological function of TDP-43, role of TDP-43 in transcription and gene regulation, TDP-43 associations with mitochondrial pathobiology and offer a summary of overall new developments in the area of TDP-43 pathobiology discovered in ALS, FTD, and other neurological disorders such as AD, Parkinson disease (PD), and LBD. Besides, we deliberate on the wealth of knowledge gained through high-throughput genomics, imaging, and biomarker studies of these diseases that support our view that these neurological disorders can be considered as a several facets of a spectrum of disease mechanisms, described as TDP-43 proteinopathies. To conclude, we comment on the significance of new developments in TDP-43-related investigation to the clinical practice of neurology and its role in drug development and precision medicine.

TDP-43 physiological and pathophysiological function

TDP-43 is a small ubiquitously expressed major pathological protein located on the chromosome 1, encoded by the TAR DNA-binding protein (TARDBP) gene having 96% similarity among mice and human.[23] The physiological and pathophysiological function of TDP-43 is varied and it plays major roles in the regulation of mitochondrial pathophysiology, gene expression, RNA processing, and stabilization of messenger RNA, including transcription, splicing, and transport[3,24–26] as illustrated in Figs. 1.2 and 1.3. Besides its role in gene regulatory mechanisms, TDP-43 is also involved in apoptosis, cell division, and microRNA biogenesis.

TDP-43 and mitochondria

TDP-43: mitochondrial association, function, and its role in quality control

Studies by several groups around the globe have confirmed the link of TDP-43 with mitochondria (Fig. 1.2). In the past, several efforts have been taken to explore TDP-43 subcellular organelle target(s) and first report of mitochondrial association with exogenously expressed wild-type (WT) or mutant TDP-43 (mtTDP-43) was reported in mitochondrial-enriched portions from NSC-34 cell lines.[27] Later on TDP-43 dynamics and localization, including aggregate formation in the inner membrane of mitochondria (IMM), was demonstrated in HEK293 cells, mouse, and human brain and spinal cords (SCs).[28–31] Recent work demonstrated the presence of TDP-43 in the matrix of mitochondria and its interaction with proteins critical for the mitochondrial electron transport chain (ETC) complex, e.g., its binding with mitochondrial genome-encoded mRNA, demonstrating a very distinct and direct role of TDP-43 in regulating mitochondrial function and physiology.[32,33]

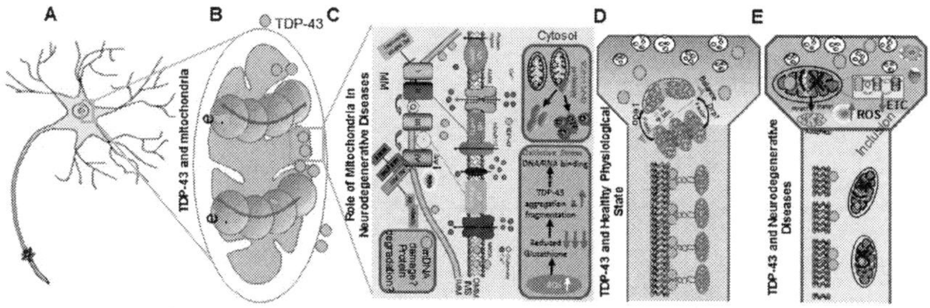

FIGURE 1.2 TDP-43 and its role in mitochondria physiological and pathophysiological health. (A) A sketch of neuron showing the nucleus, cell soma, dendrites, and neurites. (B) TDP-43 and its association with mitochondria showing TDP-43 in inner mitochondrial membrane (IMM), intermembrane space (IMS), and on the wall of cristae and outer mitochondrial membrane (OMM). (C) Toxic insults lead to mutations in mitochondrial DNA which produce reactive oxygen species (ROS) and impair calcium homeostasis and accelerate oxidative stress (OS) build up, reduced glutathione ultimately trigger TDP-43 aggregation and fragmentation hasten deleterious vicious cycle of mitochondrial damage and cell death in neurological diseases such ALS, FTD, AD, PD, and dementia. (D) Mitochondrial fission and fusion dynamics go through constant fission/fusion, and trafficking and maintain a healthy balance between in this physiological processes which are important for enabling the mitochondria to work normally in a healthy physiological state. (E) In diseased conditions, due to OS buildup and ROS formation in mitochondria, the dynamic balance between fusion/fission of mitochondria is impaired. This disturbed balance leads to TDP-43 aggregates formation and depletion. In addition, abnormal changes in mitochondria morphology, its internal shape and structure, OS formation and phosphorylation phenomenon can be noted in TDP-43 proteinopathy. *Part of the Figure adapted and modified with permission from Gao J, Wang L, Yan T, Perry G, Wang X. TDP-43 proteinopathy and mitochondrial abnormalities in neurodegeneration.* Mol Cell Neurosci *2019;**100**:103396.*

Mitochondrial oxidative phosphorylation system (OXPHOS) deficits, such as reductions in complex I–IV activity, mitochondrial membrane potential (Ψm), and upregulation of mitochondrial uncoupling protein 2 (UCP2) family have been shown in NSC-34 lines overexpressing WT or mtTDP-43.[34–36] Furthermore, mitochondria-linked TDP-43 is highly phosphorylated in ALS and FTD patient-derived induced pluripotent stem cells (iPSCs) neurons and/or fibroblasts, emphasizing the important role played by TDP-43 in post-translational modifications.[30,37] In addition, common mitochondria quality control mechanisms involve mitophagy[38] and aberrant alterations in mitophagy-associated genes are closely linked with several neurological diseases including ALS, FTD, AD, PD, and dementia.[39,40]

TDP-43: mitochondrial ion homeostasis, fission and fusion dynamics, and trafficking

Several studies demonstrated mitochondrial fission/fusion dynamics as a vital components of mitochondrial physiology, together with mitochondrial respiratory chain complex assembly and respiratory system efficiency,[41] mitochondrial ATP bioenergetics,[42] mitochondrial Ca^{2+} dynamics and homeostasis,[43–45]

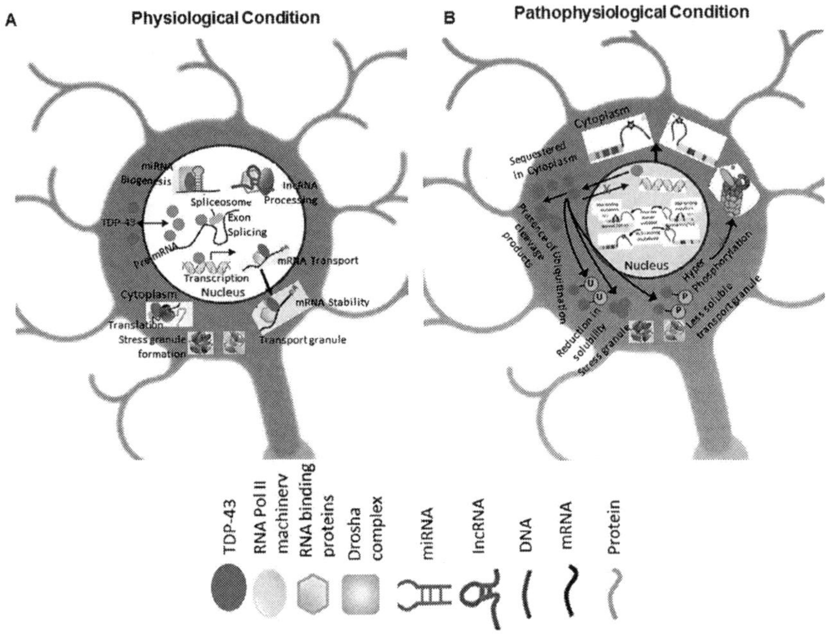

FIGURE 1.3 Physiological and pathophysiological function and molecular interplay of TDP-43. (A) In healthy physiological environments, TDP-43 is generally located in the cell nucleus, but continuous shuttling between the cytoplasm and the nucleus is quite common. TDP-43 has shown numerous physiological and gene regulatory function including its role at transcriptional level, exon splicing, miRNA biogenesis, mRNA transport, IncRNA processing, and stabilization of mRNA. In cytoplasm TDP-43 regulates stress granule formation, transport of ribonucleoprotein (RNP), translation of stress granules, as well as other physiological processes. (B) Under neuropathological disorders, TDP-43 shuttles from the nucleus to the cytoplasm and where its function is impaired and it undergoes hyper phosphorylation and/or ubiquitination. In addition, presence of cleaved TDP-43 fragments and less soluble TDP-43 inclusions are prone to aggregation in diseased conditions. These changes might be in response to various changes occurring both at the cellular and gene levels, such as loss-of-function or gain-of-function mechanisms due to toxic accumulation.

intraorganellar Ca^{2+} waves,[46,47] and control of mitochondrial morphology by reactive oxygen species formation.[48] Fragmented, damaged mitochondria, with impaired inner membrane structures[49] and mitochondrial ETC, have been shown to play dominant, early pathological roles in ALS.[50–52] Recently, mitochondria aggregation in transgenic mice overexpressing WT-TDP-43 and damaged mitochondrial cristae in ALS-TDP-43 were reported in NSC34 MNs.[27,34,53–55] ALS patient-derived fibroblasts, carrying mtTDP-43, have also been shown to damage and trigger swollen mitochondria, reinforcing previous findings.[56,57]

Mitochondrial fission/fusion are regulated by dynamin-like protein 1 (DLP1)[58] and its recruiting factors such as Fis1, Mff, MiD49, and MiD51[59]; mitofusin (Mfn) 1, mitofusin (Mfn) 2, and optic atrophy protein 1(OAP1).[32,38,60]

Cytoplasmic TDP-43 inclusion proteinopathy is implicated in patients with AD-type dementia and corticobasal degeneration.[32,61,62] In normal physiological conditions as well as during pathophysiological states, at the sites of bioenergetics requirements, mitochondria are transported to required sites by motor–adaptor complexes.[63] Overexpression of WT TDP-43 in cultured primary MNs leads to changes in anterograde and retrograde transport of mitochondria in both axons and dendrites,[29] including a decrease of mitochondrial trafficking in both axons and dendrites, recapitulating what occurs in TDP-43 overexpression,[29] suggesting the possible involvement of common downstream pathways for TDP-43-mediated mitochondrial transport. In addition, through microtubule network cytoplasmic TDP-43 regulates RNA granules formation.[64] Therefore, in spite of lack of direct evidence, TDP-43-associated mitochondrial transport, via cytoskeleton transport, is interesting to study in order to gain a better insight into TDP-43-induced mitochondrial transport defects.

TDP-43 and gene regulation

TDP-43: splicing and regulation of gene expression

Expression of TDP-43 is found in most tissues, is well-conserved between invertebrates and vertebrates, and contains two tandem RNA recognition motifs, RNA-recognition motif 1 (RRM1) and 2 (RRM2).[65,66] Due to its ability to bind RNA, including mRNA and pre-mRNA, TDP-43 has a prominent role in posttranscriptional RNA processing, RNA stability, RNA splicing, exon splicing and skipping, transportation, translation, and regulation.[67] Its main splicing regulatory elements discovered to date are a (UG)m(U)n regulatory region adjacent to 3'ss, a Pyridine-rich enhancer region close to 5'ss, and 2 complex regulatory exonic regions inside the exon.[68,69] Gly-rich C-terminal region is important for TDP-43 inhibitory activity and its interaction to heterogeneous ribonucleoproteins (hnRNPs) splicing inhibitory properties.[70] Furthermore, deletion of TDP-43 from cell lines, using small interfering RNA, has a great impact in inducing splicing alterations because of mutations in exon nine regulatory elements, which can be useful for therapeutic targeting purpose.[71]

TDP-43 inhibits the assembly of TATA box of the HIV-1 long terminal repeat region, both in the absence and/or presence of the Tat protein by binding to −18/+28 poly pyr-rich region of the trans activation response (TAR) DNA element.[72] TDP-43 has been involved in the exon splicing of several genes, e.g., cystic fibrosis transmembrane conductance regulator (CFTCR),[4,68,71] apolipoprotein A2,[69] and spinal motor neuron.[70] The comprehensive role TDP-43 plays in the regulation of RNA expression and splicing was demonstrated by recent studies where TDP-43 was shown to be bound to noncoding RNAs, such as metastasis-associated lung adenocarcinoma transcript 1

(MALAT1), a lncRNA gene class, and nuclear-enriched abundant transcript 1 (NEAT1), a novel nuclear long virus inducible noncoding RNA that encodes proteins crucial for regulation of neuronal development and survival and are implicated in various neurologic disease including FTD.[73]

TDP-43 also shows a functional role in stabilization of human neurofilament light chains (hNFL) messenger RNA (mRNA)[26] and modulation of cyclin-dependent kinase 6 (CDK6) gene expression.[74] Several new findings indicate exciting links between neuronal cell cycle gene expression, neuronal cell death, and TDP-43, observed during TDP-43 pathogenesis. This notion is braced by the findings of TDP-43-associated regulation of CDK6 in diseased condition.[74] New reports emerging to indicate that cells are delicately sensitive to TDP-43 expression levels and maintain a tight balance by which they regulate TDP-43 expression that have been demonstrated in transgenic (Tg) mouse models and cell culture.[75] This regulation happens via direct binding of the TDP-43 protein to its own 3′UTR sequence, and thereby accounts for the autoregulation of TDP-43 mRNA volatility and degradation via feedback loops, which are critical for the maintenance of prerequisite levels of TDP-43.[76] Thus, genetic, epigenetic, and environmental elements that perturb the TDP-43 autoregulatory feedback loop and associated complex network may contribute to physiological condition, neuronal dysfunction, and disease pathogenesis.

TDP-43: RNA processing and gene transcription

TDP-43 is a vital protein for normal cellular function and imbalance in its nucleus–cytoplasm distribution and aggregation leads to the misregulation of RNA metabolism, cellular homeostasis, and cytotoxicity. While shuttling between nucleus and cytoplasm, under healthy physiological situation, TDP-43 assists in regulation of numerous RNA processing steps, such as splicing, stabilization, trafficking, depletion, miRNA production,[77] cryptic exons splicing,[78,79] axonal RNA transport,[64] microRNA biogenesis[80,81] through its DNA/RNA-binding ability. Additionally, TDP-43 affects RNAs that encode proteins involved in autophagy, cellular protein homeostasis, and clearance pathways.[162] In pathological conditions such as ALS, FTD, and other neurological diseases, TDP-43 is mislocalized to the cytoplasm and forms stress granules and inclusion bodies both in neuronal and glial cells.[82,83] Roughly 25% AD cases also have these inclusions and stress granules.[84] In SH-SY5Y cells, 102 TDP-43 targets were identified for UG repeat and pyr-rich TDP-43-binding sequences, mostly within introns, while ~25% were outside introns. More than 100 up- or downregulated genes, mostly involved in neuroinflammation, were identified in HeLa cells when TDP-43 is depleted by small interfering RNA (siRNA). In addition, ~900 genes linked to endomembrane organization were aberrantly spliced due to the deficiency of TDP-43.[85] TDP-43 has also been involved in helping the mRNA stability of the hNFL, preventing its degradation by binding the 3′UTR of the hNFL transcript.[26] Alterations in hNFL profile may lead to neurodegeneration in several diseases such as ALS and FTD and suggest a crucial link concerning the neurofilament clump formations.

TDP-43 as a potential therapeutic biomarker

The TDP-43 pathology spectrum covers several neurodegenerative diseases. Diseases characterized by TDP-43 aberration as a prime disease hallmark includes FTD-TDP caused by Ubiquitin (+) neuronal aggregates formation (FTD-U), ALS, mixed ALS and FTD-TDP phenotypes, Perry syndrome, and FTD–inclusion body myopathy–Paget syndrome.[13,86–91] Diseases characterized by TDP-43 as a secondary pathophysiological hallmarks includes AD, PD, and its linked neurological disorders such as PD with brain dementia and LBD with AD[92] and rare disorders such as Guam ALS-PD.[93–96] All of these scientific breakthrough confirmed aberrant accumulation of TDP-43 in motor cortex (MCX), hippocampus, and SC of individuals diagnosed with any of these devastating disorders. The results of these work demonstrate degenerated neurons with TDP-43 accumulation in the cytoplasm, comprise ubiquitin (+) and hyper-phosphorylated protein which, subsequently, is cleaved and form C-terminal fragments, while healthy neurons retain TDP-43 predominantly in the nucleus compartment.[12,13] TDP-43 buildup is also a hallmark feature of FTD-U in white matter, Guam parkinsonism-dementia, FTD-MND mixed phenotype, and hippocampal sclerosis.[14,97–102]

TDP-43 and ALS/FTD: a tale of two disease

TDP-43 mutations are responsible for roughly 1% of all ALS cases,[103] and cytoplasmic buildup of TDP-43 in ALS is found both in neuronal and glial cell populations of the MCX, brain stem, and SC.[88,104] However, in FTD, TDP-43 presence is widespread and inclusions are detected in the neocortex, prefrontal cortex, and dentate gyrus of hippocampal cells.[13,90,105] Dozens of mouse models have been created in an endeavor to untangle the pathophysiologic contributions and disease mechanism of TDP-43 in protein aggregates, protein homeostasis, autophagy, and clearance pathways linked to ALS and FTD.

Two prominent mechanism suggested to explain TDP-43-mediated neurodegeneration are (i) loss-of-function due to downregulation in TDP-43 expression, leading to a loss of TDP-43's normal cellular functions, supporting the notion that loss of function could contribute to ALS-FTD pathology[106,107] and (ii) gain-of-function because these aggregates sequester RNA-binding proteins, which leads to splicing defects, altered nucleocytoplasmic transport and changes in TDP-43-mediated RNA metabolism.[108–110] Support for the hypothesis that gain-of-function leads to impaired autoregulation was suggested in TDP-43Q331K/Q331K, TDP-43M337V, and TDP-43G298S knock-in mice.[111,112]

In addition, endogenous TDP-43 mutations in mice demonstrate the gain-of-splicing functions that further strengthen this hypothesis.[113]

TDP-43 and Alzheimer's disease

Abnormal TDP-43 immunoreactivity and its potential clinical implications in AD have been explored using several techniques including genomics, imaging, biochemical and behavioral studies.[14,62,114–117] Presence of TDP-43 inclusions occurs in ~25%–30% of sporadic AD[14,62,115] and ~14% in familial AD.[117] TDP-43 seems to develop in a stereotypical style in AD following a modified 6-stage scheme. It starts with deposition in amygdala, and eventually in cortex and hippocampus.[114,118–122] The spread of neuropil tangles and neurofibrillary tangles of TDP-43 pathology overlaps with tau pathology but not with amyloid plaques in AD.[123] Nevertheless, a recent study concludes that hippocampal atrophy appears to be bigger in AD with TDP-43 pathology.[114] In conclusion, since TDP-43 pathobiology in AD have wide spectrum, and some of the disease hallmark are not distinguishable from FTD-TDP and ALS-TDP, including existence of neuronal cytoplasmic and intranuclear inclusions, and therefore more biochemical, genomics, and bioimaging markers are needed to conclude that tau and TDP-43 have a strong interaction.

TDP-43, Lewy body dementia, and age-related tauopathy

The 6-stage topography scheme of TDP-43 inclusion and α-synuclein distribution are some of the main characteristics of LBD,[96] and age-related tauopathy.[124,125] The frequency of TDP-43 in LBD and age-related tauopathy is ~29%, much lower than the AD where it ranges from 40% to 75%.[96,126,127] Besides, most of the cases of age-related tauopathy (~75%) with TDP-43 occurred at stage 1 and TDP inclusion does not goes beyond stage 3.[119,127] In LBD and age-related dementia, the bulk of the TDP-43 inclusions were perivascular and greatly differ from the characteristic neuronal inclusions or neurite tangles witnessed in AD.[128] These characteristics pathological differences might be the main reason for the absence of any association between clinical characteristics and TDP-43 in LBD/age-related dementia compared to AD.[129]

TDP-43 and other neurodegenerative disorders

In addition to TDP-43-positive protein inclusions in ALS, FTD, AD, HD, and LBD, TDP-43 abnormalities have also been reported in chronic traumatic encephalopathy (CTE)[130] and traumatic brain injury (TBI).[131] McKee and colleague investigated 12 cases of CTE-diagnosed athletic patients and 10 of these athletes showed extensive TDP-43 proteinopathy.[130,132] In CTE patients with severe TDP-43 pathology, the architecture of TDP-43 manifestation is similar to FTD-TDP.[130,133]

Several preclinical studies, such as cortical contusion injury (CCI) or fluid percussion injury TBI models in rodents, have examined TDP-43 pathology and discovered several alterations in TDP-43, including proteolysis, cleavage, and mislocalization of TDP-43 into TDP-25 and TDP-35 fragments, cytoplasmic TDP-43 granules formation and phosphorylation, leading to neurodegeneration.[131,134,135] Moreover, in a weight-drop model of TBI, along with neuronal damage, cleaved TDP-43 fragments are also increased in astrocytes and other glial cells.[136] In a CCI model of repetitive mild TBI, there is a consistent surge in TDP-43 expression and aggregation in the cortex and hippocampus of rodents.[137] Rodent's models of blast TBI exhibited significant increase in TDP-43 in the brain as well as classical TDP-43 fragmentation and mislocalization, a unique trademark of TDP-43 proteinopathy.[138,139] The evidence available so far on TDP-43 pathobiology following CTE and TBI are inadequate due to several factors and more investigation of TDP-43 pathology in various CTE, TBI, and blast TBI patients is required to begin to adduce a causal relation between TBI and TDP-43 pathology.

TDP-43 and novel therapy

TDP-43 proteinopathy is a shared disease hallmark of several neurological diseases such as ALS, FTD, AD, and dementia and very few options for clinical interventions are currently available.[140] Evidences in Tg mice are emerging to suggest that by inhibiting the TDP-43 mitochondrial localization it is possible to enhance neuronal survival and behavioral deficits.[30] Accordingly, TDP-43 mitochondrial localization, dynamics, and associated lead molecules and biomarkers would be a unique therapeutic target drugs able to ameliorate neurodegeneration in these diseases. In addition, mitochondria and associated TDP-43 are critical targets of Aβ, tau, and α-synuclein; therefore, mitochondria is the converging point of broad range of proteinopathies-related neurological disease and consequently further investigation might shed light on how TDP-43 associated with mitochondria play a role in disease development. Recent report indicate that accurate measurement of plasma TDP-43 levels such as TDP-43 Simoa assay (Quanterix) permit the TDP-43 pathology indexing using full-length protein calibrator and antibodies against AA 203–209 and the C-terminal region.[141] This assay helps in the diagnosis of diverse forms of dementia as well as to differentiate similar symptoms such as TDP-43 proteinopathy and tauopathy due to its ability to identify both full-length and pathological, truncated forms of the protein. In addition, efforts are underway to develop TDP-43-based biomarkers in plasma and cerebrospinal fluid for primary and secondary TDP-43 proteinopathies[142–144] which might assist in diagnosis and design of clinical trials of disease-modifying therapies.

In past decade, significant priority has been diverted toward finding and repurposing small molecular weight compounds able to target pathological features of TDP-43. Some of the existing compounds used for clinical trials are shown to be effective on TDP-43. Proteome analyses of SW480-ADH and BE-M17/iPSC cells treated with Vitamin D and aryl hydrocarbon receptor (AHR) agonists have been shown to increase TDP-43 expression levels.[145,146] Treatment with Vorinostat (suberoylanilide hydroxamic acid, SAHA), a class of histone deacetylase (HDAC) inhibitors, was demonstrated to downregulate cytosolic TDP-43 in PGRN-deficient cells, compared to control cells.[147] Additionally, reduction of 25 kDa (TDP-25) and 35 kDa (TDP-35) cleaved phosphorylated TDP-43 by proteases in ALS/FTD may serve as a good strategy to monitor reduced, deleterious effects of TDP-43 pathobiology. Indeed, D-sorbitol, commonly used as an exogenous stressor, and c-JNK inhibitors, salvage deleterious axonal trafficking and TDP-43 aggregates in mtTDP-43 iPSCs.[148] Riluzole, a glutamatergic neurotransmission inhibitor is the drug approved by the US Food and Drug Administration for ALS treatment and it has been shown to inhibit TDP-43 self-interaction and inclusion formation.[149] Rolipram, a selective PDE4 inhibitor, has been reported to reduce neuronal cytoplasmic TDP-43 levels in the neurons of Tg mice.[150] Antiepileptic drug, Valproate, was reported to lessen the deleterious effects of TDP-43 in cell lines.[151]

Furthermore, the four-part TDP-43 staging pathology progresses in a defined pattern[152] and also pinpoints possible drug targets. For example, treatments that stop or reduce TDP-43 spread may halt disease in the very beginning. Disease pathology initiates with the seeded sequestration of TDP-43 and, therefore, the most important therapeutic aim for the TDP-43 proteinopathies would be to hamper the buildup of proteinaceous inclusions and aggregates, support their quick disposal, or check/antagonize their connections with normal TDP-43.

TDP-43 and its relevance to clinical practice

It is important for clinicians to understand how the recent findings of TDP-43 as a critical disease-linked protein in FTD-TDP, ALS-TDP, AD, PD, dementia and other neurodegenerative disorders affect the forthcoming diagnosis, prognosis, treatment, and clinical practice of patients with TDP-43 proteinopathies. Individual patients having TDP-43 aberrations can share clinical symptoms within MND, FTD, ALS-FTD, AD, dementia, and several other neurological disorders or combinations of the above.[11,153] For clinicians and doctors, treating the sporadic ALS patients and ~90% of FTD patients comprised of FTD-tau or FTD-TDP, which have pivotal TDP-43 pathology, initiating trials addressing therapeutic drugs gap/limitations is of great importance.[154] TDP-43 most likely shares disease etiology with FTD-U patient groups that, until recently, were considered to encompass a blend of several disease entities. This will provide great help in estimations of heterogeneity during clinical trials designs to estimate appropriate number of patients needed to achieve power to assess significant drugs effect size that is clinically meaningful.

A staging system to categorize TDP-43 pathobiology of autopsy brain has been suggested for several diseases including ALS, FTD, and AD.[119,127,152,155] They anticipate that in the future this staging scheme will be able to classify TDP-43-associated disease, which will help researchers, doctors, and pathologists categorize pathology progression at autopsy, based on spread patterns in different regions of the brain.[152] A recent study suggests that misfolded Aβ, α-synuclein, and tau spreads from cell to cell in a seeded and self-templating, predictable fashion, and this in vitro and in vivo study predicts similar seeding pattern of disease development for TDP-43 pathology.[156] Based on the study by Brettschneider and colleagues, these researchers defined four different stages of patterns of TDP-43 aggregates spread and seeding in the brains and SC of ALS[152] which was later modified. Stage 1: TDP-43 aggregates accumulate mainly in MCX, brainstem, and SC; stage 2: inclusions started visible in the brain area such as precerebellar nuclei, red nucleus, as well as prefrontal cortex; stage 3: disease pathology extended to striatum, and in stage 4: this spread went further to temporal lobe indicating that pathology begin in the MCX and radiate downward to the SC via axons with caveats that autopsy brains provides only a static snapshot and therefore accurate guarantee of directionality is not certain[157]. Therefore, evaluating the staging pattern of TDP-43 pathology progression will definitely provide a potential treatment avenue to block TDP-43 spread, which might halt disease progression at an initial disease stage. While some initial studies have established a relationship among TDP-43, CTE, TBI, head trauma, and other neurological diseases, further studies are needed to confirm these claims.

The decisive role of the clinical pathologist in the management and etiology of TDP-43 pathology is crucial. Therefore, it's the clinician's and researchers task to obtain a patient's sample of both fresh and fixed tissues in a consistent manner, with standard protocols employed across hospitals for executing diagnostic tests, whether in-house or when sending the brain and/or collected samples to a referral neuropathologic or diagnostic center. Specimens should be procured in such a way so as to vigorously preserve tissue physiognomies. This standardized protocol will help in precise neuropathologic diagnosis and classification so that molecular-genetic testing for mutations in genes, especially in familial cases, can be performed beyond TDP-43 pathology. However, ethically challenging issues must be weighed in molecular-genetic clinical diagnostics of TDP 43 patients and the implications for living relatives and, in general, such testing is limited and occasionally not available.

Discussions and perspectives

The hallmark features of neurological disease are TDP-43 proteinopathy, TDP-43 isoforms, and its secondary modifications, including phosphorylation. Forthcoming studies are urgently needed to explore whether and how kinases and ubiquitin ligases play a major role in TDP-43 posttranslational modifications. Processes such as microRNA generation or the mitosis–meiosis dynamics which contribute to disease mechanisms need more attention. More importantly, scientists should focus more on understanding and deciphering the role of TDP-43 in RNA splicing, gene and transcriptional regulation, and mRNA stability, which will provide additional clues about how this protein plays a regulatory role in disease pathophysiology. For example, annotated genomic sequences such as promoters/enhancers or mRNA-associated splicing regulated events provides a wealth of previously unexplored data-mining opportunities and clue about precision-drug targets. Another key finding needing attention and further investigation is the observation that TDP-43 is a specific caspase substrate in Jurkat T lymphocytes, which are involved in initiation and execution of apoptosis and resultant cleavage products.[158]

Beyond implications in neurodegeneration, TDP-43 is a part of a 19 protein unit assemblage, called the Microprocessor complex Drosha/DiGeorge syndrome chromosomal region 8 (DGCR8). DGCR8 Microprocessor is one of the central molecular machineries that takes part in the generation of small (22 nt), noncoding micro RNAs (miRNAs) which play a major role in the control of gene expression in mammalian system, including humans, chimpanzee, and rodents.[159,160] The function and role played by the proteins associated with Drosha/DGCR8 complex is not clear, and, yet, there is sufficient evidence that indicates that TDP-43 may be a likely target protein, although the role needs more examination to understand its functional properties in Microprocessor complex.

Availability of iPSCs, organoids, and efficient siRNA reagents provides us a framework for selective knock-down of TDP-43 expression which will allow us to check for functionality of several potential drug targets on a large scale. Lately, new and emerging technologies, such as high-throughput next generation sequencing (HT-NGS), have established that a single RNA-binding protein can change several hundreds of alternatively spliced transcripts and its variants.[161] Therefore, approaches such as HT-NGS combined with spatial transcriptomics, HiPlex RNAScope, and BASEscope will be necessary to understand the complex role of TDP-43 in neurological diseases as these tools provide both spatial and temporal resolution at the single base level. In addition, although ~1000 of RNAs are intricately linked with TDP-43, experiments in *Drosophila melanogaster*, and its integration with unbiased systems biology and classical genetics, advocates that it is within our reach to identify precise metabolic pathways pertinent to TDP-43-dependent phenotypes. Accordingly, a clever and futuristic therapeutic approach may require not only interfering with the mechanisms of cytoplasmic aggregate formation but also targeting the definite downstream cellular pathways in which TDP-43 normally functions.

As the therapies moves closer to the clinic, our expectation is that the drug design based on molecular targets will be able to slow or reverse TDP-43 pathophysiology in diseases such as ALS, FTD, AD, dementia, and several other neurological disorders discussed in this chapter, all of which are fatal and impose severe burdens to society, both economically and emotionally. In a nutshell, I think TDP-43 associated brain disorders are the major healthcare threat to our economy and our society. It's a ticking time bomb and therefore greater public recognition is critical in encouraging prevention and early intervention,

List of abbreviations

AD Alzheimer's disease
ALS amyotrophic lateral sclerosis
APP amyloid precursor protein
C9orf72 chromosome 9 open reading frame 72
CCI concussion injury
CFTR cystic fibrosis transmembrane conductance regulator
CTE chronic traumatic encephalopathy
DGCR8 DiGeorge syndrome chromosomal region 8
FTD frontotemporal dementia
FTD-tau frontotemporal dementia with tau-positive inclusions
FTD-TDP frontotemporal dementia with TDP-43-positive inclusions
FTDU frontotemporal dementia with ubiquitin-positive inclusions
GRN granulin gene
hNFL neurofilament light chains
hnRNPs heterogeneous ribonucleoproteins
HTS high-throughput sequencing
IMM inner mitochondrial membrane
IMS intermembrane space
iPSCs induced pluripotent stem cells
LBD Lewy body dementia
MCX motor cortex
Mfn mitofusin
MN motor neuron
MND motor neuron disease
mRNA messenger RNA
mtTDP-43 mutant TDP-43
OMM outer mitochondrial membrane
OS oxidative stress
OXPHOS oxidative phosphorylation system
PD Parkinson's disease
PINK1 PTEN-induced kinase 1
ROS reactive oxygen species
siRNA small interfering RNA

TARDBP transactive response DNA-binding protein gene
TBI traumatic brain injury
TDP-43 transactive response DNA-binding protein 43
Tg transgenic
VDAC1 voltage-gated dependent anion channel 1
WT wild type

Conflicts of interests

None of the authors has any conflict of interest to disclose. The author declares that the research was conducted without any commercial or financial relationships that could be construed as a potential conflict of interest. The views expressed in this article are those of the author and do not reflect the official policy or position of the Icahn School of Medicine at Mount Sinai.

Author contributions

M K Jaiswal conceived of, drafted, edited, and approved the book chapter for publication.

All aspects of this article have been solely carried out by Dr. Jaiswal.

Acknowledgments

The author would like to thank Professor Drs. Bernhard U. Keller, Ananta paine and Kenjiro Seki for collaboration and valuable discussions.

References

[1] Ayala YM, Zago P, D'Ambrogio A, Xu YF, Petrucelli L, Buratti E, Baralle FE. Structural determinants of the cellular localization and shuttling of TDP-43. *J Cell Sci* 2008b;**121**:3778–85.

[2] Winton MJ, Igaz LM, Wong MM, Kwong LK, Trojanowski JQ, Lee VM. Disturbance of nuclear and cytoplasmic TAR DNA-binding protein (TDP-43) induces disease-like redistribution, sequestration, and aggregate formation. *J Biol Chem* 2008;**283**:13302–9.

[3] Buratti E, Baralle FE. Characterization and functional implications of the RNA binding properties of nuclear factor TDP-43, a novel splicing regulator of CFTR exon 9. *J Biol Chem* 2001;**276**:36337–43.

[4] Buratti E, Dork T, Zuccato E, Pagani F, Romano M, Baralle FE. Nuclear factor TDP-43 and SR proteins promote in vitro and in vivo CFTR exon 9 skipping. *EMBO J* 2001;**20**:1774–84.

[5] Cuevas-Diaz Duran R, Wei H, Wu JQ. Single-cell RNA-sequencing of the brain. *Clin Transl Med* 2017;**6**:20.

[6] Pick A. On the relationship between senile cerebral atrophy and aphasia [German]. *Prag Med Wochenschr* 1892;**17**:165–7.

[7] Alzheimer A. On peculiar disease cases of later age (In German). *Z Ges Neurol Psychiat* 1911;**4**:356–85.

[8] Forman MS, Farmer J, Johnson JK, Clark CM, Arnold SE, Coslett HB, Chatterjee A, Hurtig HI, Karlawish JH, Rosen HJ, et al. Frontotemporal dementia: clinicopathological correlations. *Ann Neurol* 2006;**59**:952–62.

[9] McKhann GM, Albert MS, Grossman M, Miller B, Dickson D, Trojanowski JQ. Clinical and pathological diagnosis of frontotemporal dementia: report of the work group on frontotemporal dementia and pick's disease. *Arch Neurol* 2001;**58**:1803–9.

[10] Baker M, Mackenzie IR, Pickering-Brown SM, Gass J, Rademakers R, Lindholm C, Snowden J, Adamson J, Sadovnick AD, Rollinson S, et al. Mutations in progranulin cause tau-negative frontotemporal dementia linked to chromosome 17. *Nature* 2006;**442**:916–9.

[11] Cruts M, Gijselinck I, van der Zee J, Engelborghs S, Wils H, Pirici D, Rademakers R, Vandenberghe R, Dermaut B, Martin JJ, et al. Null mutations in progranulin cause ubiquitin-positive frontotemporal dementia linked to chromosome 17q21. *Nature* 2006;**442**:920–4.

[12] Arai T, Hasegawa M, Akiyama H, Ikeda K, Nonaka T, Mori H, Mann D, Tsuchiya K, Yoshida M, Hashizume Y, et al. TDP-43 is a component of ubiquitin-positive tau-negative inclusions in frontotemporal lobar degeneration and amyotrophic lateral sclerosis. *Biochem. Biophys Res Commun* 2006;**351**:602–11.

[13] Neumann M, Sampathu DM, Kwong LK, Truax AC, Micsenyi MC, Chou TT, Bruce J, Schuck T, Grossman M, Clark CM, et al. Ubiquitinated TDP-43 in frontotemporal lobar degeneration and amyotrophic lateral sclerosis. *Science* 2006;**314**:130133.

[14] Amador-Ortiz C, Lin WL, Ahmed Z, Personett D, Davies P, Duara R, GraffRadford NR, Hutton ML, Dickson DW. TDP-43 immunoreactivity in hippocampal sclerosis and Alzheimer's disease. *Ann Neurol* 2007;**61**:435–45.

[15] Brouwers N, Bettens K, Gijselinck I, Engelborghs S, Pickut BA, Van Miegroet H, Montoya AG, Mattheijssens M, Peeters K, De Deyn PP, et al. Contribution of TARDBP to Alzheimer's disease genetic etiology. *J Alzheim Dis* 2010;**21**:423–30.

[16] Higashi S, Iseki E, Yamamoto R, Minegishi M, Hino H, Fujisawa K, Togo T, Katsuse O, Uchikado H, Furukawa Y, et al. Concurrence of TDP-43, tau and alpha-synuclein pathology in brains of Alzheimer's disease and dementia with Lewy bodies. *Brain Res* 2007;**1184**:284–94.

[17] Nakashima-Yasuda H, Uryu K, Robinson J, Xie SX, Hurtig H, Duda JE, Arnold SE, Siderowf A, Grossman M, Leverenz JB, et al. Co-morbidity of TDP-43 proteinopathy in Lewy body related diseases. *Acta Neuropathol* 2007;**114**:221–9.

[18] Fujishiro H, Uchikado H, Arai T, Hasegawa M, Akiyama H, Yokota O, Tsuchiya K, Togo T, Iseki E, Hirayasu Y. Accumulation of phosphorylated TDP43 in brains of patients with argyrophilic grain disease. *Acta Neuropathol* 2009;**117**:151158.

[19] DeJesus-Hernandez M, Mackenzie IR, Boeve BF, Boxer AL, Baker M, Rutherford NJ, Nicholson AM, Finch NA, Flynn H, Adamson J, et al. Expanded GGGGCC hexanucleotide repeat in noncoding region of C9ORF72 causes chromosome 9p-linked FTD and ALS. *Neuron* 2011;**72**:245–56.

[20] Renton AE, Majounie E, Waite A, Simon-Sanchez J, Rollinson S, Gibbs JR, Schymick JC, Laaksovirta H, van Swieten JC, Myllykangas L, et al. A hexanucleotide repeat expansion in C9ORF72 is the cause of chromosome 9p21-linked ALS-FTD. *Neuron* 2011;**72**:257–68.

[21] Hodges JR, Davies RR, Xuereb JH, Casey B, Broe M, Bak TH, Kril JJ, Halliday GM. Clinicopathological correlates in frontotemporal dementia. *Ann Neurol* 2004;**56**:399–406.

[22] Talbot K, Ansorge O. Recent advances in the genetics of amyotrophic lateral sclerosis and frontotemporal dementia: common pathways in neurodegenerative disease. *Hum Mol Genet* 2006;**15 Spec**(2):R182–7.

[23] Xu Y-F, Prudencio M, Hubbard JM, Tong J, Whitelaw EC, Jansen-West K, Stetler C, Cao X, Song J, Zhang Y-J. The pathological phenotypes of human TDP-43 transgenic mouse models are independent of downregulation of mouse Tdp-43. *PLoS One* 2013;**8**:e69864.

[24] Fiesel FC, Voigt A, Weber SS, Van den Haute C, Waldenmaier A, Gorner K, Walter M, Anderson ML, Kern JV, Rasse TM, et al. Knockdown of transactive response DNA-binding protein (TDP-43) downregulates histone deacetylase 6. *EMBO J* 2010;**29**:209–21.

[25] Lagier-Tourenne C, Polymenidou M, Cleveland DW. TDP-43 and FUS/TLS: emerging roles in RNA processing and neurodegeneration. *Hum Mol Genet* 2010;**19**:R46–64.

[26] Strong MJ, Volkening K, Hammond R, Yang W, Strong W, Leystra-Lantz C, Shoesmith C. TDP43 is a human low molecular weight neurofilament (hNFL) mRNA-binding protein. *Mol Cell Neurosci* 2007;**35**:320–7.

[27] Hong K, Li Y, Duan W, Guo Y, Jiang H, Li W, Li C. Full-length TDP-43 and its C-terminal fragments activate mitophagy in NSC34 cell line. *Neurosci Lett* 2012;**530**:144–9.

[28] Stribl C, Samara A, Trümbach D, Peis R, Neumann M, Fuchs H, Gailus-Durner V, Hrabě de Angelis M, Rathkolb B, Wolf E, et al. Mitochondrial dysfunction and decrease in body weight of a transgenic knock-in mouse model for TDP43. *J Biol Chem* 2014;**289**:10769–84.

[29] Wang W, Li L, Lin WL, Dickson DW, Petrucelli L, Zhang T, Wang X. The ALS disease-associated mutant TDP-43 impairs mitochondrial dynamics and function in motor neurons. *Hum Mol Genet* 2013;**22**:4706–19.

[30] Wang W, Wang L, Lu J, Siedlak SL, Fujioka H, Liang J, Jiang S, Ma X, Jiang Z, da Rocha EL, et al. The inhibition of TDP-43 mitochondrial localization blocks its neuronal toxicity. *Nat Med* 2016;**22**:869–78.

[31] Xu Y-F, Zhang Y-J, Lin W-L, Cao X, Stetler C, Dickson DW, Lewis J, Petrucelli L. Expression of mutant TDP-43 induces neuronal dysfunction in transgenic mice. *Mol Neurodegener* 2011;**6**:73.

[32] Davis SA, Itaman S, Khalid-Janney CM, Sherard JA, Dowell JA, Cairns NJ, Gitcho MA. TDP-43 interacts with mitochondrial proteins critical for mitophagy and mitochondrial dynamics. *Neurosci Lett* 2018a;**678**:8–15.

[33] Termsarasab P, Thammongkolchai T, Gao J, Wang L, Liang J, Wang X. Cytoplasmic mislocalization and mitochondrial colocalization of TDP-43 are common features between normal aged and young mice. *Exp. Biol. Med.* 2020 1535370220914253.

[34] Lu J, Duan W, Guo Y, Jiang H, Li Z, Huang J, Hong K, Li C. Mitochondrial dysfunction in human TDP-43 transfected NSC34 cell lines and the protective effect of dimethoxy curcumin. *Brain Res. Bull.* 2012;**89**:185–90.

[35] Moujalled D, Grubman A, Acevedo K, Yang S, Ke YD, Moujalled DM, Duncan C, Caragounis A, Perera ND, Turner BJ, et al. TDP-43 mutations causing amyotrophic lateral sclerosis are associated with altered expression of RNAbinding protein hnRNP K and affect the Nrf2 antioxidant pathway. *Hum Mol Genet* 2017;**26**:1732–46.

[36] Szelechowski M, Amoedo N, Obre E, Léger C, Allard L, Bonneu M, Claverol S, Lacombe D, Oliet S, Chevallier S, et al. Metabolic reprogramming in amyotrophic lateral sclerosis. *Sci Rep* 2018;**8**:3953.

[37] Genin EC, Bannwarth S, Lespinasse F, Ortega-Vila B, Fragaki K, Itoh K, Villa E, Lacas-Gervais S, Jokela M, Auranen M, et al. Loss of MICOS complex integrity and mitochondrial damage, but not TDP-43 mitochondrial localisation, are likely associated with severity of CHCHD10-related diseases. *Neurobiol Dis* 2018;**119**:159–71.

[38] Chen H, Chan DC. Mitochondrial dynamics–fusion, fission, movement, and mitophagy–in neurodegenerative diseases. *Hum Mol Genet* 2009;**18**:R169–76.

[39] Pickrell AM, Youle RJ. The roles of PINK1, parkin, and mitochondrial fidelity in Parkinson's disease. *Neuron* 2015;**85**:257–73.

[40] Youle RJ, Narendra DP. Mechanisms of mitophagy. *Nat Rev Mol Cell Biol* 2011;**12**:9–14.

[41] Cogliati S, Frezza C, Soriano ME, Varanita T, Quintana-Cabrera R, Corrado M, Cipolat S, Costa V, Casarin A, Gomes LC, et al. Mitochondrial cristae shape determines respiratory chain supercomplexes assembly and respiratory efficiency. *Cell* 2013;**155**:160–71.

[42] Benard G, Bellance N, James D, Parrone P, Fernandez H, Letellier T, Rossignol R. Mitochondrial bioenergetics and structural network organization. *J Cell Sci* 2007;**120**:838–48.

[43] Frieden M, James D, Castelbou C, Danckaert A, Martinou JC, Demaurex N. $Ca^{(2+)}$ homeostasis during mitochondrial fragmentation and perinuclear clustering induced by hFis1. *J Biol Chem* 2004;**279**:22704–14.

[44] Jaiswal MK, Keller BU. Cu/Zn superoxide dismutase typical for familial amyotrophic lateral sclerosis increases the vulnerability of mitochondria and perturbs Ca^{2+} homeostasis in SOD1G93A mice. *Mol Pharmacol* 2009;**75**:478–89.

[45] Jaiswal MK, Zech WD, Goos M, Leutbecher C, Ferri A, Zippelius A, Carri MT, Nau R, Keller BU. Impairment of mitochondrial calcium handling in a mtSOD1 cell culture model of motoneuron disease. *BMC Neurosci.* 2009;**10**:64.

[46] Jaiswal MK. Motoneuron specific calcium dysregulation and perturbed cellular calcium homestasis in amyotrophic lateral sclerosis: recent advances gained from genetically modified animals and cell culture models. In: *Motor neuron diseases: causes, classification and treatments in Bradley J Turner and Julie B Atkin*. New York): Nova Publishers; 2012. p. 87–114.

[47] Szabadkai G, Simoni AM, Chami M, Wieckowski MR, Youle RJ, Rizzuto R. Drp-1-dependent division of the mitochondrial network blocks intraorganellar Ca^{2+} waves and protects against Ca^{2+}-mediated apoptosis. *Mol. Cell* 2004;**16**:59–68.

[48] Yu T, Robotham JL, Yoon Y. Increased production of reactive oxygen species in hyperglycemic conditions requires dynamic change of mitochondrial morphology. *Proc Natl Acad Sci USA* 2006;**103**:2653–8.

[49] Sasaki S, Horie Y, Iwata M. Mitochondrial alterations in dorsal root ganglion cells in sporadic amyotrophic lateral sclerosis. *Acta Neuropathol* 2007;**114**:633639.

[50] Jaiswal MK. Calcium, mitochondria, and the pathogenesis of ALS: the good, the bad, and the ugly. *Front Cell Neurosci* 2013;**7**:199.

[51] Jaiswal MK. Selective vulnerability of motoneuron and perturbed mitochondrial calcium homeostasis in amyotrophic lateral sclerosis: implications for motoneurons specific calcium dysregulation. *Mol. Cel Therap* 2014;**2**:26.

[52] Jaiswal MK. *The role of mitochondria, oxidative stress and altered calcium homeostasis in amyotrophic lateral sclerosis: from current developments in the laboratory to clinical treatments*. Frontiers Media SA; 2017.

[53] Gao J, Wang L, Yan T, Perry G, Wang X. TDP-43 proteinopathy and mitochondrial abnormalities in neurodegeneration. *Mol Cell Neurosci* 2019;**100**:103396.

[54] Salvatori I, Ferri A, Scaricamazza S, Giovannelli I, Serrano A, Rossi S, D'Ambrosi N, Cozzolino M, Giulio AD, Moreno S, et al. Differential toxicity of TAR DNA-binding protein 43 isoforms depends on their submitochondrial localization in neuronal cells. *J Neurochem* 2018;**146**:585–97.

[55] Xu YF, Gendron TF, Zhang YJ, Lin WL, D'Alton S, Sheng H, Casey MC, Tong J, Knight J, Yu X, et al. Wild-type human TDP-43 expression causes TDP-43 phosphorylation, mitochondrial aggregation, motor deficits, and early mortality in transgenic mice. *J Neurosci* 2010;**30**:10851–9.

[56] Gautam M, Jara JH, Kocak N, Rylaarsdam LE, Kim KD, Bigio EH, Hande Ozdinler P. Mitochondria, ER, and nuclear membrane defects reveal early mechanisms for upper motor neuron vulnerability with respect to TDP-43 pathology. *Acta Neuropathol* 2019;**137**:47−69.

[57] Onesto E, Colombrita C, Gumina V, Borghi MO, Dusi S, Doretti A, Fagiolari G, Invernizzi F, Moggio M, Tiranti V, et al. Gene-specific mitochondria dysfunctions in human TARDBP and C9ORF72 fibroblasts. *Acta Neuropathol Commun* 2016;**4**:47.

[58] Palmer CS, Elgass KD, Parton RG, Osellame LD, Stojanovski D, Ryan MT. Adaptor proteins MiD49 and MiD51 can act independently of Mff and Fis1 in Drp1 recruitment and are specific for mitochondrial fission. *J Biol Chem* 2013;**288**:27584−93.

[59] Loson OC, Song Z, Chen H, Chan DC. Fis1, Mff, MiD49, and MiD51 mediate Drp1 recruitment in mitochondrial fission. *Mol Biol Cell* 2013;**24**:659667.

[60] Detmer SA, Chan DC. Functions and dysfunctions of mitochondrial dynamics. *Nat Rev Mol Cell Biol* 2007;**8**:870−9.

[61] James BD, Wilson RS, Boyle PA, Trojanowski JQ, Bennett DA, Schneider JA. TDP-43 stage, mixed pathologies, and clinical Alzheimer's-type dementia. *Brain J Neurol* 2016;**139**:2983−93.

[62] Uryu K, Nakashima-Yasuda H, Forman MS, Kwong LK, Clark CM, Grossman M, Miller BL, Kretzschmar HA, Lee VM, Trojanowski JQ, et al. Concomitant TAR-DNA-binding protein 43 pathology is present in Alzheimer disease and corticobasal degeneration but not in other tauopathies. *J Neuropathol Exp Neurol* 2008;**67**:555−64.

[63] Sheng ZH, Cai Q. Mitochondrial transport in neurons: impact on synaptic homeostasis and neurodegeneration. *Nat Rev Neurosci* 2012;**13**:77−93.

[64] Alami NH, Smith RB, Carrasco MA, Williams LA, Winborn CS, Han SSW, Kiskinis E, Winborn B, Freibaum BD, Kanagaraj A, et al. Axonal transport of TDP-43 mRNA granules is impaired by ALS-causing mutations. *Neuron* 2014;**81**:536−43.

[65] Ayala YM, Pantano S, D'Ambrogio A, Buratti E, Brindisi A, Marchetti C, Romano M, Baralle FE. Human, Drosophila, and C.elegans TDP43: nucleic acid binding properties and splicing regulatory function. *J Mol Biol* 2005;**348**:575−88.

[66] Cléry A, Blatter M, Allain FH. RNA recognition motifs: boring? Not quite. *Curr Opin Struct Biol* 2008;**18**:290−8.

[67] Lee EB, Lee VMY, Trojanowski JQ. Gains or losses: molecular mechanisms of TDP43-mediated neurodegeneration. *Nat Rev Neurosci* 2012;**13**:38−50.

[68] Buratti E, Brindisi A, Pagani F, Baralle FE. Nuclear factor TDP-43 binds to the polymorphic TG repeats in CFTR intron 8 and causes skipping of exon 9: a functional link with disease penetrance. *Am J Hum Genet* 2004;**74**:13221325.

[69] Mercado PA, Ayala YM, Romano M, Buratti E, Baralle FE. Depletion of TDP 43 overrides the need for exonic and intronic splicing enhancers in the human apoA-II gene. *Nucleic Acids Res* 2005;**33**:6000−10.

[70] Wang HY, Wang IF, Bose J, Shen CK. Structural diversity and functional implications of the eukaryotic TDP gene family. *Genomics* 2004;**83**:130−9.

[71] Ayala YM, Pagani F, Baralle FE. TDP43 depletion rescues aberrant CFTR exon 9 skipping. *FEBS Lett* 2006;**580**:1339−44.

[72] Ou SH, Wu F, Harrich D, Garcia-Martinez LF, Gaynor RB. Cloning and characterization of a novel cellular protein, TDP-43, that binds to human immunodeficiency virus type 1 TAR DNA sequence motifs. *J Virol* 1995;**69**:3584−96.

[73] Tollervey JR, Curk T, Rogelj B, Briese M, Cereda M, Kayikci M, König J, Hortobágyi T, Nishimura AL, Zupunski V, et al. Characterizing the RNA targets and position-dependent splicing regulation by TDP-43. *Nat Neurosci* 2011;**14**:452−8.

[74] Ayala YM, Misteli T, Baralle FE. TDP-43 regulates retinoblastoma protein phosphorylation through the repression of cyclin-dependent kinase 6 expression. *Proc Natl Acad Sci USA* 2008a;**105**:3785–9.

[75] Igaz LM, Kwong LK, Lee EB, Chen-Plotkin A, Swanson E, Unger T, Malunda J, Xu Y, Winton MJ, Trojanowski JQ, et al. Dysregulation of the ALSassociated gene TDP-43 leads to neuronal death and degeneration in mice. *J Clin Investig* 2011;**121**:726–38.

[76] Ayala YM, De Conti L, Avendaño-Vázquez SE, Dhir A, Romano M, D'Ambrogio A, Tollervey J, Ule J, Baralle M, Buratti E, et al. TDP-43 regulates its mRNA levels through a negative feedback loop. *EMBO J* 2011;**30**:277288.

[77] Pesiridis GS, Tripathy K, Tanik S, Trojanowski JQ, Lee VM. A "twohit" hypothesis for inclusion formation by carboxyl-terminal fragments of TDP-43 protein linked to RNA depletion and impaired microtubule-dependent transport. *J Biol Chem* 2011;**286**:18845–55.

[78] Humphrey J, Emmett W, Fratta P, Isaacs AM, Plagnol V. Quantitative analysis of cryptic splicing associated with TDP-43 depletion. *BMC Med Genom* 2017;**10**:38.

[79] Ling JP, Pletnikova O, Troncoso JC, Wong PC. TDP-43 repression of nonconserved cryptic exons is compromised in ALS-FTD. *Science* 2015;**349**:650–5.

[80] Buratti E, De Conti L, Stuani C, Romano M, Baralle M, Baralle F. Nuclear factor TDP-43 can affect selected microRNA levels. *FEBS J* 2010;**277**:2268–81.

[81] King IN, Yartseva V, Salas D, Kumar A, Heidersbach A, Ando DM, Stallings NR, Elliott JL, Srivastava D, Ivey KN. The RNA-binding protein TDP43 selectively disrupts microRNA-1/206 incorporation into the RNA-induced silencing complex. *J Biol Chem* 2014;**289**:14263–71.

[82] Aulas A, Vande Velde C. Alterations in stress granule dynamics driven by TDP-43 and FUS: a link to pathological inclusions in ALS? *Front Cell Neurosci* 2015;**9**:423.

[83] McGurk L, Gomes E, Guo L, Mojsilovic-Petrovic J, Tran V, Kalb RG, Shorter J, Bonini NM. Poly(ADP-Ribose) prevents pathological phase separation of TDP-43 by promoting liquid demixing and stress granule localization. *Mol Cell* 2018;**71**:703–717.e709.

[84] Ash PEA, Vanderweyde TE, Youmans KL, Apicco DJ, Wolozin B. Pathological stress granules in Alzheimer's disease. *Brain Res* 2014;**1584**:52–8.

[85] Shiga A, Ishihara T, Miyashita A, Kuwabara M, Kato T, Watanabe N, Yamahira A, Kondo C, Yokoseki A, Takahashi M, et al. Alteration of POLDIP3 splicing associated with loss of function of TDP-43 in tissues affected with ALS. *PLoS One* 2012;**7**:e43120.

[86] Cairns NJ, Neumann M, Bigio EH, Holm IE, Troost D, Hatanpaa KJ, Foong C, White 3rd CL, Schneider JA, Kretzschmar HA, et al. TDP-43 in familial and sporadic frontotemporal lobar degeneration with ubiquitin inclusions. *Am J Pathol* 2007;**171**:227–40.

[87] Leverenz JB, Yu CE, Montine TJ, Steinbart E, Bekris LM, Zabetian C, Kwong LK, Lee VM, Schellenberg GD, Bird TD. A novel progranulin mutation associated with variable clinical presentation and tau, TDP43 and alphasynuclein pathology. *Brain J Neurol* 2007;**130**:1360–74.

[88] Mackenzie IR, Rademakers R, Neumann M. TDP-43 and FUS in amyotrophic lateral sclerosis and frontotemporal dementia. *Lancet Neurol* 2010;**9**:9951007.

[89] Mishima T, Koga S, Lin WL, Kasanuki K, Castanedes-Casey M, Wszolek ZK, Oh SJ, Tsuboi Y, Dickson DW. Perry syndrome: a distinctive type of TDP-43 proteinopathy. *J Neuropathol Exp Neurol* 2017;**76**:676682.

[90] Tan RH, Yang Y, Kim WS, Dobson-Stone C, Kwok JB, Kiernan MC, Halliday GM. Distinct TDP-43 inclusion morphologies in frontotemporal lobar degeneration with and without amyotrophic lateral sclerosis. *Acta Neuropathol Commun* 2017;**5**:76.

[91] Weihl CC, Pestronk A, Kimonis VE. Valosin-containing protein disease: inclusion body myopathy with Paget's disease of the bone and fronto-temporal dementia. *Neuromuscul Disord* 2009;**19**:308–15.

[92] Tauffenberger A, Chitramuthu BP, Bateman A, Bennett HP, Parker JA. Reduction of polyglutamine toxicity by TDP-43, FUS and progranulin in Huntington's disease models. *Hum Mol Genet* 2013;**22**:782–94.

[93] Geser F, Winton MJ, Kwong LK, Xu Y, Xie SX, Igaz LM, Garruto RM, Perl DP, Galasko D, Lee VM, et al. Pathological TDP-43 in parkinsonismdementia complex and amyotrophic lateral sclerosis of Guam. *Acta Neuropathol* 2008;**115**:133–45.

[94] Chang XL, Tan MS, Tan L, Yu JT. The role of TDP-43 in Alzheimer's disease. *Mol Neurobiol* 2016;**53**:3349–59.

[95] Dickson DW. Parkinson's disease and parkinsonism: neuropathology. *Cold Spring Harb Perspect Med* 2012;**2**.

[96] McAleese KE, Walker L, Erskine D, Thomas AJ, McKeith IG, Attems J. TDP-43 pathology in Alzheimer's disease, dementia with Lewy bodies and ageing. *Brain Pathol* 2017;**27**:472–9.

[97] Hasegawa M, Arai T, Akiyama H, Nonaka T, Mori H, Hashimoto T, Yamazaki M, Oyanagi K. TDP-43 is deposited in the Guam parkinsonism-dementia complex brains. *Brain J Neurol* 2007;**130**:1386–94.

[98] Mackenzie IR, Bigio EH, Ince PG, Geser F, Neumann M, Cairns NJ, Kwong LK, Forman MS, Ravits J, Stewart H, et al. Pathological TDP-43 distinguishes sporadic amyotrophic lateral sclerosis from amyotrophic lateral sclerosis with SOD1 mutations. *Ann Neurol* 2007;**61**:427–34.

[99] Neumann M, Kwong LK, Truax AC, Vanmassenhove B, Kretzschmar HA, Van Deerlin VM, Clark CM, Grossman M, Miller BL, Trojanowski JQ, et al. TDP-43-positive white matter pathology in frontotemporal lobar degeneration with ubiquitin-positive inclusions. *J Neuropathol Exp Neurol* 2007a;**66**:177–83.

[100] Neumann M, Mackenzie IR, Cairns NJ, Boyer PJ, Markesbery WR, Smith CD, Taylor JP, Kretzschmar HA, Kimonis VE, Forman MS. TDP-43 in the ubiquitin pathology of frontotemporal dementia with VCP gene mutations. *J Neuropathol Exp Neurol* 2007b;**66**:152–7.

[101] Seelaar H, Schelhaas HJ, Azmani A, Kusters B, Rosso S, Majoor-Krakauer D, de Rijk MC, Rizzu P, ten Brummelhuis M, van Doorn PA, et al. TDP-43 pathology in familial frontotemporal dementia and motor neuron disease without Progranulin mutations. *Brain J Neurol* 2007;**130**:1375–85.

[102] Tan CF, Eguchi H, Tagawa A, Onodera O, Iwasaki T, Tsujino A, Nishizawa M, Kakita A, Takahashi H. TDP-43 immunoreactivity in neuronal inclusions in familial amyotrophic lateral sclerosis with or without SOD1 gene mutation. *Acta Neuropathol* 2007;**113**:535–42.

[103] Sreedharan J, Blair IP, Tripathi VB, Hu X, Vance C, Rogelj B, Ackerley S, Durnall JC, Williams KL, Buratti E, et al. TDP-43 mutations in familial and sporadic amyotrophic lateral sclerosis. *Science* 2008;**319**:1668–72.

[104] Bodansky A, Kim JM, Tempest L, Velagapudi A, Libby R, Ravits J. TDP-43 and ubiquitinated cytoplasmic aggregates in sporadic ALS are low frequency and widely distributed in the lower motor neuron columns independent of disease spread. *Amyotroph Lateral Scler* 2010;**11**:321–7.

[105] Davidson Y, Kelley T, Mackenzie IR, Pickering-Brown S, Du Plessis D, Neary D, Snowden JS, Mann DM. Ubiquitinated pathological lesions in frontotemporal lobar degeneration contain the TAR DNA-binding protein, TDP-43. *Acta Neuropathol* 2007;**113**:521–33.

[106] Mitra J, Guerrero EN, Hegde PM, Liachko NF, Wang H, Vasquez V, Gao J, Pandey A, Taylor JP, Kraemer BC, et al. Motor neuron disease-associated loss of nuclear TDP-43 is linked to DNA double-strand break repair defects. *Proc Natl Acad Sci USA* 2019;**116**:4696–705.

[107] Roczniak-Ferguson A, Ferguson SM. Pleiotropic requirements for human TDP-43 in the regulation of cell and organelle homeostasis. *Life Sci Alliance* 2019;**2**.

[108] Buratti E, Baralle FE. TDP-43: gumming up neurons through protein and protein-RNA interactions. *Trends Biochem Sci* 2012;**37**:237–47.

[109] Hergesheimer RC, Chami AA, de Assis DR, Vourc'h P, Andres CR, Corcia P, Lanznaster D, Blasco H. The debated toxic role of aggregated TDP-43 in amyotrophic lateral sclerosis: a resolution in sight? *Brain J Neurol* 2019;**142**:1176–94.

[110] Khalfallah Y, Kuta R, Grasmuck C, Prat A, Durham HD, Vande Velde C. TDP-43 regulation of stress granule dynamics in neurodegenerative disease-relevant cell types. *Sci Rep* 2018;**8**:7551.

[111] Ebstein SY, Yagudayeva I, Shneider NA. Mutant TDP-43 causes early-stage dose-dependent motor neuron degeneration in a TARDBP Knockin mouse model of ALS. *Cell Reports* 2019;**26**:364–373.e364.

[112] White MA, Kim E, Duffy A, Adalbert R, Phillips BU, Peters OM, Stephenson J, Yang S, Massenzio F, Lin Z, et al. TDP-43 gains function due to perturbed autoregulation in a Tardbp knock-in mouse model of ALS-FTD. *Nat Neurosci* 2018;**21**:552–63.

[113] Fratta P, Sivakumar P, Humphrey J, Lo K, Ricketts T, Oliveira H, Brito-Armas JM, Kalmar B, Ule A, Yu Y, et al. Mice with endogenous TDP-43 mutations exhibit gain of splicing function and characteristics of amyotrophic lateral sclerosis. *EMBO J* 2018;**37**.

[114] Josephs KA, Whitwell JL, Knopman DS, Hu WT, Stroh DA, Baker M, Rademakers R, Boeve BF, Parisi JE, Smith GE, et al. Abnormal TDP-43 immunoreactivity in AD modifies clinicopathologic and radiologic phenotype. *Neurology* 2008;**70**:1850–7.

[115] Kadokura A, Yamazaki T, Lemere CA, Takatama M, Okamoto K. Regional distribution of TDP-43 inclusions in Alzheimer disease (AD) brains: their relation to AD common pathology. *Neuropathology* 2009;**29**:566–73.

[116] King A, Sweeney F, Bodi I, Troakes C, Maekawa S, Al-Sarraj S. Abnormal TDP-43 expression is identified in the neocortex in cases of dementia pugilistica, but is mainly confined to the limbic system when identified in high and moderate stages of Alzheimer's disease. *Neuropathology* 2010;**30**:408–19.

[117] Lippa CF, Rosso AL, Stutzbach LD, Neumann M, Lee VM, Trojanowski JQ. Transactive response DNA-binding protein 43 burden in familial Alzheimer disease and Down syndrome. *Arch Neurol* 2009;**66**:1483–8.

[118] Hu WT, Josephs KA, Knopman DS, Boeve BF, Dickson DW, Petersen RC, Parisi JE. Temporal lobar predominance of TDP-43 neuronal cytoplasmic inclusions in Alzheimer disease. *Acta Neuropathol* 2008;**116**:215–20.

[119] Josephs KA, Murray ME, Whitwell JL, Tosakulwong N, Weigand SD, Petrucelli L, Liesinger AM, Petersen RC, Parisi JE, Dickson DW. Updated TDP-43 in Alzheimer's disease staging scheme. *Acta Neuropathol* 2016;**131**:571–85.

[120] Nag S, Yu L, Boyle PA, Leurgans SE, Bennett DA, Schneider JA. TDP-43 pathology in anterior temporal pole cortex in aging and Alzheimer's disease. *Acta Neuropathol Commun* 2018;**6**:33.

[121] Wilson AC, Dugger BN, Dickson DW, Wang DS. TDP-43 in aging and Alzheimer's disease - a review. *Int J Clin Exp Pathol* 2011;**4**:147–55.

[122] Yokota O, Davidson Y, Arai T, Hasegawa M, Akiyama H, Ishizu H, Terada S, Sikkink S, Pickering-Brown S, Mann DM. Effect of topographical distribution of alpha-synuclein pathology on TDP-43 accumulation in Lewy body disease. *Acta Neuropathol* 2010;**120**:789–801.

[123] Duyckaerts C, Delatour B, Potier MC. Classification and basic pathology of Alzheimer disease. *Acta Neuropathol* 2009;**118**:5–36.

[124] Josephs KA, Murray ME, Tosakulwong N, Weigand SD, Knopman DS, Petersen RC, Jack Jr CR, Whitwell JL, Dickson DW. Brain atrophy in primary age-related tauopathy is linked to transactive response DNA-binding protein of 43 kDa. *Alzheimers Dement* 2019;**15**:799806.

[125] Josephs KA, Murray ME, Tosakulwong N, Whitwell JL, Knopman DS, Machulda MM, Weigand SD, Boeve BF, Kantarci K, Petrucelli L, et al. Tau aggregation influences cognition and hippocampal atrophy in the absence of betaamyloid: a clinico-imaging-pathological study of primary age-related tauopathy (PART). *Acta Neuropathol* 2017;**133**:705–15.

[126] Arai T, Mackenzie IR, Hasegawa M, Nonoka T, Niizato K, Tsuchiya K, Iritani S, Onaya M, Akiyama H. Phosphorylated TDP-43 in Alzheimer's disease and dementia with Lewy bodies. *Acta Neuropathol* 2009;**117**:125–36.

[127] Josephs KA, Murray ME, Whitwell JL, Parisi JE, Petrucelli L, Jack CR, Petersen RC, Dickson DW. Staging TDP-43 pathology in Alzheimer's disease. *Acta Neuropathol* 2014a;**127**:441–50.

[128] Lin WL, Castanedes-Casey M, Dickson DW. Transactivation response DNA-binding protein 43 microvasculopathy in frontotemporal degeneration and familial Lewy body disease. *J Neuropathol Exp Neurol* 2009;**68**:11671176.

[129] Josephs KA, Whitwell JL, Weigand SD, Murray ME, Tosakulwong N, Liesinger AM, Petrucelli L, Senjem ML, Knopman DS, Boeve BF, et al. TDP-43 is a key player in the clinical features associated with Alzheimer's disease. *Acta Neuropathol* 2014b;**127**:811–24.

[130] McKee AC, Gavett BE, Stern RA, Nowinski CJ, Cantu RC, Kowall NW, Perl DP, Hedley-Whyte ET, Price B, Sullivan C, et al. TDP-43 proteinopathy and motor neuron disease in chronic traumatic encephalopathy. *J Neuropathol Exp Neurol* 2010;**69**:918–29.

131. Tan XL, Sun M, Brady RD, Liu S, Llanos R, Cheung S, et al. Transactive response DNA-binding protein 43 abnormalities after traumatic brain injury. *J Neurotrauma* 2018;**36**(1). https://doi.org/10.1089/neu.2017.5491. https://www.liebertpub.com/doi/abs/10.1089/neu.2017.5491.

[132] McKee AC, Stern RA, Nowinski CJ, Stein TD, Alvarez VE, Daneshvar DH, Lee HS, Wojtowicz SM, Hall G, Baugh CM, et al. The spectrum of disease in chronic traumatic encephalopathy. *Brain J Neurol* 2013;**136**:43–64.

[133] McKee AC, Stein TD, Kiernan PT, Alvarez VE. The neuropathology of chronic traumatic encephalopathy. *Brain Pathol* 2015;**25**:350–64.

[134] Wang HK, Lee YC, Huang CY, Liliang PC, Lu K, Chen HJ, Li YC, Tsai KJ. Traumatic brain injury causes frontotemporal dementia and TDP-43 proteolysis. *Neuroscience* 2015a;**300**:94–103.

[135] Wright DK, Liu S, van der Poel C, McDonald SJ, Brady RD, Taylor L, Yang L, Gardner AJ, Ordidge R, O'Brien TJ, et al. Traumatic brain injury results in cellular, structural and functional changes Resembling motor neuron disease. *Cerebr Cortex* 2017;**27**:4503–15.

[136] Huang CY, Lee YC, Li PC, Liliang PC, Lu K, Wang KW, Chang LC, Shiu LY, Chen MF, Sun YT, et al. TDP-43 proteolysis is associated with astrocyte reactivity after traumatic brain injury in rodents. *J Neuroimmunol* 2017;**313**:61–8.

[137] Zhang J, Teng Z, Song Y, Hu M, Chen C. Inhibition of monoacylglycerol lipase prevents chronic traumatic encephalopathy-like neuropathology in a mouse model of repetitive mild closed head injury. *J Cerebr Blood Flow Metabol* 2015;**35**:706.

[138] Heyburn L, Abutarboush R, Goodrich S, Urioste R, Batuure A, Statz J, Wilder D, Ahlers ST, Long JB, Sajja V. Repeated low-level blast overpressure leads to endovascular disruption and alterations in TDP-43 and Piezo2 in a rat model of blast TBI. *Front Neurol* 2019;**10**:766.

[139] Yang Z, Lin F, Robertson CS, Wang KK. Dual vulnerability of TDP-43 to calpain and caspase-3 proteolysis after neurotoxic conditions and traumatic brain injury. *J Cerebr Blood Flow Metabol* 2014;**34**:1444–52.

[140] Zinman L, Cudkowicz M. Emerging targets and treatments in amyotrophic lateral sclerosis. *Lancet Neurol* 2011;**10**:481–90.

[141] Rachelle S, Martha F, Carolin H, Amanda H, Ione W, Katrina D, Jason W, Jonathan S, Henrik Z, Jonathan R. THUR 037 Identification of biofluid markers of TDP-43 pathology. *J Neurol Neurosurg Psychiatr* 2018;**89**:A5–6.

[142] Foulds P, McAuley E, Gibbons L, Davidson Y, Pickering-Brown SM, Neary D, Snowden JS, Allsop D, Mann DM. TDP-43 protein in plasma may index TDP-43 brain pathology in Alzheimer's disease and frontotemporal lobar degeneration. *Acta Neuropathol* 2008;**116**:141–6.

[143] Kasai T, Tokuda T, Ishigami N, Sasayama H, Foulds P, Mitchell DJ, Mann DM, Allsop D, Nakagawa M. Increased TDP-43 protein in cerebrospinal fluid of patients with amyotrophic lateral sclerosis. *Acta Neuropathol* 2009;**117**:55–62.

[144] Steinacker P, Hendrich C, Sperfeld AD, Jesse S, von Arnim CA, Lehnert S, Pabst A, Uttner I, Tumani H, Lee VM, et al. TDP-43 in cerebrospinal fluid of patients with frontotemporal lobar degeneration and amyotrophic lateral sclerosis. *Arch Neurol* 2008;**65**:1481–7.

[145] Ash PEA, Stanford EA, Al Abdulatif A, Ramirez-Cardenas A, Ballance HI, Boudeau S, Jeh A, Murithi JM, Tripodis Y, Murphy GJ, et al. Dioxins and related environmental contaminants increase TDP-43 levels. *Mol Neurodegener* 2017;**12**:35.

[146] Cristobo I, Larriba MJ, Ríos Vdl, García F, Muñoz A, Casal JI. Proteomic analysis of 1α,25-Dihydroxyvitamin D3 action on human colon cancer cells reveals a link to splicing regulation. *J Proteom* 2011;**75**:384–97.

[147] Alquezar C, Esteras N, Encarnación Adl, Moreno F, de Munain AL, MartínRequero Á. Increasing progranulin levels and blockade of the ERK1/2 pathway: Upstream and downstream strategies for the treatment of progranulin deficient frontotemporal dementia. *Eur Neuropsychopharmacol* 2015;**25**:386–403.

[148] Kreiter N, Pal A, Lojewski X, Corcia P, Naujock M, Reinhardt P, Sterneckert J, Petri S, Wegner F, Storch A, et al. Age-dependent neurodegeneration and organelle transport deficiencies in mutant TDP43 patient-derived neurons are independent of TDP43 aggregation. *Neurobiol Dis* 2018;**115**:167–81.

[149] Oberstadt M, Stieler J, Simpong DL, Römuß U, Urban N, Schaefer M, Arendt T, Holzer M. TDP-43 self interaction is modulated by redox-active compounds Auranofin, Chelerythrine and Riluzole. *Sci Rep* 2018;**8**:2248.

[150] Khosravi B, LaClair KD, Riemenschneider H, Zhou Q, Frottin F, Mareljic N, Czuppa M, Farny D, Hartmann H, Michaelsen M, et al. Cell-to-cell transmission of C9orf72 poly-(Gly-Ala) triggers key features of ALS/FTD. *EMBO J* 2020;**39**:e102811.

[151] Wang X, Ma M, Teng J, Che X, Zhang W, Feng S, Zhou S, Zhang Y, Wu E, Ding X. Valproate attenuates 25-kDa C-terminal fragment of TDP-43Induced neuronal toxicity via suppressing endoplasmic reticulum stress and activating autophagy. *Int J Biol Sci* 2015b;**11**:752–61.

[152] Brettschneider J, Del Tredici K, Toledo JB, Robinson JL, Irwin DJ, Grossman M, Suh E, Van Deerlin VM, Wood EM, Baek Y, et al. Stages of pTDP-43 pathology in amyotrophic lateral sclerosis. *Ann Neurol* 2013;**74**:20–38.

[153] Benajiba L, Le Ber I, Camuzat A, Lacoste M, Thomas-Anterion C, Couratier P, Legallic S, Salachas F, Hannequin D, Decousus M, et al. TARDBP mutations in motoneuron disease with frontotemporal lobar degeneration. *Ann Neurol* 2009;**65**:470–3.

[154] Mackenzie IR, Neumann M, Bigio EH, Cairns NJ, Alafuzoff I, Kril J, Kovacs GG, Ghetti B, Halliday G, Holm IE, et al. Nomenclature for neuropathologic subtypes of frontotemporal lobar degeneration: consensus recommendations. *Acta Neuropathol* 2009;**117**:15–8.

[155] Brettschneider J, Del Tredici K, Irwin DJ, Grossman M, Robinson JL, Toledo JB, Fang L, Van Deerlin VM, Ludolph AC, Lee VM, et al. Sequential distribution of pTDP-43 pathology in behavioral variant frontotemporal dementia (bvFTD). *Acta Neuropathol* 2014;**127**:423–39.

[156] Nonaka T, Hasegawa M. Prion-like properties of assembled TDP-43. *Curr Opin Neurobiol* 2019;**61**:23–8.

[157] Braak H, Brettschneider J, Ludolph AC, Lee VM, Trojanowski JQ, Del Tredici K. Amyotrophic lateral sclerosis–a model of corticofugal axonal spread. *Nat Rev Neurol* 2013;**9**:708–14.

[158] Van Damme P, Martens L, Van Damme J, Hugelier K, Staes A, Vandekerckhove J, Gevaert K. Caspase-specific and nonspecific in vivo protein processing during Fas-induced apoptosis. *Nat Methods* 2005;**2**:771–7.

[159] Gregory RI, Yan KP, Amuthan G, Chendrimada T, Doratotaj B, Cooch N, Shiekhattar R. The Microprocessor complex mediates the genesis of microRNAs. *Nature* 2004;**432**:235–40.

[160] Shivdasani RA. MicroRNAs: regulators of gene expression and cell differentiation. *Blood* 2006;**108**:3646–53.

[161] Licatalosi DD, Mele A, Fak JJ, Ule J, Kayikci M, Chi SW, Clark TA, Schweitzer AC, Blume JE, Wang X, et al. HITS-CLIP yields genome-wide insights into brain alternative RNA processing. *Nature* 2008;**456**:464–9.

[162] Autophagy and Its Impact on Neurodegenerative Diseases: New Roles for TDP-43 and C9orf72. *Front Mol Neurosci* 2017;**10**(170). https://doi.org/10.3389/fnmol.2017.00170.

Chapter 2

Structural dissection of TDP-43: insights into physiological and pathological self-assembly

Vijay Kumar
Amity Institute of Neuropsychology & Neurosciences, Amity University, Noida, Uttar Pradesh, India

Introduction

In 2006, TAR DNA-binding protein of 43 kDa (TDP-43) was first shown to be present in the intraneuronal inclusion bodies of patients with amyotrophic lateral sclerosis (ALS) and frontotemporal lobar degeneration (FTLD).[1,2] TDP-43-positive cytoplasmic inclusions have been found in 97% ALS and 45% FTLD patients.[3−5] Simultaneously, it has been also reported to be associated with other neurodegenerative diseases like Alzeimers Disease (AD), Parkinsons Disease (PD), Huntington Disease (HD), and limbic-predominant age-related TDP-43 encephalopathy[6−12] (commonly called as TDP-43 proteinopathies). ALS and FTLD is an adult-onset incurable neurodegenerative disease with symptoms included motor and cognitive defects.[13−15] These diseases display characteristic polyubiquitinated and hyperphosphorylated cytoplasmic inclusions with misfolded aggregates of TDP-43.[16−18] In ALS, TDP-43 aggregates are present in the upper and lower motor neurons of the brain, mainly brainstem, and spinal cord as well as in the frontal cortex and hippocampus in some of the patients.[1,19] In FTLD patients, the aggregates are predominantly present in the frontal and temporal cortex and hippocampus as well as in the brainstem and spinal cord motoneurons.[2,20]

TDP-43 is ubiquitously expressed protein[21] and belongs to the classic heterogeneous ribonucleoprotein (hnRNP) family.[22] Structurally, TDP-43 contains an N-terminal domain (NTD, residues: 1−76) with a distinct fold and has been shown to form a dimer or a higher-order oligomer.[5,23−25] TDP-43 has two highly conserved RNA-binding domains (RRM1, residues: 106−176 and RRM2, residues: 191−259) required for binding sequence-specific RNA and DNA.[26−28] The C-terminal domain (CTD, residues: 274−414) is highly

TDP-43 and Neurodegeneration. https://doi.org/10.1016/B978-0-12-820066-7.00001-1

unstructured and harbors the majority of ALS-causing mutations, contains a glycine-rich prion-like region or low complexity domain.[29] It is largely involved in mediating protein–protein interactions[30] and formation of stress granules (SGs),[31] potentially via its ability to phase separate.[32,33] Besides, a nuclear localization signal (NLS, residues: 82–98) and a nuclear export signal (NES, residues: 239–250) are present.[34] Fig. 2.1 depicts the structural organization of full-length TDP-43.

TDP-43 plays key roles in preserving cellular function and survival and involve in myriad of functions including micro-RNA processing,[35] RNA transcription, translation, splicing, and stabilization;[36–38] cell division,[39] apoptosis,[40] axonal transport;[41] and embryo development.[42] Similar to other RNA-binding proteins, TDP-43 has a beads on-a string architecture comprising various independent structural and functional domains. Despite being a relatively small protein (43 kDa), three-dimensional structure determination of full-length TDP-43 protein is extremely difficult because of the presence of a significant portion of intrinsically disordered regions (IDRs). TDP-43 is estimated to contain 15%–30% of IDRs in the full protein and almost 70% of IDRs in the C-terminal portion of the protein.[43] However, several three-dimensional structures of the NTD, RNA recognition motifs (RRMs), and short fragments of the CTD have become available in recent years. A total of 34 structures of TDP-43 has been deposited in Protein Data Bank (PDB) till July 2020 (Table 2.1).

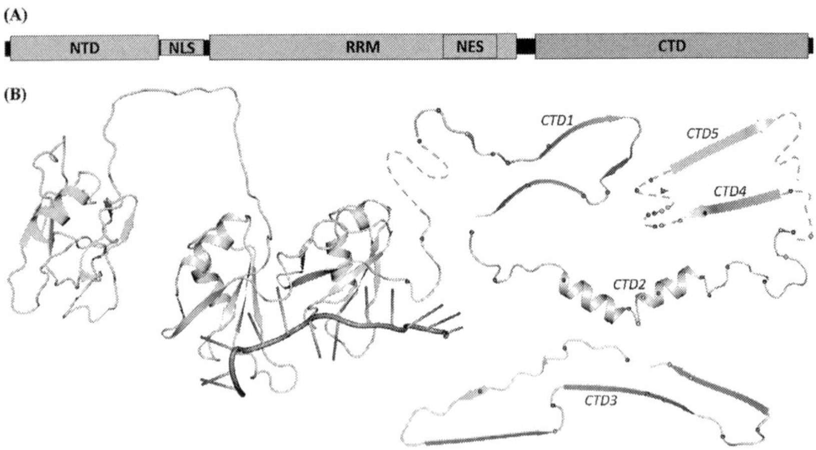

FIGURE 2.1 TDP-43 sequence and structure representation. (A) TDP-43 domain map with relative lengths of the N-terminal domain (NTD, turquoise), RNA-recognition motifs (light blue), and the C-terminal domain (gray). (B) Representative three-dimensional structures of TDP-43 regions. NTD (PDB id: 5MRG), RRM1-RRM2 with RNA (PDB id: 4BS2), CTD segments: CTD1 (PDB id: 6N3C), CTD2 (PDB id: 2N3X), CTD3 (PDB id: 6N3A), CTD4 (PDB id: 5WIA), CTD5 (PDB id: 5WIQ). *Adapted and modified from François-Moutal L, et al. Structural insights into TDP-43 and effects of post-translational modifications.* Front Mol Neurosci *2019;**12**(301).*

TABLE 2.1 Structures of TDP-43 available in Protein Data Bank (PDB) database.

Domain	Residues	PDB	Experiment	State	References
NTD	1–77	2N4P	NMR	Monomer	[51]
	2–79	5MDI	X-ray (2.1 Å)	Dimer	[45]
	1–102	5MRG	NMR	Monomer	[97]
	1–77	5X4F	NMR	Monomer	[98]
	1–80	6B1G	NMR	Dimer	[52]
	1–80	6T4B	X-ray (2.55 Å)	Dimer	[53]
RRM1	96–185	2CQG	NMR	Monomer	
	103–179	4IUF	X-ray (2.75 Å)	Monomer	[99]
	103–168	4Y00	X-ray (3.00 Å)	Monomer	[67]
	103–180	4Y0F	X-ray (2.65 Å)	Monomer	[67]
RRM2	193–267	1WF0	NMR	Monomer	
	192–261	3D2W	X-ray (1.65 Å)	Homo-2-mer	[54]
tRRM	102–269	4BS2	NMR	Monomer	[26]
RRM2 peptides	248–253	5W50	X-ray (1.4 Å)	Homo-12-mer	[73]
	247–257	5W52	Electron crystallography (1.4 Å)	Homo-10-mer	
	247–257	5W7V	Electron microscopy (3.8 Å)	Triple-helical fibril	

Continued

TABLE 2.1 Structures of TDP-43 available in Protein Data Bank (PDB) database.—cont'd

Domain	Residues	PDB	Experiment	State	References
CTD	307–348	2N2C	NMR	Monomer	
	311–360	2N3X	NMR	Monomer	[81]
	311–360	2N4G	NMR	Monomer	
	311–360	2N4H	NMR	Monomer	
	312–317	5WHN	X-ray (1.1 Å)	Homo-15-mer	[85]
	370–375	5WHP	X-ray (1.0 Å)	Homo-10-mer	
	396–402	5WIA	X-ray (1.0 Å)	Homo-10-mer	
	312–317	5WIQ	X-ray (1.25 Å)	Homo-10-mer	
	300–306	5WKD	X-ray (1.8 Å)	Homo-10-mer	
	328–333	6CB9	X-ray (1.1 Å)	Homo-10-mer	
	321–326	6CEW	X-ray (1.2 Å)	Homo-12-mer	
	312–317	5WKB	Electron crystallography (1.0 Å)	Homo-10-mer	
	312–317	6CF4	Electron crystallography (0.75 Å)	Homo-10-mer	
	333–343	6CFH	Electron crystallography (1.5 Å)	Homo-20-mer	
	311–360	6N3A	Electron microscopy (3.3 Å)	Homo-10-mer	[86]
	312–352	6N3B	Electron microscopy (3.8 Å)	Homo-10-mer	
	288–314	6N3C	Electron microscopy (3.3 Å)	Homo-20-mer	
	312–347	6N37	Electron microscopy (3.8 Å)	Homo-10-mer	

Thus, given the involvement of TDP-43 to the multiple physiological functions and its importance in neurodegeneration, we here in this chapter elucidate the structure and dynamics of TDP-43 in quest of understanding its physiological and pathophysiological functions for developing therapeutic interventions.

Structure overview of the N-terminal domain

NTD assembles into several forms ranging from dimer to higher-order oligomers both in vitro and in vivo. Recently, it has been shown that NTD mediates reversible oligomerization of TDP-43 leading to the formation of physiological TDP-43 oligomers[44–46] involved in splicing activity,[47] and pathological oligomerization through liquid–liquid phase separation (LLPS) and SG formation.[48–50] To date, six NTD structures, including three monomeric and three dimeric forms, have been submitted to the PDB (Table 2.1).

The monomeric NTD folds into a compact structure which consists of five β-sheets (β2-β1-β5-β3-β4) and single α-helix, with an additional β-hairpin present between β4- and β5-strands, similar to ubiquitin and Dix domains[51] (Fig. 2.2). The dimeric NTD structures showed a head-to-tail configuration where negatively charged tail of each domain is packed against the positively charged head region.[45,52,53] As shown by Afroz et al.,[45] the positively charged head region are made by residues Asn45 (β3), Ser48 (β3-β4 loop), Arg52, Arg55 (β4), and Asn76 (β5). Similarly, the negatively charged tail region contains the residues Glu17 (β2), Gln34 (α1), and the loop residues Glu14 and Glu21. Both positively charged head region interacts with negatively charged tail region and results in charge complementation and is one of the prominent factors in stabilization of dimeric NTD. Besides, H-bonds formed by Glu17 (β2), Met51, and Gly53 (β4) along with Asn45 and Ser48 contribute to dimer formation. This interaction is further strengthened by salt-bridge interaction

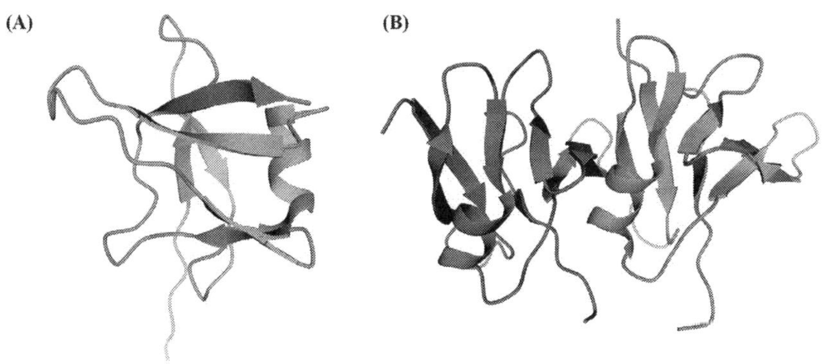

FIGURE 2.2 TDP-43's NTD structure. NMR structure of (A) monomeric NTD (PDB id: 2N4P), (B) dimeric NTD (PDB id: 6B1G).

formed by Arg55-Glu21and Arg52-Glu17. The hydrophobic interactions at the interface are formed by met51, Arg52, and Arg55 (Molecule 1) and Ile16 and Pro19 (Molecule 2). Finally, Gln34 (α-helix) of Molecule 1 forms H-bonds with Leu 41 (β3) and Asn76 (β5) of Molecule 2.

Though these three structures are very much similar to each other, the dimeric interfaces between the molecules differ slightly, likely due to crystallization conditions. As reported by Wright et al. in their recent structure,[53] a sulfate ion binding to Arg52 and Arg55 binding at the interface increases the distance between the monomers. The sulfation binding results in the rearrangement of side chains of Arg52 and Arg55 which weakens the salt-bridge binding between Arg52-Glu34 and Arg55-Glu21 reported in earlier structure.[45] Interestingly, these structures suggest that NTD alone can form a dynamic solenoid structure, but the significance of full-length TDP-43 polymerization is still unknown.

Structure overview of the RNA recognition motifs

TDP-43 RRMs cover regions of amino acids ranging from residues 106−177 (RRM1) to residues 192−259 (RRM2). These RRMs contain two highly conserved segments denoted as ribonucleoprotein (RNP) domain, RNP1 and RNP2, consisting of eight and six amino acids, respectively.[54] RRM1 has ^{X}KGFGFVRF^{Y} as RNP1 and ^{X}LIVLGL^{Y} as RNP2. While RRM2 has ^{X}RAFAFVTF^{Y} as RNP1 and ^{X}VFVGRC^{Y} as RNP2 required for nucleic acid binding. Till now, four three-dimensional structures of RRM1, two three-dimensional structures of RRM2, and one three-dimensional structure of the tandem RRMs are available in PDB (Table 2.1). Both RRM1 and RRM2 of TDP-43 adopt a canonical RRM fold, i.e., a five-stranded β-sheet stacked against two α-helices (β1α1β2β3α2β4) and an extra β-hairpin (β3′β3″) present between α2 and β4 (Fig. 2.3). The conserved RNP1 and RNP2 segments are located in β1 and β3 and form stacking interactions with the bases and sugar rings of single-stranded RNA and DNA.[26] The structure of (AUG)12

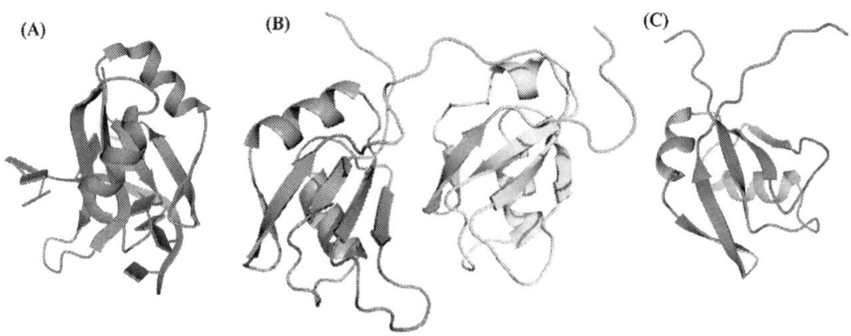

FIGURE 2.3 TDP-43's RRM structure. (A) RRM1structure with RNA (PDB id: 4IUF), (B) Tethered RRM structure (PDB id:4BS2) with RRM1 colored in orange and RRM2 in yellow. (C) RRM2 structure (PDB id: 1WF0).

RNA binds to tandem RRMs revealed that the binding region includes the canonical binding site on the β-sheet surfaces of RRMs, including both β-hairpins as well as the linker between the RRMs and C-terminal of RRM2.[26] Surprisingly, the TDP-43 tandem RRMs bind RNA in the 5′ to 3′ direction from RRM1 to RRM2, as opposite to typical tandem RRMs of the proteins including Sex-lethal,[55] PABP,[56] HuD,[57] Nucleolin,[58] Hrp1,[59] and U2AF65.[60] The other unique difference seen in RRMs is that these typical tandem RRMs bring the C-terminal β2-strand of RRM2 in closer proximity to the β4-strand of the N-terminal RRM1 (β4 β2 type of arrangement), and thus the partially helical linker (11–14 residues) bridge only two β-strands and thus adopts a compact conformation. However, in TDP-43 tandem RRMs, the orientation of RRM are reversed (β2 β4 type) and thus the 14 amino acids linker needs to bridge four β-strands instead of two and thus forms an extended conformation.[61]

TDP-43 preferentially binds to (TG/UG) sequences as demonstrated by many studies.[54,62–64] RRM1 is known to be sufficient and essential for RNA binding with nanomolecular range, while RRM2 is essential for providing the stability.[54,64–66]

An ALS-linked mutation in the RRM1 domain, D169G, does not affect RNA binding, though it results in a loss of H-bonding with Thr115 that leads to the conformational change of TDP-43. The crystal structure of D169G RRM1 in complex with single-stranded DNA, 5′-GTTGAGCGTT-3′, indicates a minor difference in the β-turn containing the D169G mutation.[67]

Amyloidogenic core

Interestingly, the residues 247–257 (DLIIKGISVHI) in RRM2 have the potential to form amyloidogenic fibrils[68,69] and are surface exposed in the partially folded intermediates of RRM2 conformation,[66,70,71] responsible for misfolding and aggregation of TDP-43.[61,72] A recent structural study of this peptide demonstrated the presence of amyloid polymorphism characterized by different backbone conformations and five different classes of steric zippers.[73] The Cryo-EM structures of the peptide demonstrated eight different classes of fibrils including cylinders, twists, sheets, and more predominant three-start helix. The three-start helix structure appeared with a left-handed twist composed of 27 parallel β-sheets, with 9 β-sheets per asymmetric unit, revealing a threefold screw symmetry. The microelectron diffraction (MicroED) structure of this amyloidogenic peptide revealed a parallel, face-to-back class 2 steric zipper assembly (Fig. 2.4).

It has been shown that both the truncated RRM2 and peptides from the β3- and β5-strands of RRM2 can assemble into amyloid fibrils. Several studies along with ours indicate that RRM2 samples several partially folded intermediate states.[65,66,70,71] These potential unfolding intermediates showed the characteristic features implicated in misfolding and aggregation of the protein.

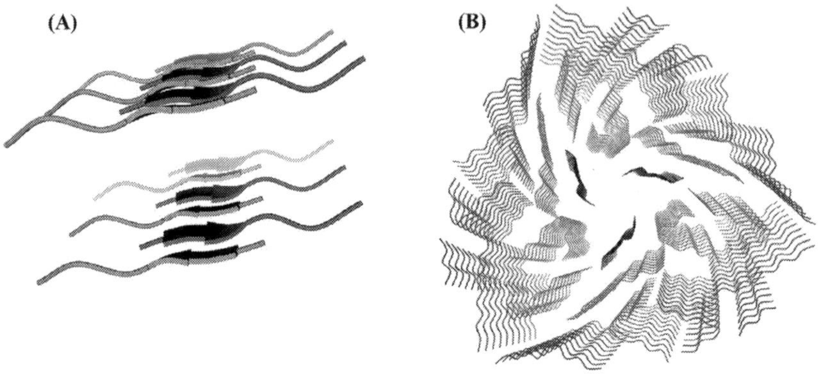

FIGURE 2.4 Atomic structures of RRM2 amyloidogenic peptides. (A) X-ray structure of ^{248}LIIKGI^{253}. (B) Cryo-EM structure of the ^{247}DLIIKGISVHI^{257} fibril.

These include decreased β-structural content, decreased intramolecular Ile, Leu, and Val contacts within the β-strands, and increased solvent exposure of aggregation-prone residues. These results thus clearly suggest that the RRM2 misfolding leads to the alteration of native TDP-43 into pathological aggregates responsible for neurodegenerative diseases.

Structure overview of the C-terminal domain

TDP-43 CTD is the most unstructured domain and harbors about 50 disease-linked mutations.[74] Many studies have suggested the involvement of this region in the aberrant misfolding and aggregation of TDP-43.[33,75–78] This domain has been studied widely, but due to its highly disordered nature, the structural studies of this domain are mainly done in fragments of CTD. The CTD was initially suggested to behave as a prion-like domain and required for aggregation.[29] However, further studies have indicated that the smaller fragments of CTD are sufficient for aggregation and neurodegenerative studies. Guo et al.[79] showed amyloid fibrils formation from 286 to 331 which is neurotoxic to primary cortical neurons. Another study identified that the residues 311–360 undergo a helix to β-sheet transition which is responsible for aggregation and cytoplasmic inclusions.[80,81] Also, Mompean et al.[82] demonstrated that segments 342–366 showed transitions from a random coil to β-sheet and thus act as a minimum toxic core residue present in TDP-43. These studies illustrated that there are several short segments of CTD responsible for amyloid fibrils formation and aggregation. These segments or peptides of CTD are shown to form stable secondary structures that are observed through crystallography, electron microscopy, and MicroED (Table 2.1). CTD forms two Gly/aromatic/Ser (GaroS)-rich regions and an amyloidogenic core consisting of a hydrophobic region and Q/N-rich region.[83]

The GaroS regions (GaroS1: 273–317 and GaroS2: 368–414) have been shown to interact with RNA granules, undergo LLPS, and contribute to the formation of SGs.[32] The structure of the hydrophobic region (residues 318–340) adopts an omega-loop helical structure (Ω-loop-helix) in a lipid-like environment as shown by Lim et al.[84] The helical nature of this hydrophobic region was further confirmed by Jiang et al.,[81] where they showed a helix-turn-helix structure in a nonlipid environment which gets disrupted by ALS-causing mutations, G335D and Q343R. This hydrophobic region can also adopt cross-β sheet architecture as shown by Mompean et al.[83] and Jiang et al.[80]

A recent study by Guenther et al.[85] showed that segments of CTD can form a steric-zipper β sheet and a kinked, labile structure, called as LARKS (low-complexity aromatic-rich kinked segments). These LARKS-forming segments stacked into kinked β-sheets and form reversible associations (Velcro-like) between CTD to form SGs or other membrane-less organelles. The authors in this study revealed crystal structures of 10 peptides of CTD through X-ray diffraction and MicroED (Fig. 2.5). They showed that six of the peptides including ^{300}GNNQGSN^{306}, ^{321}AMMAAA^{326}, ^{328}AALQSS^{333}, ^{333}SWGMMGMLASQ^{343}, ^{370}GNNSYS^{375}, and ^{396}GFNGGFG^{402} form steric zippers consisting in-register β-sheets. Out of these six segments, four segments (300–306, 328–333, 370–375, and 396–402) form parallel β-sheets, while the other two segments (321–328 and 333–343) form antiparallel sheets. Interestingly, the crystal structures of the segment of 312–317 of the CTD revealed a kinked, in-register, parallel sheets. The kink occurs at Phe 316 which has a left-handed α-helix Ramachandran angle value. The kinked backbone makes the two phenylalanines of the peptide facing in the same side of the sheet and finally creating two different interfaces.

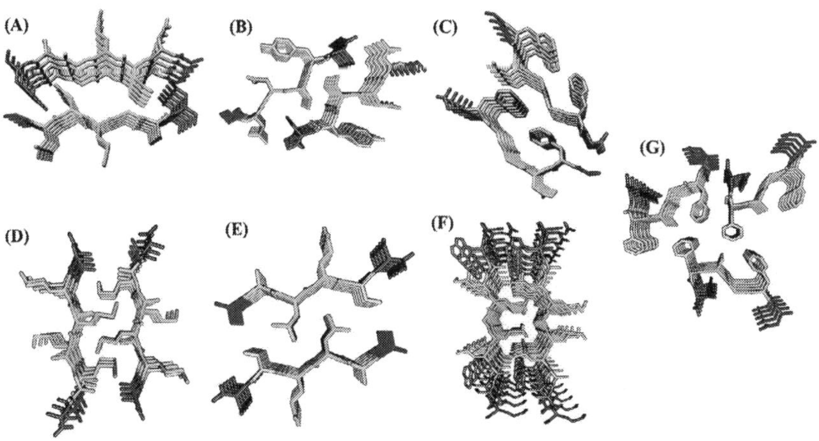

FIGURE 2.5 Structures of the low complexity domain (LCD) of CTD. Six steric-zipper structures are shown for the peptides, (A) ^{300}GNNQGSN^{306} (PDB id: 5WKD), (B) ^{370}GNNSYS^{375} (PDB id: 5WIA), (C) ^{396}GFNGGFG^{402} (PDB id: 5WIQ), (D) ^{321}AMMAAA^{326} (PDB id: 6CEW), (E) ^{328}AALQSS^{333} (PDB id: 6CB9), and (F) ^{333}SWGMMGMLASQ^{343} (PDB id: 6CFH). (G) The ^{312}NFGAFS^{317} (PDB id: 5WHN) segments form a kinked β-sheet.

The intrinsically disordered CTD contains two evolutionarily conserved segments, a hydrophobic region (321–340) and a Q/N-rich region (341–363), suggesting their functional importance. Very recently, Cao et al.[86] showed the first cryo-EM structures of these conserved regions named as "SegA." The structure analysis revealed the amyloid polymorphism of the segment SegA. The authors showed the presence of three polymorphs (as sym, asym, and slow) in which the hydrophobic region adopts a primarily identical "dagger-shaped" fold (Fig. 2.6). Contrastingly, the adjoining Q/N-rich region adopts many different conformations, where the 341–347 regions predominantly being exposed to the solvent in the two polymorphic structures considered as "pathologically relevant" (SegA-sym and SegA-asym) (Fig. 2.6A and B). This is different from the structural model proposed by Mompean and coworkers, who showed that TDP-43 residues 341–357 form β-hairpins and assemble with a β-turn configuration to form amyloid.[82,87]

The authors also show the structure of another polymorph, SegA-slow which was quite different from two of the polymorphs. In this polymorph, the Q/N-rich region adopts β-hairpin conformation where residues 341–348 are always buried in the fibril core (Fig. 2.6C). The β-hairpin in this model showed a β-arc configuration, where two of the β-strands form two different β-sheets. However, Mompean et al.[87] demonstrated that the β-hairpins are in the β-turn configuration, where the two β-strands make a single β-sheet. Therefore, hairpins with β-arc are predominant configuration present in the hydrophobic amyloidogenic region, whereas the Q/N-rich amyloid fibrils have hairpins with β-turn. Also, Kametani et al.[88] has showed that the Met, Ser, and

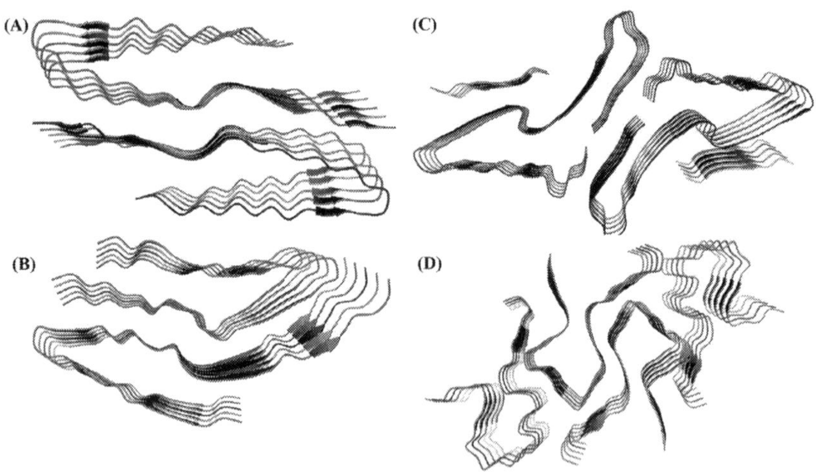

FIGURE 2.6 Cryo-EM structures of TDP-43 polymorphic fibrils. Schematic of full-length TDP-43's SegA (residues 311–360) and SegB (residues 286–331). (A) SegA-sym conformation (PDB id: 6N37), (B) SegA-asym conformation (PDB id: 6N3B), (C) SegA-slow conformation (PDB id: 6N3A), and (D) conformation of SegB A315E (PDB id: 6N3C).

Q/N residues of the dagger-shaped hydrophobic region were oxidized, phosphorylated, and deamidated, respectively, in ALS brain samples. This finding clearly indicates the solvent-exposure of these residues in bonafide TDP-43 fibrils in vivo.

These structures thus present a key part of the TDP-43 amyloid foldome, indicating the role of the hydrophobic region. Nevertheless, the pathological conformations attained by Q/N-rich region might be well relevant.

Interdomain interactions in TDP-43

The study of interdomain interactions is key to understand the physiological and pathophysiological functions of TDP-43. A number of structural and functional studies characterized the interactions between the N-and C-termini of TDP-43.[45,89] Song and coworkers through NMR elegantly demonstrated the interdomain interactions coordinated by disordered CTD.[90] They showed that the unfolded form of NTD (at pH 5.0) interacts with RRMs in TDP-43 construct (residues: 1−269) and inhibits the oligomerization of isolated NTD. Similarly, the authors reported the interdomain interactions between RRMs and CTD, i.e., TDP-43 (residues: 102−414) were coordinated by the CTD residues, Gly300-Gln343. These residues were shown to adopt helical structure and constitute the hydrophobic segment which also interacts with membrane.[84] Further, the authors went to reveal that the CTD interacts with both NTD and RRMs in full-length TDP-43 construct, except for the Gly residues. The nine key residues in RRMs showed interactions with prion-like domains Thr115, Asp119, Gly141, Arg151, Asp169, Gly170 and Arg171 in RRM1, and Met202 and Gly252 in RRM2. Interestingly, many of these residues are involved in nucleic acid binding.[26,91]

TDP-43 interdomain interactions have low affinity (μM) which results in the "open" and "closed" states of TDP-43. These conformational dynamics as suggested by authors are essential for TDP-43 to perform its physiological functions, i.e., binding to nucleic acids or other proteins, and/or self-assembly to form functional and reversible oligomer as well as to undergo pathological amyloid/fibrils formation.[84,92]

Structure overview of TDP-43 fibrils

The structural features of full-length TDP-43 fibrils that confer toxic functions are still unknown. Recently, the structure and dynamics of three TDP-43 constructs, including full-length TDP-43, TDP-35, and TDP-16, in their amyloid-fibrillar forms has been reported using magic-angle spinning SSNMR.[93] TDP-43 and TDP-35 amyloid fibrils showed broad SSNMR signals, representing structural polymorphism. The structural polymorphisms in many amyloid fibrils have been observed previously by SSNMR.[94,95] The chemical shift profiles indicated a predominant β-sheet-rich conformation for the TDP-43 and TDP-35 constructs, similar to the conformation largely observed in many amyloid structures.[93] Remarkably, SSNMR fingerprints of both TDP-35 and TDP-43 are superimposable, suggesting the similar conformation of the amyloid core in these fragments. However, TDP-16 amyloid fibrils showed a different SSNMR fingerprint indicating less pronounced β-sheet-rich structural polymorphism. The authors also showed that the GaroS1, the hydrophobic, and the Q/N-rich segments were involved in the formation of common rigid core of these TDP-43 constructs. However, TDP-16 amyloid core contains GaroS2 segment (G380-I383 and A391-G396), suggesting a different molecular organization of the amyloid core structures in these three different TDP-43 constructs.[93]

Structure models of full-length TDP-43

Recently, Wright et al.[53] presented a full-length model of tryptophan-free TDP-43 (denoted as TDP-43_{WtoA}). Accordingly, the monomeric model represents a dynamic system with a predominantly disordered CTD, but a relatively compact N-terminal region comprising NTD and RRMs. The compact N-terminal region has little contacts between the structural domains except between NTD and RRM2 (Fig. 2.7A).

The structure model of full-length TDP-43 obtained using the Robetta server[96] is shown in Fig. 2.7B. Robetta provides the ab initio and comparative models of protein. Here, the comparative models are built from known PDB templates and the final models are generated using the RosettaCM program.[96] The generated model is mainly composed of four lobes, two of which are represented by RRM1 and RRM2 domains, and the other two containing compact NTD and flexible CTD. This structure model is clearly depicting the unspecific oligomerization of TDP-43 due to presence of flexible and disordered regions in CTD as well as the linker between the domains.

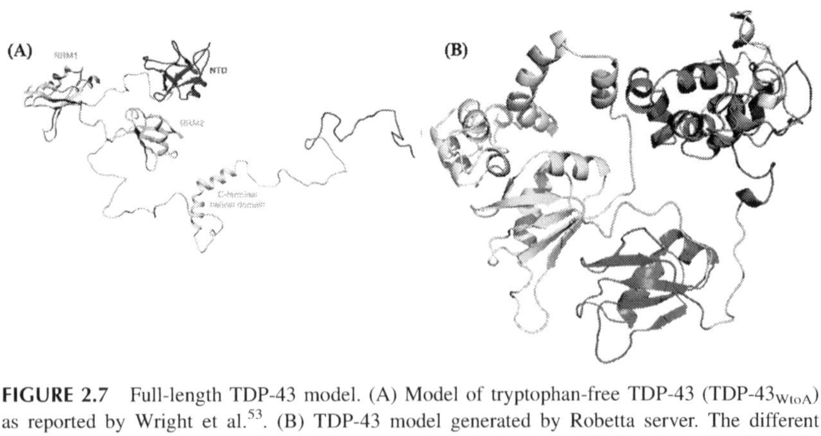

FIGURE 2.7 Full-length TDP-43 model. (A) Model of tryptophan-free TDP-43 (TDP-43_{WtoA}) as reported by Wright et al.[53]. (B) TDP-43 model generated by Robetta server. The different regions are color coded as follows: NTD (cyan), RRM1 (yellow), RRM2 (blue), GaroS1 (red), hydrophobic segment (magenta), Q/N region (gray), and GaroS2 (orange).

Conclusion

The pathophysiology of TDP-43 proteinopathies includes mislocalization, misfolding, aggregation, and segregation of RNA in SGs. All these processes are driven or modified by the effects on the three-dimensional structure of TDP-43. The three-dimensional structure of full-length TDP-43 is not available because of the presence of largely disordered C-terminal tail and flexible linkers between the domains. To understand the structure–function relationship of protein, it is necessary to understand the interdomain interactions between the protein, though it is largely unknown in TDP-43. TDP-43 forms a physiological oligomer mediated by the interactions of globular NTD and self-associating CTD. In spite of various insights into the interactions of individual domains of TDP-43, we still lack an inclusive picture of how full-length protein is arranged in respect of the relative positioning and interaction between domains, as well as the structures of physiological oligomers and pathological aggregates. As our knowledge of the three-dimensional structural repertoire of TDP-43 increases, it will provide insight into the understanding of structure and function of TDP-43 in both physiological and pathological contexts and thus increases the avenues for therapeutic intervention.

References

1. Arai T, et al. TDP-43 is a component of ubiquitin-positive tau-negative inclusions in frontotemporal lobar degeneration and amyotrophic lateral sclerosis. *Biochem Biophys Res Commun* 2006;**351**(3):602–11.
2. Neumann M, et al. Ubiquitinated TDP-43 in frontotemporal lobar degeneration and amyotrophic lateral sclerosis. *Science* 2006;**314**(5796):130–3.
3. Ling SC, Polymenidou M, Cleveland DW. Converging mechanisms in ALS and FTD: disrupted RNA and protein homeostasis. *Neuron* 2013;**79**(3):416–38.
4. Geser F, Lee VM, Trojanowski JQ. Amyotrophic lateral sclerosis and frontotemporal lobar degeneration: a spectrum of TDP-43 proteinopathies. *Neuropathology* 2010;**30**(2):103–12.
5. Kumar V, et al. Delineating the relationship between amyotrophic lateral sclerosis and frontotemporal dementia: sequence and structure-based predictions. *Biochim Biophys Acta* 2016;**1862**(9):1742–54.
6. Chanson JB, et al. TDP43-positive intraneuronal inclusions in a patient with motor neuron disease and Parkinson's disease. *Neurodegener Dis* 2010;**7**(4):260–4.
7. Rohn TT. Caspase-cleaved TAR DNA-binding protein-43 is a major pathological finding in Alzheimer's disease. *Brain Res* 2008;**1228**:189–98.
8. Schwab C, et al. Colocalization of transactivation-responsive DNA-binding protein 43 and huntingtin in inclusions of Huntington disease. *J Neuropathol Exp Neurol* 2008;**67**(12):1159–65.
9. Chen-Plotkin AS, Lee VM, Trojanowski JQ. TAR DNA-binding protein 43 in neurodegenerative disease. *Nat Rev Neurol* 2010;**6**(4):211–20.
10. Nelson PT, et al. Limbic-predominant age-related TDP-43 encephalopathy (LATE): consensus working group report. *Brain* 2019;**142**(6):1503–27.

11. Baloh RH. TDP-43: the relationship between protein aggregation and neurodegeneration in amyotrophic lateral sclerosis and frontotemporal lobar degeneration. *FEBS J* 2011;**278**(19):3539–49.
12. Nakashima-Yasuda H, et al. Co-morbidity of TDP-43 proteinopathy in Lewy body related diseases. *Acta Neuropathol* 2007;**114**(3):221–9.
13. Robberecht W, Philips T. The changing scene of amyotrophic lateral sclerosis. *Nat Rev Neurosci* 2013;**14**(4):248–64.
14. Liscic RM, et al. ALS and FTLD: two faces of TDP-43 proteinopathy. *Eur J Neurol* 2008;**15**(8):772–80.
15. Kumar V, et al. Therapeutic progress in amyotrophic lateral sclerosis-beginning to learning. *Eur J Med Chem* 2016;**121**:903–17.
16. Polymenidou M, Cleveland DW. The seeds of neurodegeneration: prion-like spreading in ALS. *Cell* 2011;**147**(3):498–508.
17. Mackenzie IR, et al. Nomenclature for neuropathologic subtypes of frontotemporal lobar degeneration: consensus recommendations. *Acta Neuropathol* 2009;**117**(1):15–8.
18. Mackenzie IR, et al. Pathological TDP-43 distinguishes sporadic amyotrophic lateral sclerosis from amyotrophic lateral sclerosis with SOD1 mutations. *Ann Neurol* 2007;**61**(5):427–34.
19. Mackenzie IR, et al. Nomenclature and nosology for neuropathologic subtypes of frontotemporal lobar degeneration: an update. *Acta Neuropathol* 2010;**119**(1):1–4.
20. Tan RH, et al. TDP-43 proteinopathies: pathological identification of brain regions differentiating clinical phenotypes. *Brain* 2015;**138**(Pt 10):3110–22.
21. Wang HY, et al. Structural diversity and functional implications of the eukaryotic TDP gene family. *Genomics* 2004;**83**(1):130–9.
22. Romano M, et al. Evolutionarily conserved heterogeneous nuclear ribonucleoprotein (hnRNP) A/B proteins functionally interact with human and Drosophila TAR DNA-binding protein 43 (TDP-43). *J Biol Chem* 2014;**289**(10):7121–30.
23. Sun Y, Chakrabartty A. Phase to phase with TDP-43. *Biochemistry* 2017;**56**(6):809–23.
24. Prasad A, et al. Molecular mechanisms of TDP-43 misfolding and pathology in amyotrophic lateral sclerosis. *Front Mol Neurosci* 2019;**12**:25.
25. Cohen TJ, Lee VM, Trojanowski JQ. TDP-43 functions and pathogenic mechanisms implicated in TDP-43 proteinopathies. *Trends Mol Med* 2011;**17**(11):659–67.
26. Lukavsky PJ, et al. Molecular basis of UG-rich RNA recognition by the human splicing factor TDP-43. *Nat Struct Mol Biol* 2013;**20**(12):1443–9.
27. Austin JA, et al. Disease causing mutants of TDP-43 nucleic acid binding domains are resistant to aggregation and have increased stability and half-life. *Proc Natl Acad Sci U S A* 2014;**111**(11):4309–14.
28. Tollervey JR, et al. Characterizing the RNA targets and position-dependent splicing regulation by TDP-43. *Nat Neurosci* 2011;**14**(4):452–8.
29. Cushman M, et al. Prion-like disorders: blurring the divide between transmissibility and infectivity. *J Cell Sci* 2010;**123**(Pt 8):1191–201.
30. Buratti E, et al. TDP-43 binds heterogeneous nuclear ribonucleoprotein A/B through its C-terminal tail: an important region for the inhibition of cystic fibrosis transmembrane conductance regulator exon 9 splicing. *J Biol Chem* 2005;**280**(45):37572–84.
31. Dewey CM, et al. TDP-43 is directed to stress granules by sorbitol, a novel physiological osmotic and oxidative stressor. *Mol Cell Biol* 2011;**31**(5):1098–108.
32. Kato M, et al. Cell-free formation of RNA granules: low complexity sequence domains form dynamic fibers within hydrogels. *Cell* 2012;**149**(4):753–67.

33. Molliex A, et al. Phase separation by low complexity domains promotes stress granule assembly and drives pathological fibrillization. *Cell* 2015;**163**(1):123–33.
34. Ayala YM, et al. Structural determinants of the cellular localization and shuttling of TDP-43. *J Cell Sci* 2008;**121**(Pt 22):3778–85.
35. Gregory RI, et al. The Microprocessor complex mediates the genesis of microRNAs. *Nature* 2004;**432**(7014):235–40.
36. Ayala YM, et al. Human, Drosophila, and *C.elegans* TDP43: nucleic acid binding properties and splicing regulatory function. *J Mol Biol* 2005;**348**(3):575–88.
37. Lagier-Tourenne C, Polymenidou M, Cleveland DW. TDP-43 and FUS/TLS: emerging roles in RNA processing and neurodegeneration. *Hum Mol Genet* 2010;**19**(R1):R46–64.
38. Casafont I, et al. TDP-43 localizes in mRNA transcription and processing sites in mammalian neurons. *J Struct Biol* 2009;**167**(3):235–41.
39. Yamashita M, et al. Distinct pathways leading to TDP-43-induced cellular dysfunctions. *Hum Mol Genet* 2014;**23**(16):4345–56.
40. Vogt MA, et al. TDP-43 induces p53-mediated cell death of cortical progenitors and immature neurons. *Sci Rep* 2018;**8**(1):8097.
41. Alami NH, et al. Axonal transport of TDP-43 mRNA granules is impaired by ALS-causing mutations. *Neuron* 2014;**81**(3):536–43.
42. Sephton CF, et al. TDP-43 is a developmentally regulated protein essential for early embryonic development. *J Biol Chem* 2010;**285**(9):6826–34.
43. Meszaros B, Erdos G, Dosztanyi Z. IUPred2A: context-dependent prediction of protein disorder as a function of redox state and protein binding. *Nucleic Acids Res* 2018;**46**(W1):W329–37.
44. Shiina Y, et al. TDP-43 dimerizes in human cells in culture. *Cell Mol Neurobiol* 2010;**30**(4):641–52.
45. Afroz T, et al. Functional and dynamic polymerization of the ALS-linked protein TDP-43 antagonizes its pathologic aggregation. *Nat Commun* 2017;**8**(1):45.
46. Zhang YJ, et al. The dual functions of the extreme N-terminus of TDP-43 in regulating its biological activity and inclusion formation. *Hum Mol Genet* 2013;**22**(15):3112–22.
47. Chang CK, et al. The N-terminus of TDP-43 promotes its oligomerization and enhances DNA binding affinity. *Biochem Biophys Res Commun* 2012;**425**(2):219–24.
48. Sasaguri H, et al. The extreme N-terminus of TDP-43 mediates the cytoplasmic aggregation of TDP-43 and associated toxicity in vivo. *Brain Res* 2016;**1647**:57–64.
49. Romano V, et al. The structural integrity of TDP-43 N-terminus is required for efficient aggregate entrapment and consequent loss of protein function. *Prion* 2015;**9**(1):1–9.
50. Suzuki H, et al. Nuclear TDP-43 causes neuronal toxicity by escaping from the inhibitory regulation by hnRNPs. *Hum Mol Genet* 2015;**24**(6):1513–27.
51. Mompean M, et al. The TDP-43 N-terminal domain structure at high resolution. *FEBS J* 2016;**283**(7):1242–60.
52. Wang A, et al. A single N-terminal phosphomimic disrupts TDP-43 polymerization, phase separation, and RNA splicing. *EMBO J* 2018;**37**(5).
53. Wright GSA, et al. Purification and structural characterization of aggregation-prone human TDP-43 involved in neurodegenerative diseases. *iScience* 2020;**23**(6):101159.
54. Kuo PH, et al. Structural insights into TDP-43 in nucleic-acid binding and domain interactions. *Nucleic Acids Res* 2009;**37**(6):1799–808.
55. Handa N, et al. Structural basis for recognition of the tra mRNA precursor by the Sex-lethal protein. *Nature* 1999;**398**(6728):579–85.

56. Deo RC, et al. Recognition of polyadenylate RNA by the poly(A)-binding protein. *Cell* 1999;**98**(6):835–45.
57. Wang X, Tanaka Hall TM. Structural basis for recognition of AU-rich element RNA by the HuD protein. *Nat Struct Biol* 2001;**8**(2):141–5.
58. François-Moutal L, et al. Structural insights into TDP-43 and effects of post-translational modifications. *Front Mol Neurosci* 2019;**12**(301).
59. Perez-Canadillas JM. Grabbing the message: structural basis of mRNA 3'UTR recognition by Hrp1. *EMBO J* 2006;**25**(13):3167–78.
60. Mackereth CD, et al. Multi-domain conformational selection underlies pre-mRNA splicing regulation by U2AF. *Nature* 2011;**475**(7356):408–11.
61. Shodai A, et al. Aberrant assembly of RNA recognition motif 1 links to pathogenic conversion of TAR DNA-binding protein of 43 kDa (TDP-43). *J Biol Chem* 2013;**288**(21):14886–905.
62. Sephton CF, et al. Identification of neuronal RNA targets of TDP-43-containing ribonucleoprotein complexes. *J Biol Chem* 2011;**286**(2):1204–15.
63. Colombrita C, et al. TDP-43 and FUS RNA-binding proteins bind distinct sets of cytoplasmic messenger RNAs and differently regulate their post-transcriptional fate in motoneuron-like cells. *J Biol Chem* 2012;**287**(19):15635–47.
64. Furukawa Y, et al. A molecular mechanism realizing sequence-specific recognition of nucleic acids by TDP-43. *Sci Rep* 2016;**6**:20576.
65. Prakash A, et al. Comparative analysis of thermal unfolding simulations of RNA recognition motifs (RRMs) of TAR DNA-binding protein 43 (TDP-43). *J Biomol Struct Dyn* 2019;**37**(1):178–94.
66. Prakash A, et al. Structural heterogeneity in RNA recognition motif 2 (RRM2) of TAR DNA-binding protein 43 (TDP-43): clue to amyotrophic lateral sclerosis. *J Biomol Struct Dyn* 2020:1–11.
67. Chiang CH, et al. Structural analysis of disease-related TDP-43 D169G mutation: linking enhanced stability and caspase cleavage efficiency to protein accumulation. *Sci Rep* 2016;**6**:21581.
68. Saini A, Chauhan VS. Delineation of the core aggregation sequences of TDP-43 C-terminal fragment. *Chembiochem* 2011;**12**(16):2495–501.
69. Kumar V, et al. Exploring the aggregation-prone regions from structural domains of human TDP-43. *Biochim Biophys Acta Protein Proteonomics* 2019;**1867**(3):286–96.
70. Tavella D, Zitzewitz JA, Massi F. Characterization of TDP-43 RRM2 partially folded states and their significance to ALS pathogenesis. *Biophys J* 2018;**115**(9):1673–80.
71. Mackness BC, et al. Folding of the RNA recognition motif (RRM) domains of the amyotrophic lateral sclerosis (ALS)-linked protein TDP-43 reveals an intermediate state. *J Biol Chem* 2014;**289**(12):8264–76.
72. Yang C, et al. The C-terminal TDP-43 fragments have a high aggregation propensity and harm neurons by a dominant-negative mechanism. *PLoS One* 2010;**5**(12):e15878.
73. Guenther EL, et al. Atomic-level evidence for packing and positional amyloid polymorphism by segment from TDP-43 RRM2. *Nat Struct Mol Biol* 2018;**25**(4):311–9.
74. Buratti E, Baralle FE. The molecular links between TDP-43 dysfunction and neurodegeneration. *Adv Genet* 2009;**66**:1–34.
75. Conicella AE, et al. ALS mutations disrupt phase separation mediated by alpha-helical structure in the TDP-43 low-complexity C-terminal domain. *Structure* 2016;**24**(9):1537–49.
76. Mompean M, et al. Electrostatic Repulsion Governs TDP-43 C-terminal domain aggregation. *PLoS Biol* 2016;**14**(4):e1002447.

77. Budini M, et al. Cellular model of TAR DNA-binding protein 43 (TDP-43) aggregation based on its C-terminal Gln/Asn-rich region. *J Biol Chem* 2012;**287**(10):7512–25.
78. Chen AK, et al. Induction of amyloid fibrils by the C-terminal fragments of TDP-43 in amyotrophic lateral sclerosis. *J Am Chem Soc* 2010;**132**(4):1186–7.
79. Guo W, et al. An ALS-associated mutation affecting TDP-43 enhances protein aggregation, fibril formation and neurotoxicity. *Nat Struct Mol Biol* 2011;**18**(7):822–30.
80. Jiang LL, et al. Structural transformation of the amyloidogenic core region of TDP-43 protein initiates its aggregation and cytoplasmic inclusion. *J Biol Chem* 2013;**288**(27):19614–24.
81. Jiang LL, et al. Two mutations G335D and Q343R within the amyloidogenic core region of TDP-43 influence its aggregation and inclusion formation. *Sci Rep* 2016;**6**:23928.
82. Mompean M, et al. Structural characterization of the minimal segment of TDP-43 competent for aggregation. *Arch Biochem Biophys* 2014;**545**:53–62.
83. Mompean M, et al. An amyloid-like pathological conformation of TDP-43 is stabilized by hypercooperative hydrogen bonds. *Front Mol Neurosci* 2016;**9**:125.
84. Lim L, et al. ALS-causing mutations significantly perturb the self-assembly and interaction with nucleic acid of the intrinsically disordered prion-like domain of TDP-43. *PLoS Biol* 2016;**14**(1):e1002338.
85. Guenther EL, et al. Atomic structures of TDP-43 LCD segments and insights into reversible or pathogenic aggregation. *Nat Struct Mol Biol* 2018;**25**(6):463–71.
86. Cao Q, et al. Cryo-EM structures of four polymorphic TDP-43 amyloid cores. *Nat Struct Mol Biol* 2019;**26**(7):619–27.
87. Mompean M, et al. Structural evidence of amyloid fibril formation in the putative aggregation domain of TDP-43. *J Phys Chem Lett* 2015;**6**(13):2608–15.
88. Kametani F, et al. Mass spectrometric analysis of accumulated TDP-43 in amyotrophic lateral sclerosis brains. *Sci Rep* 2016;**6**:23281.
89. Wang IF, et al. The self-interaction of native TDP-43 C terminus inhibits its degradation and contributes to early proteinopathies. *Nat Commun* 2012;**3**:766.
90. Wei Y, et al. Inter-domain interactions of TDP-43 as decoded by NMR. *Biochem Biophys Res Commun* 2016;**473**(2):614–9.
91. Lua S, et al. Structural, stability, dynamic and binding properties of the ALS-causing T46I mutant of the hVAPB MSP domain as revealed by NMR and MD simulations. *PLoS One* 2011;**6**(11):e27072.
92. Qin H, et al. TDP-43 N terminus encodes a novel ubiquitin-like fold and its unfolded form in equilibrium that can be shifted by binding to ssDNA. *Proc Natl Acad Sci U S A* 2014;**111**(52):18619–24.
93. Shenoy J, et al. Structural dissection of amyloid aggregates of TDP-43 and its C-terminal fragments TDP-35 and TDP-16. *FEBS J* 2020;**287**(12):2449–67.
94. Tycko R. Physical and structural basis for polymorphism in amyloid fibrils. *Protein Sci* 2014;**23**(11):1528–39.
95. Tycko R, Wickner RB. Molecular structures of amyloid and prion fibrils: consensus versus controversy. *Acc Chem Res* 2013;**46**(7):1487–96.
96. Raman S, et al. Structure prediction for CASP8 with all-atom refinement using Rosetta. *Proteins* 2009;**77**(Suppl. 9):89–99.

97. Mompean M, et al. Point mutations in the N-terminal domain of transactive response DNA-binding protein 43 kDa (TDP-43) compromise its stability, dimerization, and functions. *J Biol Chem* 2017;**292**(28):11992–2006.
98. Jiang LL, et al. The N-terminal dimerization is required for TDP-43 splicing activity. *Sci Rep* 2017;**7**(1):6196.
99. Kuo PH, et al. The crystal structure of TDP-43 RRM1-DNA complex reveals the specific recognition for UG- and TG-rich nucleic acids. *Nucleic Acids Res* 2014;**42**(7):4712–22.

Chapter 3

Posttranslational modifications of TDP-43

Samir Rahman[1,2]

[1]*Friedman Brain Institute, Icahn School of Medicine at Mount Sinai, New York, NY, United States;* [2]*Department of Genetics and Genomic Science and Institute for Multiscale Biology, Icahn School of Medicine at Mount Sinai, New York, NY, United States*

Introduction

TAR DNA-binding protein of 43 kDa (TDP-43) is a multifunctional 414 aa RNA/DNA-binding protein involved in RNA metabolism that is evolutionarily conserved and ubiquitously expressed. It is a member of the family of heterogeneous nuclear RNA ribonucleoproteins (hnRNPs) and contains two RNA recognition motifs (RRM1/RRM2) flanked by an N- and glycine-rich C-terminal domain (CTD) that mediate protein–protein interactions. Under normal physiological conditions, TDP-43 is localized predominantly in the nucleus, but in addition to harboring an N-terminal nuclear localization signal (NLS) it also possesses a nuclear export sequence (NES) within RRM2 allowing it to shuttle between the nucleus and cytoplasm. TDP-43 was initially discovered as a protein that binds to the HIV-1 transactivation response element (TAR), repressing HIV-1 gene expression.[1] In addition to transcriptional repression, it was also shown to mediate alternative splicing at the cystic fibrosis transmembrane conductance regulator (CFTR) locus.[2,3]

TDP-43 has been characterized as a major component of cytoplasmic and nuclear aggregates present in most amyotrophic lateral sclerosis (ALS) and in approximately half of TDP-43 frontotemporal lobar degeneration (FTLD) patients.[4,5] Its activity is misregulated in neurogenerative disease due to the accumulation of aberrant posttranslational modifications (PTMs). These aberrant PTMs can affect many properties: stability, aggregation, subcellular localization, function, protein–protein or protein–RNA interaction, and clearance by the proteasome/autophagic response.[6] The most common PTMs shown to affect TDP-43 function in ALS/FTLD are the following: phosphorylation, generation of CTD fragments, ubiquitination, acetylation, cysteine oxidation, and sumoylation. Interestingly, nonneurodegenerative diseases with aberrant TDP-43 function only display a subset of the full

TDP-43 and Neurodegeneration. https://doi.org/10.1016/B978-0-12-820066-7.20002-7

repertoire of PTMs listed above. Thus, identifying specific PTMs that promote TDP-43 aggregation and toxicity in neurodegenerative disease is a promising strategy toward developing targeted therapies. It is therefore imperative to elucidate the molecular properties and associated functions of TDP-43 that are affected by aberrant PTMs specific to neurodegenerative disease.

Ubiquitination

Frontotemporal dementia (FTD) corresponds to a broad group of brain disorders that affect the frontal and temporal lobes of the brain, whereby portions of these lobes shrink. Symptoms include behavioral changes and language dysfunction. FTD is the most common cause of dementia under age 65 after Alzheimer's disease (AD). The diagnosis of FTD falls into three major categories.[7] The first category encompasses tauopathies including Pick's disease, corticobasal degeneration, progressive supranuclear palsy, and FTD with parkinsonism linked to chromosome 17 (FTDP-17). The second category is FTLD with ubiquitin-positive tau-negative neuronal cytoplasmic inclusions. This can be divided into two subtypes, one being FTLD with motor neuron disease (MND), also known as ALS with dementia, and the other is FTLD with MND-type inclusions but without MND.[8–10] The last category is FTLD without tau- or ubiquitin-positive inclusions, which therefore lacks any distinctive histologic features.[11] It has been shown that FTLD with MND-type inclusions but without MND (FTLD-MDN-type) is the most common form of FTD.[12]

There is evidence supporting the close clinical association between FTD and ALS, an adult-onset neurodegenerative disease that primarily affects upper and lower motor neurons, leading to muscle weakness and paralysis, and typically death within 3–5 years. The incidence of FTD in patients with bulbar-onset ALS is high as 48%, and conversely 50% of FTD patients meet the criteria for ALS.[13,14] Furthermore, neuropsychological testing revealed that the ALS patients show impairment of frontotemporal cognitive functions.[15] Ubiquitination of misfolded proteins that aggregate in the cytoplasm and nucleus of central nervous system neurons is a characteristic feature of neurodegenerative diseases, and is common in both FTLD and ALS.[16] The identities of the ubiquitinated misfolded disease proteins were previously unknown.

Using biochemical and immunohistochemical analyses, TDP-43 was shown to be a major component of ubiquitin-positive tau-negative neuronal cytoplasmic inclusions in FTLD-MND-type, FTLD-MND, and ALS.[4,5] Pathologic TDP-43 was only recovered from affected central nervous system regions, including hippocampus, neocortex, and spinal cord. Furthermore, in the spinal cord of ALS patients, neuronal intranuclear inclusions and glial cytoplasmic inclusions were also positive for TDP-43, but ubiquitin negative.[5] Interestingly, while TDP-43 was detectable in the nuclei of unaffected neurons, it was not detected in the nuclei of affected neurons with ubiquitin-positive inclusions.[4] This implies that in affected neurons, TDP-43 is redistributed from the nucleus, where it normally functions in processes related to transcription and splicing, to the cytoplasm where it accumulates in ubiquitin-positive inclusions. Immunoprecipitation of TDP-43 followed by coimmunoblotting with antibodies against TDP-43 and ubiquitin revealed that TDP-43 is ubiquitinated. Interestingly, a high molecular weight band of 45 kDa in western blots of TDP-43 on insoluble fractions from FTLD and ALS disappears upon dephosphorylation, showing that accumulated TDP-43 is hyperphosphorylated in both neurodegenerative disorders. The discovery of the common intracellular aggregation of ubiquitinated and hyperphosphorylated TDP-43 in FTLD-MND-type, FTLD-MND, and ALS, further supports the notion that these disorders represent a pathological spectrum of a single disease with a shared TDP-43 proteinopathy.

The ubiquitin–proteasome pathway is the primary cellular mechanism to eliminate polyubiquitinated damaged or misfolded proteins, but there is also evidence showing that the loss of autophagic degradation can lead to the accumulation of ubiquitin-positive inclusions and cause neurodegeneration in mice.[17,18] Autophagy is a homeostatic process by which cells degrade and recycle cytosolic components through lysosomes. The endosomal sorting complexes required for transport (ESCRTs) are functionally linked to autophagy, as they have been shown to recognize ubiquitinated endocytosed integral membrane proteins and sort them into the intralumenal vesicles of the multivesicular body (MVB) targeting them for subsequent degradation in the lysosome/vacuole. The ESCRT-III complex internalizes ubiquitinated cargo into MVBs. Interestingly, mutations in the ESCRT-III subunit CHMP2B were associated with FTD in a large Danish pedigree and in patients with ALS.[19,20] Human cells depleted of ESCRT subunits or expressing CHMP2B mutants accumulate ubiquitin-positive aggregates.[21] Interestingly, although depletion of the ESCRT-I subunit Tsg101 and the ESCRT-III subunit Vps24 leads to the aggregation of TDP-43, highlighting the importance of autophagic degradation in the clearance of TDP-43-positive inclusions, cells expressing CHMP2B mutants do not contain TDP-43 in the cytoplasmic ubiquitin-positive inclusions. This finding coincides with recent data showing that the ubiquitin-positive inclusions in patients from the large Danish family with the CHMP2B mutation were TDP-43 negative.[22] As this contrasts with other cases of FTLD with ubiquitin-positive inclusions, it might suggest distinct molecular pathways in the MVB-mediated autophagic clearance of TDP-43-positive versus TDP-43-negative ubiquitinated inclusions, further highlighting the complexity of this disease.

UBAP1 (ubiquitin-associated protein 1) was discovered to be a risk factor for FTLD through linkage disequilibrium analysis of chromosome 9p21, a region shown to be associated with FTLD in certain families.[23] While the exact function of the protein is unknown, the gene is likely to be a member of a family of proteins that are connected to ubiquitin and the ubiquitination pathway. *UBAP1* mRNA extracted from cases with the disease-associated haplotype showed significantly lower expression levels than control. As the SNPs within this haplotype do not fall within the open reading frame of *UBAP1* but around the gene, they may affect regulatory regions that control the expression of *UBAP1*. UBAP1 and TDP-43 were shown to colocalize in cytoplasmic inclusions in the brain of an individual with familial FTLD. Altogether, these findings suggest that UBAP1 may be implicated in the metabolism of TDP-43, whereby disruption of its expression may inhibit the ubiquitin proteasome–mediated clearance of TDP-43 from cytoplasmic inclusions, contributing to the neurodegenerative phenotype of FTLD.

Polyubiquitination of TDP-43 enhances its propensity to aggregate in the cytoplasm.[24] The protein Ubiquilin, encoded by *UBQLN*, which can harbor a mutation associated with Alzheimer's and increased neurodegeneration in Drosophila, binds with high affinity to a hyper-polyubiquitinated TDP-43 mutant, inducing stronger aggregation in the cytoplasm. It was also shown that Ubiquilin and TDP-43 colocalize with the autophagosomal marker LC3, suggesting a role for Ubiquilin in delivering TDP-43 to autophagosomes to facilitate lysosome-mediated degradation. Mutations in a related gene *UBQLN2* were shown to be associated with some ALS/dementia cases, impairing proteasome-mediated protein degradation.[25] Interestingly, ALS/dementia cases with or without UBQLN2 mutations both showed Ubiquilin 2–positive inclusions in the spinal cord and hippocampus. Interestingly, while wild-type TDP-43 is almost entirely nuclear, C-terminal fragments of TDP-43 colocalize with both wild-type and mutant Ubiquilin 2 in the cytoplasm. Ubiquilin 2 was shown to interact with overexpressed TDP-43, binding to the CTD, dramatically reducing TDP-43 levels, possibly through autophagosomal or ubiquitin proteasome–mediated degradation.[26] Ubiquilins contain a ubiquitin-associated domain (UBA) that binds to ubiquitinated proteins and a ubiquitin-like domain that binds to the proteasome, thereby playing an intermediary role in targeting proteins toward proteasomal degradation. Dominant mutations in *TARDBP*, the gene encoding TDP-43, are associated with subsets of both sporadic (no family history), comprising 90% of cases, and familial ALS cases, occurring mostly within the CTD. Mutations in TDP-43 may affect polyubiquitination levels and thereby alter TDP-43 trafficking and metabolism by factors such as Ubiquilin, exacerbating neurodegeneration. Conversely, ALS and FTD-linked mutations in UQLN2 can lead to aggregation of polyubiquitinated substrates including TDP-43. ALS linked mutations in UBQLN2 can disrupt both the proteasomal and autophagic pathways of protein clearance leading to accumulation of polyubiquitinated proteins such as TDP-43, which are sequestered away from degradation by UBQLN2 via its UBA domain.[27]

Interestingly, overexpression of both WT and mutant UBQLN2 results in the activation of the inflammation-associated transcription factor NFk-Beta as well as cytoplasmic accumulation of TDP-43 in Neuro2A cells, increasing their vulnerability to death.[28] Furthermore, mice expressing ALS and FTD-linked mutants of UBQLN2 have been shown to develop memory deficits and motor neuron dysfunction.[29,30] They show an age-dependent loss of both hippocampal and motor neurons concomitant with increased ubiquitin-positive inclusions in the brain and spinal cord, and redistribution of TDP-43 from the nucleus to ubiquitin-positive inclusions in the cytoplasm of motor neurons.

The CTD of TDP-43 is critical for nuclear localization as overexpression of a TDP-43 splice variant lacking the C-terminus (TDP-S6) results in cytosolic and nuclear aggregation of this shorter isoform in both a human cell line and mouse primary neurons, recapitulating disease pathology, whereas full-length TDP-43 is localized primarily in the nucleus.[31] TDP-S6 overexpression leads to an increase in both ubiquitin (Ub) and small Ub-like modifier-2/3 (SUMO-2/3) within the TDP insoluble fraction. Whereas SUMO-2/3 was only found to colocalize with nuclear TDP aggregates, ubiquitin colocalized with both cytoplasmic and nuclear TDP aggregates. Polyubiquitin chains on aggregated TDP were linked to ubiquitin Lys-48 and Lys-63. ALS-associated mutations within the glycine-rich CTD of TDP-43 may disrupt key amino acid residues required for nuclear localization and proper function, leading to pathological aggregation and polyubiquitination.

TDP-43 and the aggregation-prone splice variant TDP-S6 have been shown to coaggregate with other RNA-binding proteins upon overexpression in HEK293T cells.[32] TDP-S6 overexpressed in HEK293T cells and primary mouse motor neurons coaggregates with the cytoplasmic stress granule–associated proteins G3BP, PABPC1, and eIF4A1, which may be implicated in translational repression during stress. Overexpressed TDP-43 was shown to coaggregate with the nuclear paraspeckle–associated RNA/DNA-binding proteins RBM14, PSF, and NonO, involved in splicing and transcriptional repression. Four evolutionarily conserved ubiquitinated residues were found in both TDP-43 and TDP-S6, localized on or near the RNA-binding motif (RRM) 1. Mutagenesis of these ubiquitination sites inhibits TDP-43 polymerization. Therefore, ubiquitination of TDP-43 within RRM1 may induce TDP-43 self-interaction within aggregates that contain other RNA-binding proteins, the solubility of which may also be affected by TDP-43 ubiquitination. Pathological aggregation of TDP-43 with other RNA-binding proteins may alter their RNA-binding properties and thus abrogate different processes in neuronal RNA metabolism, such as transcription, splicing, and translation.

As a nuclear RNA-binding protein, TDP-43 regulates the transcription of thousands of genes, including the E3 ubiquitin ligase Parkin, whose mRNA TDP-43 binds to. Interestingly, TDP-43 increases the expression level of Parkin, which in turn ubiquitinates TDP-43.[33] Parkin generates polyubiquitin chains on TDP-43 linked to ubiquitin Lys-48 and Lys-63, which induces cytosolic accumulation of TDP-43 with Parkin. Surprisingly, although Lys-48- and Lys-63-linked ubiquitin chains were previously shown to target proteins for degradation, Parkin-mediated polyubiquitination leads to cytoplasmic accumulation of TDP-43, without any evidence of protein clearance. Interestingly, Parkin forms a multiprotein complex with both TDP-43 and HDAC6, which along with TDP-43 ubiquitination is necessary to redistribute TDP-43 from the nucleus to the cytoplasm. Thus, independent of its role in protein clearance, ubiquitination may also be involved in the loss of TDP-43's nuclear function while inducing a toxic cytoplasmic gain of function implicated in disease pathology.

In addition to Parkin, other proteins were revealed to interact with TDP-43 to control ubiquitination levels, aberrations in which lead to neurotoxicity.[34] A protein interaction screen using TDP-43 as bait against a human brain cDNA library showed interaction with the ubiquitin conjugating enzyme UBE2E3 and the ubiquitin isopeptidase Y UBPY. Ubiquitinated insoluble aggregates of TDP-43 C-terminal fragments accumulate upon proteasome inhibition suggesting rapid proteasomal turnover of aberrant ubiquitinated TDP-43. All three UBE2E family members enhance ubiquitination of TDP-43, whereas the catalytic mutant UBE2E3^{C145S} is much less efficient. Furthermore, knockdown of UBE2E3 results in reduced ubiquitination. Among a subset of the 48 known disease-associated TDP-43 mutants that are mostly linked to FTLD and ALS a particular mutant TDP-43^{K263E} was prone to excessive ubiquitination, which was further enhanced by proteasome inhibition and UBE2E3 expression. Conversely, expression of UBPY reduces ubiquitination of the mutant TDP-43^{K263E}. UBPY also deubiquitinates TDP-43 C-terminal fragments. Knockdown of UBPY results in accumulation of ubiquitinated TDP-43 aggregates in Drosophila, resulting in neurotoxicity. Thus, UBE2E and UBPY control TDP-43 ubiquitination and solubility, thereby playing important roles in neurodegeneration. Recently, a Drosophila protein named CG5445 was shown to bind to a ubiquitinated ALS-linked mutant TDP-43, suppressing its aggregation, and thereby preventing eye degeneration.[35] Therefore, cells may have mechanisms to sequester ubiquitinated aggregation prone proteins like TDP-43, maintaining them in a soluble state to mitigate their cytotoxicity until degradation.

Phosphorylation

The role of aberrant phosphorylation in neurodegenerative disease was first characterized for tau proteins in Alzheimer's. It was found that hyperphosphorylation of tau in brains of Alzheimer's patients is a consequence of defects in phosphatase activity compared to age-matched controls.[36] Pathogenic TDP-43 aggregates in FTLD-U (frontotemporal lobar degeneration with ubiquitin-positive inclusions) and ALS cases were also found to be hyperphosphorylated.[37] Phosphorylated and ubiquitinated aggregates were found in the neuroblastoma cell line SH-SY5Y similar to those found in the brains of FTLD-U and ALS.[38] Two candidate sequences for nuclear localization, residues 78–84 and 187–192, were examined. Deletion of 78–84 led to cytoplasmic accumulation of TDP-43, whereas deletion of 187–192 led to the formation of punctate structures in the nucleus. Interestingly, while wild-type TDP-43 did not show staining with ubiquitination- and phosphorylation-specific antibodies either in the presence or absence of the proteasome inhibitor MG132, both the NLS and 187–192 deletion mutants showed ubiquitin- and phosphorylation-positive inclusions upon proteasome inhibition. The 187–192 mutant showed phosphorylation-positive but ubiquitin-negative inclusions in the absence of proteasome inhibition. Upon proteasome inhibition, large inclusions were both ubiquitin and phosphorylation positive, whereas small inclusions were only phospho positive. Altogether, these results suggest that phosphorylation may represent an earlier event in TDP-43 pathology, whereas ubiquitination may represent a later-stage triggering formation of large aggregates that may be short lived due to rapid turnover by the proteasome.

The tandem ser409/410 residues in the CTD of TDP-43 were uniquely phosphorylated in ubiquitin-positive cytoplasmic inclusions in all sporadic and familial forms of TDP-43 proteinopathies but not in nuclear localized wild-type TDP-43.[39] While compact neuronal inclusions were double labeled with both the phospho-ser409/410 and ubiquitin-specific antibodies, granular "preinclusions" were mostly stained with only phosphorylation-specific antibodies, further confirming that hyperphosphorylation of TDP-43 precedes ubiquitination. Interestingly, while phosphorylated C-terminal TDP-43 fragments are abundant in the cortical brain regions of both FTLD-U and ALS, ALS spinal cord samples predominantly showed full-length phosphorylated TDP-43. It was shown in another study on a broader panel of phosphorylated sites within the CTD of TDP-43 that different clinical subtypes of FTLD-U and ALS are associated with different banding patterns of the 18–26 kDa CTD fragments isolated from the cerebral cortex.[40] In vitro phosphorylation assays showed that Casein Kinase 1 (CK1) is the kinase responsible for hyperphosphorylation of TDP-43 CTD, confirmed by another study which determined the CK1 target residues by mass spectrometry.[41]

A time course study of a neuroblastoma cell line (M17D3) conditionally expressing a GFP-tagged CTD fragment of TDP-43 (GFP-$TDP_{220\text{-}414}$) showed gradual formation of large inclusions that were both ubiquitinated and phosphorylated.[42] Inclusions were shown to be degraded primarily by the ubiquitin–proteasome system rather than the autophagosomal pathway, with the assistance of molecular chaperones Hsp70 and Hsp90, which have the ability to bind to TDP-43 CTD fragments. However, phosphorylated TDP-43 CTD fragments accumulate into insoluble aggregates making them resistant to degradation. Thus, modulating the phosphorylation state of TDP-43 CTD fragments is a promising therapeutic strategy to promote clearance of TDP-43 inclusions.

Although cytoplasmic mislocalization and aggregation of TDP-43 are hallmarks of ALS, it is not clear whether these phenotypes actually cause neurodegeneration and motor dysfunction. A study modeling TDP-43 proteinopathy in *Caenorhabditis elegans* expressing either wild-type human TDP-43 or ALS CTD mutants G290A, A315T, or M337V did not show evidence of cytoplasmic mislocalization but robust nuclear aggregates were detected.[43] Interestingly, although wild-type TDP-43 showed the propensity to accumulate in insoluble aggregates, worms expressing wild-type TDP-43 only showed mild motor deficits compared to those expressing the ALS mutants, which displayed some of the characteristics of ALS and FTLD-U such as age-induced decline in motor function, shortened lifespan, and motor neuron degeneration accompanied by hyperphosphorylation, truncation, and ubiquitination of TDP-43. It was shown that the toxicity of mutant TDP-43 was dependent on ser409/410 phosphorylation, as mutating these residues to alanines to ablate phosphorylation markedly ameliorated locomotion in worms expressing mutant TDP-43. Thus, although TDP-43 aggregation is a feature of TDP-43 proteinopathies, it may not be a driver of toxicity and neurodegeneration. Neurotoxicity may be primarily driven by phosphorylation of ser409/410, which may be increased by mutations within the CTD of TDP-43 that enhance the activity of kinases such as the previously identified CK1. This, in turn, may ultimately abrogate efficient clearance of TDP-43 inclusions, resulting in gradual neurodegeneration.

It has been shown that the 15 kDa fragment at the C-terminus of TDP-43 is sufficient to induce aggregation when overexpressed in mammalian cells, but may be modulated by additional sequences.[44] The C-terminus of TDP-43 is nonstructured, which may make it prone to aggregation, as intrinsically disordered regions in proteins are susceptible to aggregation through phase separation. In full-length TDP-43, the CTD domain may be hidden in the interior of the protein, and therefore not exposed. Interestingly, while aggregation of TDP-43 CTD fragments is accompanied by increased phosphorylation of ser409/410, phosphorylation itself may represent an early cellular attempt at preventing rather than inducing aggregation, as mutating ser409/410 to aspartic acid residues to mimic phosphorylation significantly reduced aggregates. Aggregates of TDP-43 CTD fragments were shown to be degraded by the autophagosomal or ubiquitin–proteasomal pathways through p62, an adaptor protein with a UBA domain that allows it to bind to ubiquitinated substrates. Thus, phosphorylation may play an early role in regulating aggregation, whereas ubiquitination at later stages may trigger TDP-43 degradation through the autophagosome or ubiquitin–proteasome system.

Interestingly, the solubility and phosphorylation levels of different TDP-43 CTD fragments are dependent on the site of cleavage.[45] The molecular weight range of pathogenic TDP-43 CTD fragments (20–25 kDa) indicates that the RNA-binding domain RRM2 may be a major hotspot for TDP-43 cleavage sites. Cleavage that occurs N-terminally of RRM2, leaving RRM2 intact, does not decrease solubility, whereas cleavage within the middle of RRM2 results in hyperphosphorylated insoluble aggregates. Interestingly, phosphorylation itself does not increase the propensity of TDP-43 CTD fragments to aggregate, as serine to alanine mutations to block phosphorylation does not alter insolubility, and neither does the introduction of phosphomimetic mutations at Ser 379, 403, 404, 409, and 410, which may even slightly decrease insolubility. Thus, phosphorylation is likely to be a "postaggregation" process.

Neurodegenerative diseases, including ALS, have been linked to oxidative stress and the production of reactive oxygen species (ROS).[46–52] Interestingly, inducing oxidative stress in motor neuron-like cells and primary cortical neurons through glutathione depletion produces signature characteristics of TDP-43 pathology, such as aggregation, C-terminal fragmentation, cytoplasmic mislocalization, and hyperphosphorylation.[53] A nonphosphorylable mutant TDP-43 with serine to alanine substitutions at the CTD residues 403/404 and 409/410 showed the same propensity for aggregation and cytoplasmic mislocalization under oxidative stress as phosphorylated TDP-43, confirming that phosphorylation is not mandatory for pathological insolubility and cytoplasmic distribution of TDP-43. TDP-43 shows differential cellular localization and biochemical properties depending on the magnitude of oxidative stress. In cultured cells undergoing short-term nonlethal oxidative stress induced by sodium arsenite TDP-43 interacts with stalled ribosomes in stress granules through direct binding to mRNA, potentially helping to repress translation as a protective mechanism.[54] This association was transient as it was reversible upon removal of stress. However, upon long-term sublethal injury, TDP-43 dissociates from stress granules and undergoes irreversible biochemical changes such as insolubility, disulfide bond formation, and phosphorylation. TDP-43 was shown to be phosphorylated by the sodium arsenite–induced c-jun N-terminal kinase pathway. Interestingly, the neurotoxicity of TDP-43 may be modulated by its interaction with stress granule components.[55] Genes regulating stress granule formation in Saccharomyces *cerevisiae* and *Drosophila melanogaster* were shown to have an effect on TDP-43 toxicity. Notably, TDP-43 interacts with a stress granule component, polyA-binding protein (PABP), which was also shown to accumulate abnormally in ALS spinal cord motor neurons. TDP-43 toxicity also upregulates the phosphorylation of the translation initiation factor eIF2alpha. Phosphorylation of eIF2alpha leads to global inhibition of protein synthesis, thus one of the mechanisms that TDP-43 mediates its toxicity may be the induction of prolonged translation repression, leading to cell death. A small molecule inhibitor of eIF2alpha phosphorylation successfully mitigated TDP-43 toxicity in flies and mammalian neurons.

In addition to inducing prolonged translational repression, misfolded and aggregated TDP-43 can trigger endoplasmic reticulum (ER) stress, which can then activate apoptosis.[56] Homeodomain-interacting protein kinase 2 (HIPK2) is the essential link in the IRE1alpha-ASK1-JNK pathway that mediates this apoptotic response. Specifically, two highly conserved serine and threonine residues (S359/T360) within the activation loop of the kinase domain in HIPK2 are phosphorylated in response to ER stress, and HIPK2 activation correlates well with TDP-43 proteinopathy in ALS, Furthermore, blocking HIPK2 kinase activity protects motor neurons from TDP-43 cytotoxicity. Thus, pathogenic TDP-43 may induce abnormal phosphorylation of other factors involved in proapoptotic cellular stress pathways, the pharmacological prevention of which may be a promising therapeutic strategy to alleviate neurodegeneration.

Differentiation of ALS patient–derived iPSCs (induced pluripotent stem cells) into motor neurons can model disease mechanisms at the cellular level and facilitate drug discovery.[57,58] Sporadic ALS patient iPSC-derived motor neurons showed the same intranuclear TDP-43 aggregates as found in the postmortem tissue of one of the patients from which the iPSCs were derived.[57] These intranuclear aggregates were phosphorylated, but not ubiquitinated. Thus, phosphorylation is a common feature of TDP-43 mislocalized to the cytoplasm and TDP-43 aggregates in the nucleus. These findings also provide further evidence that phosphorylation is an early disease process that may precede ubiquitination. A small molecule screen in wild-type and ALS mouse stem cell–derived motor neurons identified kenpaullone as a kinase inhibitor that substantially improves motor neuron survival.[58] Kenpaullone also improved the survival of human motor neurons derived from ALS patient iPSCs expressing mutant TDP-43. HGK kinase was shown to be directly inhibited by kenpaullone, thereby reducing phosphorylation levels of downstream factors in the HGK signaling pathway, notably the c-Jun N-terminal kinase, which was previously shown to be involved in downstream phosphorylation of TDP-43 under oxidative stress. Thus, one of the mechanisms by which kenpaullone alleviates motor neuron degeneration could be the inactivation of the c-Jun N-terminal kinase, thereby decreasing the downstream phosphorylation of TDP-43, and thus facilitating clearance of TDP-43 inclusions.

The anatomical distribution of phosphorylated TDP-43 (pTDP-43) inclusions across the central nervous system can be used to classify four distinct neuropathological stages of ALS.[59] The increasing burden of pTDP-43 pathology is associated with the sequential propagation of pTDP-43 from the motor cortex and spinal cord into the prefrontal cortex and then into the hippocampus. Interestingly, all stages of ALS are associated with pTDP-43 oligodendroglial inclusions. The pattern of pTDP-43 propagation across the different neuropathological stages is consistent with the hypothesis that pTDP-43 pathology may be transmitted along axonal pathways. Furthermore, the discovery of pTDP-43 pathology in the prefrontal cortex as part of the ongoing disease process may explain the deficits in executive cognitive functions in ALS.

pTDP-43 has been successfully used as a diagnostic marker in other studies to identify clinical subtypes of FTLD and ALS.[60,61] A variety of FTLD-TDP subtypes can be recognized based on the morphology and anatomical distribution of both neuronal inclusions and dystrophic neurites, which are abnormal neuronal processes characterized by aberrant sprouting and accumulation of various cellular organelles and signaling proteins. Type A consists of FTLD cases with moderate to numerous TDP-43-positive neuronal cytoplasmic inclusions and short dystrophic neurites predominantly located in the upper cortical layers II/III. Type B consists of cases with moderate to numerous TDP-43-positive neuronal cytoplasmic inclusions and sparse dystrophic neurites across all cortical layers. Type C consists of cases with long dystrophic neurites primarily in the upper cortices with rare neuronal cytoplasmic inclusions. Lastly, type D consists of cases with numerous neuronal intranuclear inclusions distributed across all cortical layers. When comparing tissue sections immunostained with either a pTDP-43 antibody or an antibody against nonphosphorylated TDP-43 (iTDP-43), there was increased consensus among independent observers in the subtyping of FTLD-TDP cases with pTDP-43 over iTDP-43.[60] Similarly, pTDP-43 has also been to classify different subtypes of sporadic ALS based on cerebral pTDP-43 pathology.[61] Cases were divided into two types, those without pTDP-43-positive neuronal cytoplasmic inclusions in the hippocampal dentate granule cells, and those with such inclusions. Furthermore, type 2 was further divided into two subtypes, type 2A, consisting of cases with no or few dystrophic neurites, and type 2B, in which dystrophic neurites were abundant. While type 2 was associated with cognitive impairment, type 1 was not. Intriguingly, type 2B is a distinct subtype which shows poor prognosis despite less severe motor neuron loss, and also shows more glial involvement in cortical TDP-43 pathology than the other subtypes. Altogether, these studies demonstrate that TDP-43 phosphorylation is a robust signature of both FTLD-TDP and ALS and can provide insights into the heterogeneous nature of TDP-43 proteinopathy.

An RNAi screen in an ALS model of *C. elegans* designed to monitor changes in TDP-43 phosphorylation identified homologs of the mammalian tau tubulin kinases TTBK1 and TTBK2 as factors controlling TDP-43 phosphorylation.[62] TTBK1/2 were shown to directly phosphorylate TDP-43 in an in vitro phosphorylation assay while overexpressing TTBK1/2 in cultured mammalian cells promotes hyperphosphorylation and cytoplasmic aggregation of TDP-43. Furthermore, TTBK1/2 were expressed at higher levels in FTLD cortex and were shown to colocalize with phospho-TDP-43 inclusions in ALS spinal cord. Interestingly, TTBK1 has also been implicated in Alzheimer's, as it has been shown to be highly expressed in AD frontal cortex[63] and to increase tau toxicity.[64] Thus, TTBK1/2 are attractive drug targets not only for TDP-43 proteinopathies but for other neurodegenerative disorders as well. Small molecule inhibitors have been designed to target CK1, also known to directly phosphorylate TDP-43.[65] These inhibitors prevent TDP-43 phosphorylation in mammalian cell cultures, increase lifespan in Drosophila by reducing TDP-43 toxicity, and are also predicted to cross the blood–brain barrier. Thus, pharmacological inhibition of kinases that directly phosphorylate TDP-43 is a promising therapeutic strategy for ALS.

A large proportion of familial FTLD cases with ubiquitin-positive TDP-43 inclusions (FTLD-TDP-43) are associated with mutations in the progranulin gene (GRN).[66,67] Previous work on lymphoblasts isolated from FTLD patients with a loss-of-function mutation in GRN showed an increase in cell cycle activity associated with increased levels of the cyclin-dependent kinase 6 (CDK6) and phosphorylation of retinoblastoma protein (pRb).[68] Increasing progranulin (PGRN) levels prevents overactivation of the CDK6/pRb cascade, and thus decreases cell proliferation.[69] Interestingly, cell cycle–related events are now considered to be important contributors to the pathology of neurodegenerative disorders, including Alzheimer's disease (AD),[70] Parkinson's disease (PD),[71] FTLD, and ALS.[72] Notably, the reentry of quiescent, postmitotic neurons into the cell cycle may be associated with mitotic catastrophe and cell death.[73,74] Thus, pharmacologically inhibiting cell cycle activation is a promising therapeutic strategy to prevent neurodegeneration in FTLD. Intriguingly, brain penetrant CK1 inhibitors have been shown to decrease proliferative activity of PGRN-deficient lymphoblasts by preventing phosphorylation and cytoplasmic mislocalization of TDP-43 and the overactivation of the CDK6/pRb cascade.[75] These inhibitors also played a neuroprotective role in a neuronal cell model with induced TDP-43 phosphorylation. Interestingly, TDP-43 has been shown to inhibit pRb phosphorylation by repressing CDK6 expression.[76] Removal of TDP-43 increases CDK6 mRNA and protein levels, thus increasing pRb phosphorylation, leading to misregulation of the cell cycle and apoptosis. Thus, in PGRN-deficient FTLD, TDP-43 phosphorylation may result in mislocalization to the cytoplasm, leading to a nuclear loss of function in repressing CDK6, which then leads to aberrant cell cycle regulation and apoptosis.

Casein Kinases 1 and 2 are both implicated in oligomerization of TDP-43, thereby enhancing its toxicity.[77,78] A functional study in Drosophila revealed that coexpression of the Q331K mutant TDP-43 with Doubletime (DBT), the Drosophila homolog of CK1, leads to oligomerization of TDP-43 to form high molecular weight species, coinciding with enhanced toxicity.[78] Interestingly, while DBT can induce ser409/ser410 phosphorylation of both wild-type and Q331K mutant TDP-43, only the latter shows cytoplasmic mislocalization and aggregation. TDP-43 inclusions can either take the form of amorphous aggregates or highly ordered filaments. An in vitro phosphorylation study using purified soluble TDP-43 as substrate found that CK2 can induce rapid polymerization of TDP-43 into 10 nm ordered filaments similar to the TDP-43 structures found in disease.[77] Interestingly, these filaments can be recognized by heat shock proteins, Hsp90 in particular, which either inhibit polymerization or induce depolymerization of TDP-43. Heat shock proteins may induce refolding of TDP-43 to make it more soluble and therefore more easily cleared by the proteasomal machinery. Thus, enhancing the activity of heat shock proteins is a potential therapeutic strategy to mitigate toxicity in TDP-43 proteinopathies.

The strength of activity of casein kinase 1 can determine the degree of TDP-43 cytotoxicity. It was shown that the expression of a hyperactive c-terminally truncated form of CK1 (CK1δ1-317) in the human SH-SY5Y neuroblastoma cell line promotes cytoplasmic mislocalization and aggregation of TDP-43 in ubiquitin- and p62-positive inclusions.[79] Interestingly, insoluble phosphorylated TDP-43 extracted from cells coexpressing TDP-43 and CK1δ1-317 functioned as seeds for TDP-43 aggregation in cultured cells, indicating that CK1δ1-317-mediated hyperphosphorylation can induce prion-like activity in TDP-43. Furthermore, CK1δ1-317 abrogates physiological functions of TDP-43. TDP-43 is known to mediate alternative splicing of the CFTR locus and stabilize mRNA levels of HDAC6, both of which were downregulated in cells coexpressing TDP-43 and CK1δ1-317. Thus, hyperphosphorylation of TDP-43 by increased CK1 activity can induce both a nuclear loss of function in normal RNA metabolism and a cytoplasmic toxic gain of function whereby pTDP-43 acts as a seed to induce abnormal aggregation.

Protein phosphorylation levels depend on the balance between the activities of kinases and phosphatases. Although there are abundant data on TDP-43 phosphorylation by kinases, there has been limited study on the activity of phosphatases that may dephosphorylate TDP-43 and thus reverse TDP-43 cytotoxicity. A yeast two-hybrid screen using the C-terminal of TDP-43 as bait against a fetal human brain cDNA library found the calmodulin-dependent phosphatase calcineurin, known to be highly expressed in mammalian brain, to strongly interact with TDP-43.[80] An in vitro dephosphorylation assay found that calcineurin dephosphorylates TDP-43 at the canonical disease−specific phosphosites s409/410. Furthermore, deletion of calcineurin in a *C. elegans* model of TDP-43 proteinopathy leads to excessive pTDP-43, motor dysfunction, and neurodegenerative damage. Pharmacological inhibition of calcineurin increased pTDP-43 levels in cultured human cells and calcineurin was also shown to colocalize with pTDP-43 in central nervous system tissue from FTLD-TDP and ALS patients. Another study showed that mutation in SOD1 (superoxidase dismutase 1), a gene that is mutated and is the causative factor in many cases of familial ALS, affects the ability of SOD1 to bind Zn, abrogating its interaction with calcineurin.[81] This, in turn, may inhibit the function of calcineurin, resulting in the accumulation of phosphorylated TDP-43. Apart from calcineurin, the protein phosphatase PP1 has also been implicated in TDP-43 dephosphorylation.[82] Specifically, the alpha and gamma isoforms of PP1 dephosphorylate TDP-43, suppressing the function of TDP-43 in Tau exon 10 inclusion. Altogether, these results highlight the importance of further investigation of phosphatases that interact with TDP-43, in addition to kinases, to understand the significance of TDP-43 phosphorylation in disease and develop novel therapeutic strategies.

Aberrant phosphorylation of TDP-43 may abrogate the stability of other proteins that TDP-43 normally interacts with to mediate RNA metabolism processes. Notably, TDP-43 is known to interact with Drosha, which is involved in the processing of microRNAs.[83,84] In one particular study, three different phosphomimetic mutants of TDP-43 (S379E, S403/404E, and S409/410E) were expressed in the mouse neuroblastoma Neuro2A cell line to investigate the relationship between aberrant phosphorylation of TDP-43 and Drosha stability.[85] Expression of the S409/410E TDP-43 mutant leads to Drosha instability and cytotoxicity in Neuro2A cells. Thus, aberrant phosphorylation of TDP-43 may lead to nuclear loss of function by destabilizing TDP-43 interacting proteins that are implicated in the processing of functional noncoding RNAs. This, in turn, may affect downstream regulation of genes that play a neuroprotective role.

C-terminal fragmentation

Mutations in the gene encoding the secreted growth factor progranulin are responsible for a large proportion of FTLD-U cases. Most mutations generate null alleles, leading to severe loss of progranulin protein. The biochemical link between progranulin depletion and TDP-43 pathology in FTLD-U was previously unknown. A study by Zhang and colleagues in 2007 showed that knockdown of PGRN results in caspase-mediated cleavage of TDP-43 into 35 and 25 kDa CTD fragments, similar to the fragments found in the brains of familial and sporadic FTLD-U.[86] Caspase-mediated cleavage induces cytoplasmic mislocalization and aggregation of the 35 and 25 kDa TDP-43 CTD fragments. The same group later showed that ectopic expression of the 25 kDa CTD fragment (TDP-25) leads to the accumulation of ubiquitinated and hyperphosphorylated cytoplasmic inclusions, increasing cellular toxicity.[87] Interestingly, TDP-25 was more prone to phosphorylation at the disease-associated s409/s410 sites than full-length TDP-43, although phosphorylation itself was not required for inclusion formation. TDP-25 also did not affect the function and localization of endogenous full-length TDP-43. Moreover, an antibody generated specifically to recognize TDP-25 found that it was a robust marker of pathological inclusions in TDP-43 proteinopathies. Thus, aberrant cleavage of TDP-43 may represent an early step in disease progression, resulting in cytoplasmic mislocalization due to loss of the N-terminal NLS, which is then followed by aggregation and phosphorylation. Interestingly, C-terminal fragments of TDP-43 are mostly detected in the hippocampus and cortex of brains from FTLD-U and ALS patients, whereas inclusions in spinal cord motor neurons in both diseases are primarily enriched in full-length TDP-43.[88] Thus, there may be regional differences in TDP-43 pathology throughout the central nervous system.

Interestingly, a later study failed to reproduce the findings from Zhang and colleagues that showed that PGRN knockdown triggers caspase-mediated cleavage of TDP-43, bringing into question the mechanistic link between PGRN deficiency and TDP-43 pathology.[89] siRNA-mediated knockdown of PGRN in cell culture and gene knockout in mice failed to produce the disease-associated CTFs of TDP-43. Nonetheless, inducing apoptosis leads to caspase-mediated generation of a major 35 kDA fragment and a minor 25 kDA fragment, which are both initially soluble but gradually accumulate into hyperphosphorylated aggregates. As a caspase-resistant mutant still accumulates into insoluble fractions, C-terminal fragmentation does not appear to be a prerequisite for aggregation. Moreover, as phosphorylation of s409/s410 occurs late in the process of converting soluble TDP-43 into insoluble TDP-43 species, phosphorylation is also not a prerequisite for aggregation, corroborating other studies.

Contradicting the previous findings of Zhang and colleagues, expressing CTD fragments of TDP-43 in the human neuroblastoma cell line SH-SY5Y leads to abnormal sequestration of full-length endogenous TDP-43 into cytoplasmic inclusions containing ubiquitinated and phosphorylated CTD fragments, resulting in loss of nuclear localization and function of endogenous full-length TDP-43.[90] The discrepancy in the findings may be attributed to the fact that Zhang and colleagues expressed CTD fragments of TDP-43 in HEK293 cells, whereas in the later study CTD fragments were expressed in a neuroblastoma cell line, which is more similar to disease-relevant tissue, thus potentially better able to recapitulate mechanisms of TDP-43 pathology. Nonnaka et al. also found that expressing CTD fragments of TDP-43 harboring ALS causing mutations enhances the formation of inclusions. They also determined the cleavage sites of CTD fragments extracted from FTLD-U brains through mass spectrometry. Similarly, another study found that Arg-208 was the specific cleavage site of a pathological CTD fragment purified from FTLD-U brains, and that the expression of this 208-414 fragment along with other CTD fragments leads to the accumulation of ubiquitinated and hyperphosphorylated cytoplasmic aggregates in cultured cells.[91] Furthermore, expression of CTD fragments was shown to disrupt alternative splicing of the CFTR locus, one of the key functions of TDP-43, corroborating with the findings from Nonnaka and colleagues. Another study in which full-length TDP-43 or various fragments were expressed in motor neuron derived NSC-34 cells or in primary neurons found that both N- and C-terminal fragments can aggregate, but the C-terminal fragments have a higher propensity to aggregate, and that the C-terminal end of the RRM2 domain is required, though not sufficient, for aggregation.[92] Furthermore, aggregation of the TDP-43 fragments can drive coaggregation with the full-length TDP-43, reducing its nuclear localization and function. TDP-43 fragments were also shown to impair neural differentiation. Altogether, these studies show that CTD fragments in TDP-43 proteinopathies may act through a dominant negative mechanism by trapping full-length TDP-43 in cytoplasmic inclusions, preventing it from performing its physiological nuclear RNA processing functions.

Although missense mutations in TDP-43 are associated with rare forms of familial ALS, the role of wild-type TDP-43 in the majority of familial and sporadic ALS/FTLD has been poorly characterized. Homozygous and hemizygous WT human TDP-43 expressing transgenic mice show a dose-dependent degeneration of cortical and spinal motor neurons and develop spastic paralysis characteristic of ALS.[93] Degeneration of nonmotor cortical and subcortical neurons, characteristic of FTLD, was also observed. Moreover, affected neurons in both brain and spinal cord showed accumulation of ubiquitinated and phosphorylated TDP-43 aggregates in both the nucleus and cytoplasm. Interestingly, the characteristic disease-associated 25 kDa CTD fragment was found in the soluble nuclear fraction of affected neurons, which correlated with disease progression in WT TDP-43 mice. Intriguingly, another study on transgenic mice expressing TDP-25 in neurons showed that the accumulation of soluble TDP-25 can lead to cognitive decline in the absence of TDP-43-positive inclusions and without overt neurodegeneration.[94] Thus, CTD fragments of TDP-43 may also act pathologically through a mechanism that is independent of cytoplasmic aggregation. Furthermore, as the CTD of TDP-43 is a low complexity disordered region that is inherently prone to aggregation, overexpression of wild-type TDP-43 may also lead to aggregation even in the absence of specific mutations within TDP-43.

There is much debate about which specific CTD fragment of TDP-43 is implicated in ALS/FTLD disease progression. Overexpression of the 25 kDa CTD fragment of TDP-43 in mouse and human neuroblastoma cell lines is sufficient to induce cytoplasmic mislocalization and accumulation of endogenous full-length TDP-43.[95] Inhibiting autophagy increased the accumulation of CTD fragments, whereas inducing the autophagic response through inhibition of mTOR promoted clearance of the 25 kDa CTD fragment and restored TDP-43 nuclear localization. The 25 kDa CTD fragment of TDP-43 was also shown to induce forelimb impairment when expressed in rats, mimicking the clinical phenotype of ALS.[96] Conversely, expressing CTD fragments in HeLa cells revealed that the 35 kDa fragment, but not the 25 kDa fragment, is implicated in the formation of cytoplasmic inclusions and disruption of the splicing function of TDP-43.[97] TDP-35 was shown to act as a seed to sequester endogenous full-length TDP-43 into cytoplasmic inclusions, thus inhibiting nuclear localization and function. The discrepancy between this study and the study by Caccamo et al. could be attributed to the fact that HeLa cells were used rather than neuroblastoma cells, which are biologically more similar to disease-relevant tissue. Alternatively, the difference in cleavage site of the particular 25 kDa CTD fragment used in this study may explain the difference in results. The 25 kDa CTD fragment identified from previous studies was shown to be cleaved by caspases at Asp219. However, Che et al. showed that the 220–414 fragment runs as a 20 kDa band, whereas the 184–414 fragment that they express in their experiments has a true molecular mass of 25 kDa, with cleavage likely occurring at the linker region between the two RRMs. The previously studied CTD fragment considered to be 25 kDa may actually be smaller, and cleaved within RRM2, producing a different phenotype.

Regardless of which CTD fragment is primarily associated with ALS/FTLD pathology, CTD fragments are recognized as more toxic than full-length TDP-43 and therefore more susceptible to rapid clearance.[98,99] It was shown that the p62 adaptor protein, which binds to ubiquitinated and misfolded proteins, binds to TDP-43 and is likely involved in the degradation of the TDP-43 CTD fragment of 35 kDa, and not full-length TDP-43.[98] Interestingly, p62 coimmunoprecipitated with a smaller amount of TDP-43 in FTLD-TDP brains compared to control brains, suggesting that the interaction between p62 and TDP-43 is disrupted, leading to accumulation of the 35 kDa CTD fragment of TDP-43 and increased cellular toxicity. Inducing molecular chaperone activity through pharmacological upregulation of the heat shock response or genetic upregulation of a small heat shock protein, G14207, in a Drosophila model expressing either full-length TDP-43 or the 25 kDa CTD fragment of TDP-43 suppresses the aggregation of both isoforms of TDP-43.[99] Although inhibiting aggregation of full-length TDP-43 mildly suppresses neurotoxicity in photoreceptor and motor neurons, inhibiting aggregation of TDP-25 suppresses toxicity completely and leads to clearance of TDP-25 in both neuronal subtypes. Thus, TDP-25 appears to have a higher susceptibility to molecular chaperone–mediated clearance than full-length TDP-43. The molecular chaperone HSPB8 was shown to mitigate the accumulation of both TDP-35 and TDP-25 in mammalian motor neurons and suppress toxicity associated with truncated/mislocalized TDP-43 in *D. melanogaster.*[100] Overexpressing HSP67Bc, the Drosophila orthologue of human HSPB8, counteracts the accumulation of an NLS mutant of TDP-43 that mislocalizes to the cytoplasm, thereby suppressing eye degeneration. Furthermore, overexpressing HSP67Bc also counteracts the severe neurodegeneration induced by TDP-35 accumulation.

Cleavage of TDP-43 itself may not be sufficient to induce aggregation and toxicity. Pesiridis and colleagues engineered mammalian cells to express a tobacco etch virus (TEV) protease that cleaves TDP-43 containing a TEV cleavage site in order to monitor the aggregation of CTD fragments.[101] Interestingly, regions within RRM2 are less susceptible to cleavage by TEV, probably because RRM2 is a highly structured region and considered the stability core of TDP-43, whereas regions flanking RRM2 are readily cleaved. C-terminal fragments containing RRM2 were readily transported to the cytoplasm and cleared, suggesting that other processes apart from cleavage are required to induce CTF aggregation. CTFs were shown to rapidly aggregate into cytoplasmic inclusions when RNA was depleted, when dynein-mediated microtubule transport was disrupted, or when natively misfolded CTFs were introduced into these cells. Thus, binding to RNAs or association with actively transported RNPs on the microtubule network may maintain CTFs in a solubilized form, the disruption of either of which may induce misfolding of CTFs, which then serve as seeds to sequester more CTFs in cytoplasmic aggregates. Intriguingly, cytoplasmic inclusions generated by the 35 kDa fragment of TDP-43 (TDP-35) that sequester full-length TDP-43 are actually dependent on the ability to bind RNA.[102] Unlike the 25 kDa CTD fragment of TDP-43, which only contains full-length or truncated RRM2, the 35 kDa fragment contains both RRM2 and RRM1, the latter playing a dominant role in nucleic acid binding. Mutation within RRM1 prevents the formation of TDP-35 inclusions and their RNA-assisted association with TDP-43. It is hypothesized that the RNA-dependent TDP-35 inclusions may be more dynamic in nature than the insoluble aggregates, similar to RNP granules in cells, which are relatively dynamic. Interestingly, TDP-35 was shown to form mildly toxic ubiquitin-negative inclusions in the cytoplasm, whereas TDP-25 forms sufficiently toxic ubiquitin-positive inclusions in the cytoplasm.[103] Furthermore, the coaggregation of TDP-25 with full-length TDP-43 is initiated by depletion of the RNA that binds to TDP-25. Thus, TDP-35 inclusions may be fluid and dynamic as they behave like RNP granules, and so are not recognized as highly toxic by the ubiquitin−proteasome pathway. As TDP-25 lacks RRM1, which plays a more dominant role in nucleic acid binding, it may be inherently more prone to forming insoluble aggregates due to less efficient binding to RNA, thus being susceptible to ubiquitination. Depletion of RNA can then further accelerate coaggregation with full-length TDP-43, increasing cellular toxicity.

The accumulation of both soluble and insoluble CTD fragments of TDP-43 was shown to induce pathology in CNS regions relevant to FTLD/ALS in mouse models.[104,105] Homozygous transgenic mice expressing TDP-25 (TgTDP-25+/+) were shown to develop spatial and working memory deficits as well as decreased motor function compared to the hemizygous transgenic mice (TgTDP-25+/0), showing a gene dosage−dependent effect on cognitive and motor function.[105] These behavioral deficits were correlated with increased levels of soluble TDP-25 in both the nucleus and cytosol. Furthermore, the overexpression of TDP-25 is linked to reduction in autophagy and proteasome function, thus showing that increased levels of TDP-25 can mediate toxicity by inhibiting the two major protein turnover pathways. In another study on transgenic mice, a 20 kDa CTD fragment of TDP-43 (amino acids 208−414) previously identified in FTLD-TDP brains was expressed in the forebrain, leading to insoluble aggregates in the neuronal cytoplasm and progressive neurodegeneration in the hippocampus.[104] Importantly, suppressing the expression of the 208 CTD fragment enhances its clearance, corroborating with the results from Caccamo et al. which showed that overexpression of the CTD fragment inhibits protein turnover.

The 20–25 kDa CTD fragments of TDP-43 found in patient tissues are mostly fragmented within RRM2. The mechanism by which truncation of RRM2 enhances cytoplasmic aggregation and toxicity of TDP-43 is poorly understood. A recent study performed molecular dynamics simulations on the intact RRM2 domain and a truncated variant (where residues 189–207 are deleted to mimic a cleavage site within RRM2 found in ALS patients).[106] While the intact RRM2 domain is highly structured, the truncated variant associated with ALS is highly unstable, corroborating with previous experimental results, and therefore conformationally heterogenous. The structural reorganization of the truncated variant of RRM2 increases the accessibility of hydrophobic residues, including the NES, and also alters the electrostatic surface charges increasing the propensity for protein–protein interactions. Therefore, cleavage within RRM2 may increase the cytoplasmic mislocalization of TDP-43 by making the NES more accessible, and enhance aggregation by increasing the protein–protein interaction strength of the CTD.

Other than TDP-43 pathology, another molecular abnormality found specifically in motor neurons of sporadic ALS patients is the expression of glutamine/arginine (Q/R) site-unedited GluA2, a subunit of alpha-amino-3-hydroxy-5-methyl-lisoxazole-4-proprionic acid (AMPA) receptors.[107] The expression of Q/R site-unedited GluA2 is linked to downregulation of the RNA editing enzyme ADAR2 in ALS spinal cord motor neurons.[108,109] ADAR2 catalyzes A-to-I (adenosine-to-inositol) conversion at the Q/R site of GluA2 pre-mRNA.[110] ADAR2 editing activity at the Q/R site is critical for the survival of motor neurons, as motor neurons lacking ADAR2 undergo slow death due to failure of GluA2 RNA editing, resulting in the upregulation of Ca2+ permeable AMPA receptors.[111] Yamashita et al. found that conditional ADAR2 knockout mice activate the Cal2+ dependent cysteine protease calpain through upregulation of Ca2+ permeable AMPA receptors, which results in the production of C-terminally cleaved TDP-43 fragments, leading to TDP-43 cytoplasmic mislocalization and ALS pathology.[112] Interestingly, mice in which the endogenous GluA2 alleles were replaced with an allele encoding the Q/R site-edited GluA2 expressed Ca2+ impermeable AMPA receptors, inhibiting calpain activation and the production of abnormal N-terminal TDP-43 fragments. Furthermore, the spinal cords and brains of ALS patients showed the presence of calpain-dependent TDP-43 fragments, and an ALS-linked mutant TDP-43 was more vulnerable to calpain-mediated cleavage. Interestingly, a later study by the same group showed that phosphorylated full-length TDP-43 is more resistant to calpain-mediated cleavage than endogenous full-length TDP-43.[113] Previously, calpain was shown to serially cleave TDP-43 into aggregation prone large fragments, and further into smaller fragments that are no longer aggregation prone. However, nonphosphorylated calpain cleaved fragments were later shown to be as resistant to further cleavage by calpain as phosphorylated calpain cleaved fragments. Thus, both phosphorylated and calpain cleaved TDP-43 can persist intracellularly for a long time, serving as seeds for inclusions. Altogether, these studies support the role of calpain in generating aggregation prone N-terminal fragments of TDP-43 in ALS motor neurons, which is distinct from the role of caspases in generating CTD fragments of TDP-43 in the brains of FTLD/ALS patients. Nonetheless, TDP-43 has also shown dual vulnerability to calpain and caspase 3 cleavage in both cerebrocortical neurons in culture subjected to injury and in the cortex of mouse models of traumatic brain injury (TBI).[114] Furthermore, TDP-43 and its 35 kDa fragment were found to be elevated in the cerebrospinal fluid of severe TBI patients, suggesting the use of CTD fragments of TDP-43 as biomarkers for TBI.

Rather than being the causative factor in TDP-43 pathology, TDP-43 CTD cleavage may be implicated in rapid clearance of TDP-43 to prevent the cytotoxic effects of overexpressed full-length TDP-43. It was shown that the CTF of 25 kDa is generated before the production of the CTF of 35 kDa specifically through cleavage after Asp174.[115] This initial cleavage is required for the subsequent activation of caspase-3/7 to further fragment and induce clearance of full-length TDP-43. Furthermore, cleavage at Asp174 is carried out by ER membrane bound caspase-4, which may act as a sensor to detect overloaded full-length TDP-43 at the ER membrane. Inhibiting caspase-4 delays fragmentation and clearance of TDP-43, reducing cell viability.

Cysteine oxidation

Cysteine oxidation of TDP-43 is known to promote disulfide bonds that induce oligomerization of TDP-43 in ALS/FTLD. TDP-43 contains six cysteines, two in the NTD (Cys39 and Cys50), two in RRM1 (Cys173 and Cys175), and two in RRM2 (Cys198 and Cys244), that have all been shown to undergo oxidation and form disulfide bridges. Both in vitro and cell-based assays have shown that oxidative stress induces cysteine oxidation, decreasing TDP-43 solubility.[116] Cysteine residues within or near RRM2 were implicated in the formation of intra- and intermolecular disulfide bonds, the latter leading to the production of high molecular weight oligomeric species of TDP-43, which can be disrupted by treatment with a reducing agent, showing that cysteine oxidation drives oligomerization. Furthermore, FTLD brains have increased levels of disulfide bond cross-linked TDP-43 compared to control brains, confirming the importance of cysteine oxidation in TDP-43 proteinopathy. More recently, disulfide bond cross-linked multimers of TDP-43 have been found in both the cortex and spinal cord of a mutant TDP-43 expressing mouse model of ALS/FTLD.[117] Interestingly, aggregates of TDP-43 in a transgenic *C. elegans* model expressing human TDP-43 variants in the nervous system were also shown to be associated with intermolecular disulfide bonds.[118] However, generating a cysteine-free 25 kDa fragment of TDP-43 in which the cysteine within RRM2 was mutated to an alanine did not significantly disrupt the aggregation of most of the TDP-25 protein, while blocking the formation of the disulfide-linked dimer.

It was shown that cysteine oxidation within RRM1, but not the other domains, induces protein aggregation.[119] Cys173 is preferentially oxidized, leading to conformational change, followed by oxidation of Cys175, leading to cross-linking of RRM1, and disrupting its DNA-binding ability. Although RRM2 oxidation itself was not sufficient to induce protein aggregation, it can enhance RRM1-mediated aggregation, as the RRM1-RRM2 construct tended to aggregate stronger than RRM1 alone. Cys173 and Cys175 were also required to induce insolubilization of TDP-43 in cells treated with 4-hydroxynonenal (HNE), a marker of oxidative stress that is elevated in sporadic ALS patients.[120] The free thiols of Cys173 and Cys175 are crucial for the structural integrity of RRM1, as mutating them to serines induces misfolding and self-assembly of RRM1 in a disulfide bond independent manner, impairing RNA processing activity of RRM1.[121] RRM1 C173S/C175S mutations induce TDP-43 aggregation, recapitulating the hallmarks of ALS pathology, including cytoplasmic localization, ubiquitination, and phosphorylation. TDP-43 aggregates with RRM1 C173S/C175S mutations require the CTD. Thus, misfolding of RRM1 induced by the disruption of the free thiols of cysteines 173 and 175 may occlude the RNA-binding and processing activity of RRM1, and also induce conformational changes within the CTD, increasing its propensity to aggregate.

Protein disulfide isomerase (PDI) is an ER-bound protein that catalyzes and breaks disulfide bonds, and also acts as a molecular chaperone to prevent protein misfolding and aggregation. PDI expression is increased in the cerebrospinal fluid of ALS patients manifesting intracellular aggregates of mutant SOD1.[122] As SOD1 aggregation in familial ALS is also linked to cysteine oxidation and disulfide cross-linking,[123,124] PDI expression in ALS may be activated to prevent abnormal disulfide bond formation and/or mitigate aggregation of disulfide cross-linked SOD1. Interestingly, inducing ER stress in cultured HeLa and mouse neuroblastoma Neuro2A cells by overexpressing an ALS-linked mutant TDP-43 induces PDI to interact with TDP-43.[125] PDI was also shown to interact with TDP-43 in the spinal cords of transgenic mutant TDP-43 expressing mice. Rare genetic variants in PDI have been associated with ALS, and ALS-linked PDI variants were shown to cause motor dysfunction in zebrafish.[126,127] Thus, mutations in PDI may disrupt its ability to mitigate oxidative stress–induced disulfide bond cross-linking or aggregation of TDP-43 or SOD1, exacerbating motor neuron degeneration.

The N-terminal and C-terminal regions of TDP-43 contribute to TDP-43 inclusion formation in mechanistically distinct ways.[128] Notably, Bozzo and colleagues distinguish between oligomerization and the formation of large aggregates. Mutations and molecular dynamics simulations reveal different roles for the six cysteine residues in mediating unfolding and aggregation of TDP-43. The most accessible NTD cysteine residues prime the oligomerization process, which is then propagated by the four cysteine residues within RRM1/RRM2. Either removal or oxidation of cysteines can induce partial unfolding of TDP-43. However, the large aggregates are mediated in a cysteine-independent manner by unfolding within the CTD which enhances nonspecific hydrophobic interactions. Interestingly, intermolecular disulfide bonds have been shown to induce tetramerization of the NTD.[129] Therefore, the cysteines within the NTD initiate oligomerization, which may sequentially induce conformational changes through RRM1/RRM2 down to the CTD, which then induces the formation of large insoluble aggregates. Under oxidative stress, disulfide bond cross-linking of RRM2 can prevent refolding of RRM2, which may lead to gradual unfolding and increase in disorder, seeding the aggregation process.[130] Aggregate particles generated by oxidized RRM2 were also shown to be more susceptible to proteolysis, suggesting that TDP-43 fragmentation in disease can occur after the formation of insoluble aggregates. Furthermore, cysteine oxidation was shown to promote cytoplasmic mislocalization and enhance phosphorylation of s409/s410.[131] Altogether, these studies show the complex and dynamic interplay between various PTMs of TDP-43.

Acetylation

Lysine acetylation is a major PTM involved in diverse cellular signaling pathways.[132] It has also been implicated in AD and other neurodegenerative disorders.[133–137] Acetylation of misfolded tau proteins marks neurofibrillary tangles in AD and related tauopathies and is considered a pathological marker in AD.[134–136] Interestingly, recent studies have shown that acetylation of lysines within RRM1/RRM2 of TDP-43 impair RNA binding and promote accumulation of TDP-43 in hyper-phosphorylated and ubiquitinated insoluble aggregates.[138,139] In particular, acetylation of lysine 145 (K145) within RRM1, located near the RNA-binding interface, is sufficient to drive TDP-43 pathology. Acetylation was shown to be induced by oxidative stress.[138] Interestingly, an antibody targeted against acetylated K145 (ac-K145) was able to detect TDP-43 aggregates in ALS spinal cord that were also immunoreactive for phosphorylated s409/s410. However, the ac-K145 antibody did not readily detect TDP-43 inclusions in FTLD-TDP brain. These results corroborate with earlier findings in which ALS spinal cord was primarily enriched in phosphorylated full-length TDP-43, whereas FTLD-TDP brains were primarily enriched in phosphorylated CTD fragments of TDP-43, which mostly lack the RRM1 domain that contains ac-K145.

There may be a mechanistic relationship between cysteine oxidation and lysine acetylation within RRM1. As redox sensing is typically a rapidly transduced signal, cysteine oxidation may lead to the initial loss of RNA binding due to the changed conformation of RRM1, exposing K145, which is normally protected by RNA binding. Acetylation of K145 may then further prevent TDP-43 from binding to its target RNA by neutralizing the positive charge of K145. Loss of RNA binding may enhance aggregation of TDP-43, as it has been shown that RNA binding maintains the solubility of TDP-43.[140,141]

TDP-43 aggregates induced by acetylation were targeted by p62 and Ubiquilin 2, which bind to ubiquitinated misfolded proteins to target them for autophagosomal and proteasomal degradation, respectively.[139] Molecular chaperones were also targeted to these aggregates in order to disaggregate and potentially induce refolding of TDP-43 to enhance its clearance. HDAC6 is recruited to acetylated TDP-43 and is able to deacetylate TDP-43 under normal conditions, but prolonged oxidative stress prevents deacetylation, as the acetylated lysines may be inaccessible within large toxic aggregates. Interestingly, TDP-43 is known to bind and stabilize HDAC6 mRNA.[142,143] TDP-43 was also shown to form an mRNP complex on HDAC mRNA with FUS/TLS, an hnRNP protein similar to TDP-43 that is mutated in rare cases of ALS and also forms cytosolic aggregates.[144] Therefore, under normal conditions, TDP-43 may increase HDAC6 expression by stabilizing its mRNA, which would increase TDP-43 deacetylation, and thus maintain TDP-43 in its soluble native functional state.

TDP-43 also autoregulates its own mRNA by binding to its 3′UTR.[145,146] Specifically, TDP-43 enhances alternative splicing of an intron in its 3′UTR, which mimics the presence of a premature termination codon, thus triggering degradation of TDP-43 mRNA through nonsense-mediated decay (NMD). Thus, the initial loss of RNA binding through the combination of cysteine oxidation and lysine acetylation may initiate a pathological positive feedback mechanism, whereby decreased NMD increases the pool of TDP-43 mRNAs available for translation, increasing pathogenic TDP-43 protein levels, which increases the size of insoluble aggregates. In addition to the toxic gain of function achieved by these aggregates, the loss of RNA binding may lead to massive loss of neuronal function. TDP-43 was shown to bind to RNAs in mouse adult brain from approximately 6000 genes.[145] Knockdown of TDP-43 primarily depletes RNAs with very long introns enriched in TDP-43-binding sites and that encode proteins involved in synaptic activity. TDP-43 binding to introns may stabilize these RNAs by preventing aberrant splicing events that induce NMD. Furthermore, TDP-43 was also shown to bind to pre-mRNAs in human brain to influence alternative splicing.[147] A substantial proportion of alternative mRNA isoforms regulated by TDP-43 encode proteins involved in neuronal development or have been implicated in neurological disease. Thus, a major contribution of TDP-43 proteinopathy to neurodegeneration may arise from the inability of TDP-43 to regulate the stability or splicing of mRNAs that are essential to maintain neuronal function.

Conclusion

Aberrant protein aggregation is a hallmark of multiple neurodegenerative disorders, including FTLD, ALS, AD, and PD. Different proteins have been characterized as the primary components of pathological aggregates in these diseases, such as tau in AD, α-synuclein in PD, and TDP-43 in FTLD and ALS. Although there is evidence of secondary TDP-43 pathology in other neurodegenerative disorders such as AD and PD, TDP-43 appears to be the primary driver of pathological inclusions in FTLD and ALS. Although FTLD and ALS have different clinical presentations and affect different anatomical regions of the central nervous system, at the molecular level, they represent different manifestations of the same fundamental pathology of mislocalization and aggregation of TDP-43. As TDP-43 is an evolutionarily conserved and ubiquitously expressed protein involved in general RNA metabolism processes ranging from transcription to translation regulation, the mechanistic link between functional aberrations in TDP-43 and neurodegenerative disease is poorly understood. Importantly, whether aggregation of TDP-43 is a cause or consequence of disease is not fully known.

Although mutations in TDP-43 are associated with subsets of sporadic and familial ALS and rare cases of FTLD, a majority of cases involve pathology of wild-type TDP-43. TDP-43 is inherently aggregation prone because of its low complexity CTD that may facilitate phase separation–mediated protein–protein interactions between disordered regions. Under physiological conditions, the CTD may enable interactions with multiple partners in RNP granules implicated in RNA processing activities. However, under prolonged oxidative stress induced by the accumulation of ROS during aging TDP-43 may shift from RNP granules to irreversible insoluble aggregates. Studies in flies and worms show aberrant aggregation and neurodegeneration induced by mutant TDP-43. The role of wild-type TDP-43 in progressive neurodegeneration is underestimated in these studies, as the pathology of wild-type TDP-43 may develop over longer timespans of human central nervous system development that are not recapitulated in simple model systems. Mouse models have shown that overexpression of wild-type TDP-43 can result in cognitive and motor deficits in a dosage-dependent manner.

Early studies focused on the PTMs that are acquired during the early and late phases of TDP-43 aggregation. Ubiquitination appears to be the final PTM that occurs prior to turnover by the autophagosomal or proteasomal degradation pathways. Phosphorylation occurs on granular preinclusions that are smaller than the compact ubiquitinated inclusions. Multiple studies suggest that phosphorylation may inhibit clearance of TDP-43 inclusions, thus contributing to the persistence of toxic aggregates and serving as a robust signature of all familial and sporadic TDP-43 proteinopathies. C-terminal fragmentation may play a key role in driving TDP-43 aggregation by at least two distinct mechanisms. Firstly, the loss of RRM1, the primary nucleic acid−binding domain, in CTD fragments found in ALS/FTLD brains may decrease TDP-43 solubility as RNA binding maintains TDP-43 in its native conformation. Secondly, the hydrophobic core of the CTD may be less accessible within full-length TDP-43, whereas C-terminal cleavage would increase exposure of the hydrophobic residues on the CTD, increasing nonspecific hydrophobic interactions, which were shown to be the primary driver of large aggregate formation.

Recent studies have revealed oxidative stress−mediated changes in the RNA-binding domains that could constitute early steps in the conformational changes that are required for the seeding of large insoluble aggregates. Cysteine residues, particularly within RRM1, may serve as redox sensors that induce conformational changes once oxidized, which in turn favors intra- and intermolecular disulfide bonds, the latter generating high molecular weight oligomers. Lysine acetylation near the RNA-binding interference of RRM1, potentially induced through oxidation-mediated conformational changes, may directly disrupt the electrostatic interactions between TDP-43 and its RNA targets. As mutations associated with ALS/FTLD are primarily located within the CTD, the majority of studies have been devoted to CTD phosphorylation and the pathology of C-terminal fragments. Mutations within the RRM1/RRM2 domains may have been evolutionarily selected against, as they would likely lead to embryonic lethality by disrupting the regulation of thousands of RNA targets that may be reliant on TDP-43. Nonetheless, PTMs such as cysteine oxidation and lysine acetylation, which affect the NTD and RRM1/RRM2 domains, should be studied more extensively in the future as they may induce conformational changes within the CTD increasing its aggregation propensity. In fact, RRM1-C/S mutations enhanced the inclusions of familial ALS-linked TDP-43 mutants in the CTD.[121] Therefore, mutations within the CTD could influence the dynamics of CTD unfolding in response to oxidative stress−induced conformational changes in RRM1. This mechanism could be particularly important in ALS spinal cord neurons, which are enriched in phosphorylated full-length TDP-43, suggesting a CTD fragmentation–independent molecular pathway of aggregation mediated by changes within an intact RRM1. It is more accurate to describe TDP-43 pathology as an accelerating positive feedback loop rather than a linear sequence of events. Initial deposition of aggregates through CTD-mediated mechanisms could signal changes in the NTD and RRM1/RRM2, resulting in loss of nuclear function and sequestration of TDP-43 in cytoplasmic inclusions. Mutations within the CTD of TDP-43 may fine tune the rate at which TDP-43 unfolds, therefore controlling the rate of TDP-43 aggregation, which in turn affects the rate of neurodegeneration.

ALS/FTD-associated mutations in other genes may affect the processing of different PTMs throughout the life cycle of TDP-43. ALS- and FTD-linked mutations in Ubiquilin 2, which binds to ubiqutinated TDP-43, may affect the turnover of TDP-43 aggregates in both motor neurons and the hippocampus, causing motor and cognitive deficits. Mutations in PGRN, resulting in progranulin depletion, are associated with a large fraction of FTLD cases, and have been linked to C-terminal fragmentation, cytoplasmic mislocalization, and phosphorylation of TDP-43 in FTLD brains. Rare genetic variants in PDI have been associated with ALS, and may affect the disulfide cross-linking of TDP-43 through mechanisms that remain to be identified. The most common genetic cause of both familial ALS and familial FTD is a GGGGCC (G_4C_2) hexanucleotide repeat expansion within a non-coding region of the chromosome 9 open reading frame 72 (C9orf72).[148,149] This repeat expansion was identified in most families with a combined ALS/FTD phenotype and TDP-43-based pathology.[148] A very recent study showed that a glycine–arginine repeat protein (poly(GR)) translated from the G_4C_2 repeat expanded region sequesters endogenous TDP-43 in cytoplasmic inclusions.[150] Furthermore, mice expressing poly(GR) induce the mislocalization of nucleocytoplasmic shuttling factors and nucleopore complex proteins, resulting in the aberrant cytoplasmic accumulation of endogenous TDP-43. Altogether, these studies show that the misregulation of TDP-43 in both ALS and FTD is caused by multiple factors affecting PTMs, solubility, and localization of TDP-43 and that clinical phenotypes can result from a combination of toxic gain of function of TDP-43 aggregates, as well as loss of nuclear RNA metabolism processes, affecting potentially hundreds of mRNAs implicated in neuronal function and survival. Furthermore, it remains to be elucidated how TDP-43 pathology is transmitted through axonal pathways from the motor cortex and spinal cord into the hippocampus and prefrontal cortex, causing the shared motor and cognitive deficits of the ALS/FTD spectrum (Figs. 3.1 and 3.2).

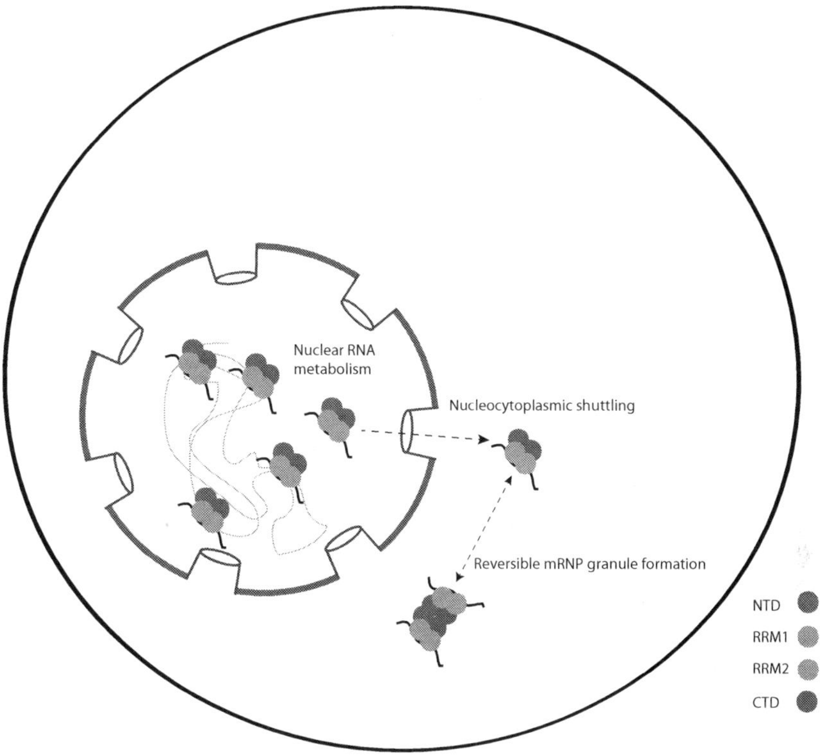

FIGURE 3.1 Localization and functions of TDP-43 in normal physiology. This figure describes the normal trafficking of full-length soluble TDP-43 bound to its target mRNAs.

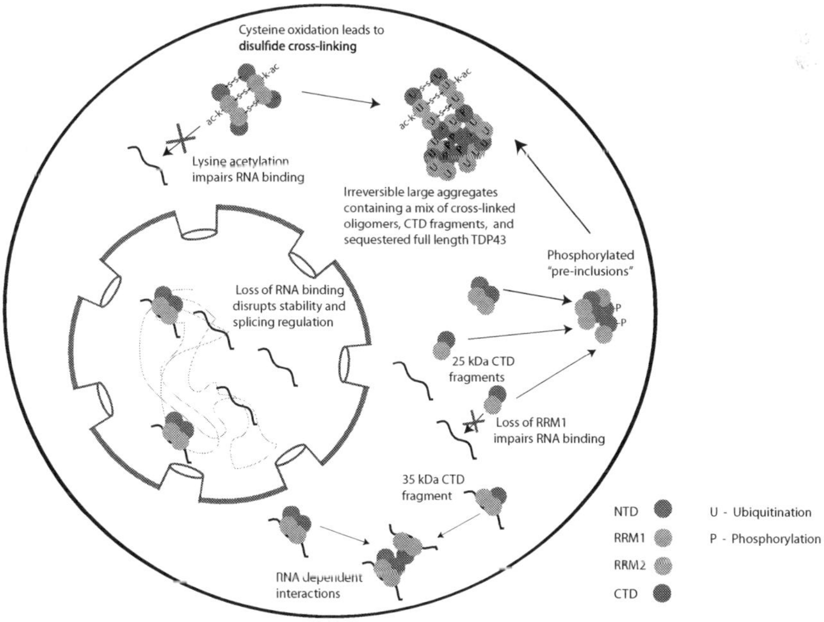

FIGURE 3.2 Posttranslational modifications of TDP-43 in ALS/FTD. This figure describes the sequence of aberrant posttranslational modifications of TDP-43 and how they sequester endogenous full-length TDP-43 into irreversible large cytoplasmic aggregates in the ALS/FTD spectrum.

Abbreviations

ALS amyotrophic lateral sclerosis
CTD C-terminal domain
CTF C-terminal fragment
FTD frontotemporal dementia
FTLD frontotemporal lobar degeneration
FTLD-TDP-43 frontotemporal lobar degeneration with TDP-43-positive inclusions
FTLD-U frontotemporal lobar degeneration with ubiquitin-positive inclusions
NES nuclear export sequence
NLS nuclear localization signal
NTD N-terminal domain
RRM1 RNA recognition motif 1
RRM2 RNA recognition motif 2
TDP-43 transactive response DNA-binding protein of 43 kDA

Acknowledgments

I would like to thank Dr. Manoj Kumar Jaiswal for giving me the opportunity to contribute to this book.

References

1. Ou SH, Wu F, Harrich D, García-Martínez LF, Gaynor RB. Cloning and characterization of a novel cellular protein, TDP-43, that binds to human immunodeficiency virus type 1 TAR DNA sequence motifs. *J Virol* 1995;**69**:3584–96.
2. Buratti E, et al. Nuclear factor TDP-43 and SR proteins promote in vitro and in vivo CFTR exon 9 skipping. *EMBO J* 2001;**20**:1774–84.
3. Buratti E, Baralle FE. Characterization and functional implications of the RNA binding properties of nuclear factor TDP-43, a novel splicing regulator of CFTR exon 9. *J Biol Chem* 2001;**276**:36337–43.
4. Neumann M, et al. Ubiquitinated TDP-43 in frontotemporal lobar degeneration and amyotrophic lateral sclerosis. *Science* 2006;**314**:130–3. 80.
5. Arai T, et al. TDP-43 is a component of ubiquitin-positive tau-negative inclusions in frontotemporal lobar degeneration and amyotrophic lateral sclerosis. *Biochem Biophys Res Commun* 2006;**351**:602–11.
6. Sambataro F, Pennuto M. Post-translational modifications and protein quality control in motor neuron and polyglutamine diseases. *Front Mol Neurosci* 2017;**10**:1–13.
7. McKhann GM. Clinical and pathological diagnosis of frontotemporal dementia: report of the work group on frontotemporal dementia and Pick's disease. *Arch Neurol* 2001;**58**:1803–9.
8. Nakano I. Temporal lobe lesions in amyotrophic lateral sclerosis with or without Dementia : a neuropathological study. *Neuropathology* 1993;**12**:215–27.
9. Wightman G, et al. Hippocampal and neocortical ubiquitin-immunoreactive inclusions in amyotrophic lateral sclerosis with dementia. *Neurosci Lett* 1992;**139**:269–74.
10. Jackson M, Lennox G, Lowe J. Motor neuron disease-inclusion dementia. *Neurodegeneration* 1996;**5**:339–50.

11. Knopman DS, Mastri AR, Frey WH, Sung JH, Rustan T. Dementia lacking distinctive histologic features. *Neurology* 1990;**40**. 251 LP − 251.
12. Lipton AM, White CL, Bigio EH. Frontotemporal lobar degeneration with motor neuron disease-type inclusions predominates in 76 cases of frontotemporal degeneration. *Acta Neuropathol* 2004;**108**:379−85.
13. Portet F, Cadilhac C, Touchon J, Camu W. Cognitive impairment in motor neuron disease with bulbar onset. *Amyotroph Lateral Scler Other Mot Neuron Disord* 2001;**2**:23−9.
14. Lomen-Hoerth C, Anderson T, Miller B. The overlap of amyotrophic lateral sclerosis and frontotemporal dementia. *Neurology* 2002;**59**:1077−9.
15. Schreiber H, et al. Cognitive function in bulbar- and spinal-onset amyotrophic lateral sclerosis: a longitudinal study in 52 patients. *J Neurol* 2005;**252**:772−81.
16. Ikeda K, Tsuchiya K. Motor neuron disease group accompanied by inclusions of unidentified protein signaled by ubiquitin. *Neuropathology* 2004;**24**:117−24.
17. Hara T, et al. Suppression of basal autophagy in neural cells causes neurodegenerative disease in mice. *Nature* 2006;**441**:885−9.
18. Komatsu M, et al. Loss of autophagy in the central nervous system causes neurodegeneration in mice. *Nature* 2006;**441**:880−4.
19. Skibinski G, et al. Mutations in the endosomal ESCRTIII-complex subunit CHMP2B in frontotemporal dementia. *Nat Genet* 2005;**37**:806−8.
20. Parkinson N, et al. ALS phenotypes with mutations in CHMP2B (charged multivesicular body protein 2B). *Neurology* 2006;**67**:1074−7.
21. Filimonenko M, et al. Functional multivesicular bodies are required for autophagic clearance of protein aggregates associated with neurodegenerative disease. *J Cell Biol* 2007;**179**:485−500.
22. Holm IE, Englund E, Mackenzie IRA, Johannsen P, Isaacs AM. A reassessment of the neuropathology of frontotemporal dementia linked to chromosome 3. *J Neuropathol Exp Neurol* 2007;**66**:884−91.
23. Rollinson S, et al. Ubiquitin associated protein 1 is a risk factor for frontotemporal lobar degeneration. *Neurobiol Aging* 2009;**30**:656−65.
24. Kim SH, et al. Potentiation of amyotrophic lateral sclerosis (ALS)-associated TDP-43 aggregation by the proteasome-targeting factor, ubiquilin 1. *J Biol Chem* 2009;**284**:8083−92.
25. Deng HX, et al. Mutations in UBQLN2 cause dominant X-linked juvenile and adult-onset ALS and ALS/dementia. *Nature* 2011;**477**:211−5.
26. Cassel JA, Reitz AB. Ubiquilin-2 (UBQLN2) binds with high affinity to the C-terminal region of TDP-43 and modulates TDP-43 levels in H4 cells: characterization of inhibition by nucleic acids and 4-aminoquinolines. *Biochim Biophys Acta−Protein Proteonomics* 2013;**1834**:964−71.
27. Osaka M, Ito D, Suzuki N. Disturbance of proteasomal and autophagic protein degradation pathways by amyotrophic lateral sclerosis-linked mutations in ubiquilin 2. *Biochem Biophys Res Commun* 2016;**472**:324−31.
28. Picher-Martel V, Dutta K, Phaneuf D, Sobue G, Julien JP. Ubiquilin-2 drives NF-κB activity and cytosolic TDP-43 aggregation in neuronal cells. *Mol Brain* 2015;**8**.
29. Ceballos-Diaz C, et al. Viral expression of ALS-linked ubiquilin-2 mutants causes inclusion pathology and behavioral deficits in mice. *Mol Neurodegener* 2015;**10**:1−13.
30. Le NTT, et al. Motor neuron disease, TDP-43 pathology, and memory deficits in mice expressing ALS-FTD-linked UBQLN2 mutations. *Proc Natl Acad Sci U S A* 2016;**113**:E7580−9.

31. Seyfried NT, et al. Multiplex SILAC analysis of a cellular TDP-43 proteinopathy model reveals protein inclusions associated with SUMOylation and diverse polyubiquitin chains. *Mol Cell Proteomics* 2010;**9**:705–18.
32. Dammer EB, et al. Coaggregation of RNA-binding proteins in a model of TDP-43 proteinopathy with selective RGG motif methylation and a role for RRM1 ubiquitination. *PLoS One* 2012;**7**.
33. Hebron ML, et al. Parkin ubiquitinates tar-DNA binding protein-43 (TDP-43) and promotes its cytosolic accumulation via interaction with histone deacetylase 6 (HDAC6). *J Biol Chem* 2013;**288**:4103–15.
34. Hans F, et al. UBE2E ubiquitin-conjugating enzymes and ubiquitin isopeptidase y regulate TDP-43 protein ubiquitination. *J Biol Chem* 2014;**289**:19164–79.
35. Uechi H, et al. Ubiquitin-binding protein CG5445 suppresses aggregation and cytotoxicity of amyotrophic lateral sclerosis-linked TDP-43 in Drosophila. *Mol Cell Biol* 2018;**38**:1–14.
36. Gong C-X, Singh TJ, Grundke-Iqbal I, Iqbal K. Phosphoprotein phosphatase activities in Alzheimer disease brain. *J Neurochem* 1993;**61**:921–7.
37. Inukai Y, et al. Abnormal phosphorylation of ser409/410 of TDP-43 in FTLD-U and ALS. *FEBS Lett* 2008;**582**:2899–904.
38. Nonaka T, et al. Phosphorylated and ubiquitinated TDP-43 pathological inclusions in ALS and FTLD-U are recapitulated in SH-SY5Y cells. *FEBS Lett* 2009;**583**:394–400.
39. Neumann M, et al. Phosphorylation of S409/410 of TDP-43 is a consistent feature in all sporadic and familial forms of TDP-43 proteinopathies. *Acta Neuropathol* 2009;**117**:137–49.
40. Hasegawa M, et al. Phosphorylated TDP-43 in frontotemporal lobar degeneration and amyotrophic lateral sclerosis. *Ann Neurol* 2008;**64**:60–70.
41. Kametani F, et al. Identification of casein kinase-1 phosphorylation sites on TDP-43. *Biochem Biophys Res Commun* 2009;**382**:405–9.
42. Zhang Y-J, et al. Phosphorylation regulates proteasomal-mediated degradation and solubility of TAR DNA binding protein-43 C-terminal fragments. *Mol Neurodegener* 2010;**5**:33.
43. Liachko NF, Guthrie CR, Kraemer BC. Phosphorylation promotes neurotoxicity in a *Caenorhabditis elegans* model of TDP-43 proteinopathy. *J Neurosci* 2010;**30**:16208–19.
44. Brady OA, Meng P, Zheng Y, Mao Y, Hu F. Regulation of TDP-43 aggregation by phosphorylation and p62/SQSTM1. *J Neurochem* 2011;**116**:248–59.
45. Furukawa Y, Kaneko K, Nukina N. Molecular properties of TAR DNA binding protein-43 fragments are dependent upon its cleavage site. *Biochim Biophys Acta - Mol Basis Dis* 2011;**1812**:1577–83.
46. Abe K, Pan LH, Watanabe M, Kato T, Itoyama Y. Induction of nitrotyrosine-like immunoreactivity in the lower motor neuron of amyotrophic lateral sclerosis. *Neurosci Lett* 1995;**199**:152–4.
47. Abe K, et al. Upregulation of protein-tyrosine nitration in the anterior horn cells of amyotrophic lateral sclerosis. *Neurol Res* 1997;**19**:124–8.
48. Beal MF, et al. Increased 3-nitrotyrosine in both sporadic and familial amyotrophic lateral sclerosis. *Ann Neurol* 1997;**42**:644–54.
49. Butterfield DA, Reed T, Newman SF, Sultana R. Roles of amyloid β-peptide-associated oxidative stress and brain protein modifications in the pathogenesis of Alzheimer's disease and mild cognitive impairment. *Free Radic Biol Med* 2007;**43**:658–77.
50. Nunomura A, et al. Neuronal RNA oxidation is a prominent feature of dementia with Lewy bodies. *Neuroreport* 2002;**13**:2035–9.

51. Lovell MA, Markesbery WR. Oxidative DNA damage in mild cognitive impairment and late-stage Alzheimer's disease. *Nucleic Acids Res* 2007;**35**:7497–504.
52. Shaw IC, Fitzmaurice PS, Mitchell JD, Lynch PG. Studies on cellular free radical protection mechanisms in the anterior horn from patients with amyotrophic lateral sclerosis. *Neurodegeneration* 1995;**4**:391–6.
53. Iguchi Y, et al. Oxidative stress induced by glutathione depletion reproduces pathological modifications of TDP-43 linked to TDP-43 proteinopathies. *Neurobiol Dis* 2012;**45**:862–70.
54. Higashi S, et al. TDP-43 associates with stalled ribosomes and contributes to cell survival during cellular stress. *J Neurochem* 2013;**126**:288–300.
55. Kim HJ, et al. Therapeutic modulation of eIF2α phosphorylation rescues TDP-43 toxicity in amyotrophic lateral sclerosis disease models. *Nat Genet* 2014;**46**:152–60.
56. Lee S, et al. Activation of HIPK2 promotes ER stress-mediated neurodegeneration in amyotrophic lateral sclerosis. *Neuron* 2016;**91**:41–55.
57. Burkhardt MF, et al. A cellular model for sporadic ALS using patient-derived induced pluripotent stem cells. *Mol Cell Neurosci* 2013;**56**:355–64.
58. Yang YM, et al. A small molecule screen in stem-cell-derived motor neurons identifies a kinase inhibitor as a candidate therapeutic for ALS. *Cell Stem Cell* 2013;**12**:713–26.
59. Brettschneider J, et al. Stages of pTDP-43 pathology in amyotrophic lateral sclerosis. *Ann Neurol* 2013;**74**:20–38.
60. Tan RH, et al. Classification of FTLD-TDP cases into pathological subtypes using antibodies against phosphorylated and non-phosphorylated TDP43. *Acta Neuropathol Commun* 2014;**2**:1–9.
61. Takeuchi R, et al. Heterogeneity of cerebral TDP-43 pathology in sporadic amyotrophic lateral sclerosis: evidence for clinico-pathologic subtypes. *Acta Neuropathol Commun* 2016;**4**:61.
62. Liachko NF, et al. The tau tubulin kinases TTBK1/2 promote accumulation of pathological TDP-43. *PLoS Genet* 2014;**10**.
63. Sato S, et al. Spatial learning impairment, enhanced CDK5/p35 activity, and downregulation of NMDA receptor expression in transgenic mice expressing tau-tubulin kinase 1. *J Neurosci* 2008;**28**:14511–21.
64. Xu J, et al. Tau-tubulin kinase 1 enhances prefibrillar tau aggregation and motor neuron degeneration in P301L FTDP-17 tau-mutant mice. *Faseb J* 2010;**24**:2904–15.
65. Salado IG, et al. Protein kinase CK-1 inhibitors as new potential drugs for amyotrophic lateral sclerosis. *J Med Chem* 2014;**57**:2755–72.
66. Baker M, et al. Mutations in progranulin cause tau-negative frontotemporal dementia linked to chromosome 17. *Nature* 2006;**442**:916–9.
67. Cruts M, et al. Null mutations in progranulin cause ubiquitin-positive frontotemporal dementia linked to chromosome 17q21. *Nature* 2006;**442**:920–4.
68. Alquezar C, et al. Alteration in cell cycle-related proteins in lymphoblasts from carriers of the c.709-1G>A PGRN mutation associated with FTLD-TDP dementia. *Neurobiol Aging* 2012;**33**. 429.e7–429.e20.
69. Alquezar C, et al. Increasing progranulin levels and blockade of the ERK1/2 pathway: upstream and downstream strategies for the treatment of progranulin deficient frontotemporal dementia. *Eur Neuropsychopharmacol* 2015;**25**:386–403.
70. Yang Y, Geldmacher DS, Herrup K. DNA replication precedes neuronal cell death in Alzheimer's disease. *J Neurosci* 2001;**21**:2661–8.

71. Höglinger GU, et al. The pRb/E2F cell-cycle pathway mediates cell death in Parkinson's disease. *Proc Natl Acad Sci U S A* 2007;**104**:3585–90.
72. Husseman JW, Nochlin D, Vincent I. Mitotic activation: a convergent mechanism for a cohort of neurodegenerative diseases. *Neurobiol Aging* 2000;**21**:815–28.
73. Copani A, et al. Activation of cell-cycle-associated proteins in neuronal death: a mandatory or dispensable path? *Trends Neurosci* 2001;**24**:25–31.
74. Herrup K, Neve R, Ackerman SL, Copani A. Divide and die: cell cycle events as triggers of nerve cell death. *J Neurosci* 2004;**24**:9232–9.
75. Alquezar C, et al. Targeting TDP-43 phosphorylation by Casein Kinase-1δ inhibitors: a novel strategy for the treatment of frontotemporal dementia. *Mol Neurodegener* 2016;**11**:1–14.
76. Ayala YM, Misteli T, Baralle FE. TDP-43 regulates retinoblastoma protein phosphorylation through the repression of cyclin-dependent kinase 6 expression. *Proc Natl Acad Sci U S A* 2008;**105**:3785–9.
77. Carlomagno Y, et al. Casein kinase II induced polymerization of soluble TDP-43 into filaments is inhibited by heat shock proteins. *PLoS One* 2014;**9**.
78. Choksi DK, et al. TDP-43 phosphorylation by casein kinase iε promotes oligomerization and enhances toxicity in vivo. *Hum Mol Genet* 2014;**23**:1025–35.
79. Nonaka T, et al. Phosphorylation of TAR DNA-binding protein of 43 kDa (TDP-43) by truncated casein kinase 1δ triggers mislocalization and accumulation of TDP-43. *J Biol Chem* 2016;**291**:5473–83.
80. Liachko NF, et al. The phosphatase calcineurin regulates pathological TDP-43 phosphorylation. *Acta Neuropathol* 2016;**132**:545–61.
81. Kim JM, et al. Impaired Cu–Zn superoxide dismutase (SOD1) and calcineurin (Cn) interaction in ALS: a Presumed consequence for TDP-43 and zinc aggregation in Tg SOD1 G93A rodent spinal cord tissue. *Neurochem Res* 2019;**44**:228–33.
82. Gu J, et al. Protein Phosphatase 1 dephosphorylates TDP-43 and suppresses its function in tau exon 10 inclusion. *FEBS Lett* 2018;**592**:402–10.
83. Gregory RI, et al. The Microprocessor complex mediates the genesis of microRNAs. *Nature* 2004;**432**:235–40.
84. Colombrita C, et al. TDP-43 and FUS RNA-binding proteins bind distinct sets of cytoplasmic messenger RNAs and differently regulate their post-transcriptional fate in motoneuron-like cells. *J Biol Chem* 2012;**287**:15635–47.
85. Kim KY, et al. A phosphomimetic mutant TDP-43 (S409/410E) induces Drosha instability and cytotoxicity in Neuro 2A cells. *Biochem Biophys Res Commun* 2015;**464**:236–43.
86. Zhang YJ, et al. Progranulin mediates caspase-dependent cleavage of TAR DNA binding protein-43. *J Neurosci* 2007;**27**:10530–4.
87. Zhang YJ, et al. Aberrant cleavage of TDP-43 enhances aggregation and cellular toxicity. *Proc Natl Acad Sci U S A* 2009;**106**:7607–12.
88. Igaz LM, et al. Enrichment of C-terminal fragments in TAR DNA-binding protein-43 cytoplasmic inclusions in brain but not in spinal cord of frontotemporal lobar degeneration and amyotrophic lateral sclerosis. *Am J Pathol* 2008;**173**:182–94.
89. Dormann D, et al. Proteolytic processing of TAR DNA binding protein-43 by caspases produces C-terminal fragments with disease defining properties independent of progranulin. *J Neurochem* 2009;**110**:1082–94.
90. Nonaka T, Kametani F, Arai T, Akiyama H, Hasegawa M. Truncation and pathogenic mutations facilitate the formation of intracellular aggregates of TDP-43. *Hum Mol Genet* 2009;**18**:3353–64.

91. Igaz LM, et al. Expression of TDP-43 C-terminal fragments in vitro recapitulates pathological features of TDP-43 proteinopathies. *J Biol Chem* 2009;**284**:8516–24.
92. Yang C, et al. The C-terminal TDP-43 fragments have a high aggregation propensity and harm neurons by a dominant-negative mechanism. *PLoS One* 2010;**5**.
93. Wils H, et al. TDP-43 transgenic mice develop spastic paralysis and neuronal inclusions characteristic of ALS and frontotemporal lobar degeneration. *Proc Natl Acad Sci U S A* 2010;**107**:3858–63.
94. Caccamo A, Majumder S, Oddo S. Cognitive decline typical of frontotemporal lobar degeneration in transgenic mice expressing the 25-kDa C-terminal fragment of TDP-43. *Am J Pathol* 2012;**180**:293–302.
95. Caccamo A, et al. Rapamycin rescues TDP-43 mislocalization and the associated low molecular mass neurofilament instability. *J Biol Chem* 2009;**284**:27416–24.
96. Dayton RD, et al. Selective forelimb impairment in rats expressing a pathological TDP-43 25 kDa C-terminal fragment to mimic amyotrophic lateral sclerosis. *Mol Ther* 2013;**21**:1324–34.
97. Che M, Jiang Y, Xie Y, Jiang L, Hu H. Aggregation of the 35-kDa fragment of TDP-43 causes formation of cytoplasmic inclusions and alteration of RNA processing. *Faseb J* 2011;**25**:2344–53.
98. Tanji K, et al. P62/sequestosome 1 binds to TDP-43 in brains with frontotemporal lobar degeneration with TDP-43 inclusions. *J Neurosci Res* 2012;**90**:2034–42.
99. Gregory JM, Barros TP, Meehan S, Dobson CM, Luheshi LM. The aggregation and neurotoxicity of TDP-43 and its als-associated 25 kDa fragment are differentially affected by molecular chaperones in drosophila. *PLoS One* 2012;**7**.
100. Crippa V, et al. The chaperone HSPB8 reduces the accumulation of truncated TDP-43 species in cells and protects against TDP-43-mediated toxicity. *Hum Mol Genet* 2016;**25**:3908–24.
101. Pesiridis GS, Tripathy K, Tanik S, Trojanowski JQ, Lee VMY. A ‘two-hit’ hypothesis for inclusion formation by carboxyl-terminal fragments of TDP-43 protein linked to RNA depletion and impaired microtubule-dependent transport. *J Biol Chem* 2011;**286**:18845–55.
102. Che MX, Jiang LL, Li HY, Jiang YJ, Hu HY. TDP-35 sequesters TDP-43 into cytoplasmic inclusions through binding with RNA. *FEBS Lett* 2015;**589**:1920–8.
103. Kitamura A, et al. Interaction of RNA with a C-terminal fragment of the amyotrophic lateral sclerosis-associated TDP43 reduces cytotoxicity. *Sci Rep* 2016;**6**:1–14.
104. Walker AK, et al. An insoluble frontotemporal lobar degeneration-associated TDP-43 C-terminal fragment causes neurodegeneration and hippocampus pathology in transgenic mice. *Hum Mol Genet* 2015;**24**:7241–54.
105. Caccamo A, et al. Reduced protein turnover mediates functional deficits in transgenic mice expressing the 25 kDa C-terminal fragment of TDP-43. *Hum Mol Genet* 2015;**24**:4625–35.
106. Morgan BR, Zitzewitz JA, Massi F. Structural rearrangement upon fragmentation of the stability core of the ALS-linked protein TDP-43. *Biophys J* 2017;**113**:540–9.
107. Kwak S, Kawahara Y. Deficient RNA editing of GluR2 and neuronal death in amyotropic lateral sclerosis. *J Mol Med* 2005;**83**:110–20.
108. Aizawa H, et al. TDP-43 pathology in sporadic ALS occurs in motor neurons lacking the RNA editing enzyme ADAR2. *Acta Neuropathol* 2010;**120**:75–84.
109. Hideyama T, et al. Profound downregulation of the RNA editing enzyme ADAR2 in ALS spinal motor neurons. *Neurobiol Dis* 2012;**45**:1121–8.
110. Higuchi M, et al. Point mutation in an AMPA receptor gene rescues lethality in mice deficient in the RNA-editing enzyme ADAR2. *Nature* 2000;**406**:78–81.

111. Hideyama T, et al. Induced loss of ADAR2 engenders slow death of motor neurons from Q/R site-unedited GluR2. *J Neurosci* 2010;**30**:11917–25.
112. Yamashita T, et al. A role for calpain-dependent cleavage of TDP-43 in amyotrophic lateral sclerosis pathology. *Nat Commun* 2012;**3**:1–13.
113. Yamashita T, Teramoto S, Kwak S. Phosphorylated TDP-43 becomes resistant to cleavage by calpain: a regulatory role for phosphorylation in TDP-43 pathology of ALS/FTLD. *Neurosci Res* 2016;**107**:63–9.
114. Yang Z, Lin F, Robertson CS, Wang KKW. Dual vulnerability of TDP-43 to calpain and caspase-3 proteolysis after neurotoxic conditions and traumatic brain injury. *J Cerebr Blood Flow Metabol* 2014;**34**:1444–52.
115. Li Q, Yokoshi M, Okada H, Kawahara Y. The cleavage pattern of TDP-43 determines its rate of clearance and cytotoxicity. *Nat Commun* 2015;**6**.
116. Cohen TJ, Hwang AW, Unger T, Trojanowski JQ, Lee VMY. Redox signalling directly regulates TDP-43 via cysteine oxidation and disulphide cross-linking. *EMBO J* 2012;**31**:1241–52.
117. Bargsted L, et al. Disulfide cross-linked multimers of TDP-43 and spinal motoneuron loss in a TDP-43A315T ALS/FTD mouse model. *Sci Rep* 2017;**7**:1–11.
118. Zhang T, Mullane PC, Periz G, Wang J. TDP-43 neurotoxicity and protein aggregation modulated by heat shock factor and insulin/IGF-1 signaling. *Hum Mol Genet* 2011;**20**:1952–65.
119. Chang CK, Chiang MH, Toh EKW, Chang CF, Huang TH. Molecular mechanism of oxidation-induced TDP-43 RRM1 aggregation and loss of function. *FEBS Lett* 2013;**587**:575–82.
120. Kabuta C, Kono K, Wada K, Kabuta T. 4-hydroxynonenal induces persistent insolubilization of TDP-43 and alters its intracellular localization. *Biochem Biophys Res Commun* 2015;**463**:82–7.
121. Shodai A, et al. Aberrant assembly of RNA recognition motif 1 links to pathogenic conversion of TAR DNA-binding protein of 43 kDa (TDP-43). *J Biol Chem* 2013;**288**:14886–905.
122. Grek C, Townsend DM. Protein disulfide isomerase superfamily in disease and the regulation of apoptosis. *Endoplasmic Reticulum Stress Dis* 2014;**1**:4–17.
123. Furukawa Y, Fu R, Deng HX, Siddique T, O'Halloran TV. Disulfide cross-linked protein represents a significant fraction of ALS-associated Cu, Zn-superoxide dismutase aggregates in spinal cords of model mice. *Proc Natl Acad Sci U S A* 2006;**103**:7148–53.
124. Karch CM, Prudencio M, Winkler DD, Hart PJ, Borchelt DR. Role of mutant SOD1 disulfide oxidation and aggregation in the pathogenesis of familial ALS. *Proc Natl Acad Sci U S A* 2009;**106**:7774–9.
125. Walker AK, et al. ALS-associated TDP-43 induces endoplasmic reticulum stress, which drives cytoplasmic TDP-43 accumulation and stress granule formation. *PLoS One* 2013;**8**:1–12.
126. Gonzalez-perez P, et al. Identification of rare protein disulfide isomerase gene variants in amyotrophic lateral sclerosis patients. *Gene* 2015;**566**:158–65.
127. Woehlbier U, et al. ALS-linked protein disulfide isomerase variants cause motor dysfunction. *EMBO J* 2016;**35**:845–65.
128. Bozzo F, et al. Structural insights into the multi-determinant aggregation of TDP-43 in motor neuron-like cells. *Neurobiol Dis* 2016;**94**:63–72.
129. Jiang LL, et al. The N-terminal dimerization is required for TDP-43 splicing activity. *Sci Rep* 2017;**7**:1–12.

130. Rabdano SO, et al. Onset of disorder and protein aggregation due to oxidation-induced intermolecular disulfide bonds: Case study of RRM2 domain from TDP-43. *Sci Rep* 2017;**7**:1–20.
131. Goh CW, et al. Chronic oxidative stress promotes GADD34-mediated phosphorylation of the TAR DNA-binding protein TDP-43, a modification linked to neurodegeneration. *J Biol Chem* 2018;**293**:163–76.
132. Choudhary C, Weinert BT, Nishida Y, Verdin E, Mann M. The growing landscape of lysine acetylation links metabolism and cell signalling. *Nat Rev Mol Cell Biol* 2014;**15**:536–50.
133. Min SW, et al. Acetylation of tau inhibits its degradation and contributes to tauopathy. *Neuron* 2010;**67**:953–66.
134. Irwin DJ, et al. Acetylated tau, a novel pathological signature in Alzheimer's disease and other tauopathies. *Brain* 2012;**135**:807–18.
135. Irwin DJ, et al. Acetylated tau neuropathology in sporadic and hereditary tauopathies. *Am J Pathol* 2013;**183**:344–51.
136. Cohen TJ, et al. The acetylation of tau inhibits its function and promotes pathological tau aggregation. *Nat Commun* 2011;**2**.
137. Cook C, et al. Acetylation of the KXGS motifs in tau is a critical determinant in modulation of tau aggregation and clearance. *Hum Mol Genet* 2014;**23**:104–16.
138. Cohen TJ, et al. An acetylation switch controls TDP-43 function and aggregation propensity. *Nat Commun* 2015;**6**.
139. Wang P, Wander CM, Yuan CX, Bereman MS, Cohen TJ. Acetylation-induced TDP-43 pathology is suppressed by an HSF1-dependent chaperone program. *Nat Commun* 2017;**8**:1–15.
140. Huang YC, et al. Inhibition of TDP-43 aggregation by nucleic acid binding. *PLoS One* 2013;**8**.
141. Sun Y, Arslan PE, Won A, Yip CM, Chakrabartty A. Binding of TDP-43 to the 3′UTR of its cognate mRNA enhances its solubility. *Biochemistry* 2014;**53**:5885–94.
142. Fiesel FC, et al. Knockdown of transactive response DNA-binding protein (TDP-43) downregulates histone deacetylase 6. *EMBO J* 2010;**29**:209–21.
143. Fiesel FC, Schurr C, Weber SS, Kahle PJ. TDP-43 knockdown impairs neurite outgrowth dependent on its target histone deacetylase 6. *Mol Neurodegener* 2011;**6**:1–10.
144. Kim SH, Shanware NP, Bowler MJ, Tibbetts RS. Amyotrophic lateral sclerosis-associated proteins TDP-43 and FUS/TLS function in a common biochemical complex to co-regulate HDAC6 mRNA. *J Biol Chem* 2010;**285**:34097–105.
145. Polymenidou M, et al. Long pre-mRNA depletion and RNA missplicing contribute to neuronal vulnerability from loss of TDP-43. *Nat Neurosci* 2011;**14**:459–68.
146. Bhardwaj A, Myers MP, Buratti E, Baralle FE. Characterizing TDP-43 interaction with its RNA targets. *Nucleic Acids Res* 2013;**41**:5062–74.
147. Tollervey JR, et al. Characterizing the RNA targets and position-dependent splicing regulation by TDP-43. *Nat Neurosci* 2011;**14**:452–8.
148. DeJesus-Hernandez M, et al. Expanded GGGGCC hexanucleotide repeat in noncoding region of C9ORF72 causes chromosome 9p-linked FTD and ALS. *Neuron* 2011;**72**:245–56.
149. Renton AE, et al. A hexanucleotide repeat expansion in C9ORF72 is the cause of chromosome 9p21-linked ALS-FTD. *Neuron* 2011;**72**:257–68.
150. Cook CN, et al. C9orf72 poly(GR) aggregation induces TDP-43 proteinopathy. *Sci Transl Med* 2020;**12**:1–12.

Chapter 4

Regulation of autophagy by TDP-43: a promising therapeutic intervention in neurodegenerative diseases

Anil Kumar Tomar, Savita Yadav
Department of Biophysics, All India Institute of Medical Sciences, New Delhi, India

Introduction

A ubiquitously expressed and encoded by the TARDBP gene, transactive response DNA-binding protein 43 (TDP-43) has a 414 amino acid long sequence. A member of the heterogeneous nuclear ribonucleoprotein (hnRNP) family, TDP-43 is a nucleotide-binding protein that binds to both DNA and RNA and consists of two RNA recognition motifs. These motifs are RRM1 (106-177aa) and RRM2 (192-259aa). Its primary function is the regulation of mRNA processing, transcription, and posttranscription. Other documented roles are stress granule (SG) response and microRNA biogenesis.[1–3] Autophagy, a process by which intracellular machinery eliminates impaired or excessive cytoplasmic material, is crucial for normal cellular development. It helps in overcoming cellular stress and any defects in associated processes are linked to various human diseases including cancers, infections, heart dysfunction, and neurological disorders. It is believed that autophagy controls the expression of TDP-43 and TDP-43 regulates the events of autophagy. Thus, there exists a feedback regulatory loop between autophagy and TDP-43, specifically in neurodegenerative diseases.

Autophagy

Autophagy means *self-eating* and it is the body's way to eat out damaged cells in order to regenerate new and healthy cells. Technically, it is a degradation pathway which is an adoptive intracellular homeostasis process. It is an evolutionarily conserved process that intracellular machinery utilizes to

TDP-43 and Neurodegeneration. https://doi.org/10.1016/B978-0-12-820066-7.00005-9

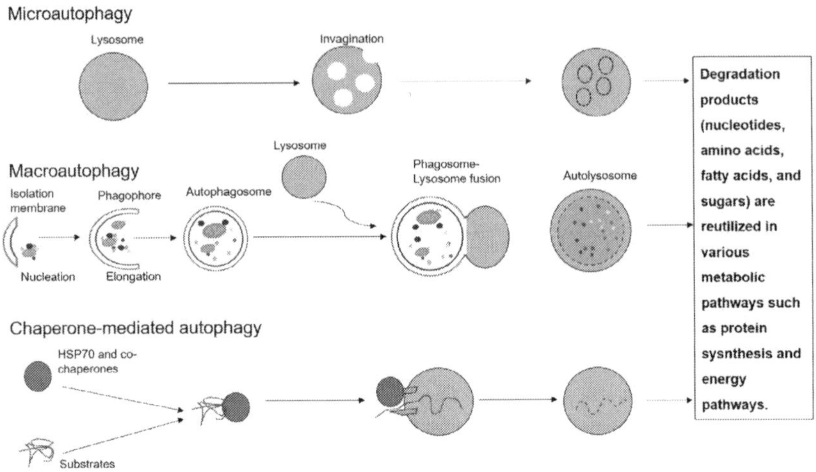

FIGURE 4.1 Different types of autophagy.[11]

eliminate impaired or excessive cytoplasmic material, such as organelles and biological macromolecules with the help of lysosomal enzymes called acidic hydrolases. Most mammalian cells have three types of autophagy, each of which performs a very specific task.[4,5] These are macroautophagy, microautophagy, and chaperone-mediated autophagy (Fig. 4.1). In the case of macroautophagy, a double membranous autophagosome develops that sequesters impaired cellular materials and subsequently delivers them to the lysosome for degradation.[6] In mammalian species, the outer membrane of autophagosome fuses with a lysosome, and hydrolytic enzymes degrade the autophagy cargo, components of which are then released into the cytosol. While in microautophagy, direct uptake of cytoplasmic contents into the lysosome takes place through invagination.[7] In chaperone-mediated autophagy, cytosolic chaperones unfold the targeted proteins and then transport them directly across the lysosomal membrane for degradation.[8,9] The breakdown products such as nucleotides, amino acids, fatty acids, and sugars are reutilized in various metabolic pathways. Though all three types of autophagy processes lead to degradation of cytoplasmic contents and then release of the breakdown products back into the cytosol so that they can be reused by the cell, macroautophagy is most explored and best studied. Autophagy is considered a cellular protective process as it is crucial for maintaining intracellular homeostasis, recycling of nutrients, and clearance of impaired cellular materials. It is well known that autophagy is crucial for normal development and helps in overcoming adverse starvation and cellular stress conditions. The sequential steps of the autophagy mechanism are (i) formation of autophagosome, (ii) cargo selection, (iii) autophagosome and lysosome fusion, and (iv) degradation; and disruption at any stage affects the whole process as each step is crucial for the success of autophagy process.[4] Defects in autophagy have been associated with various human diseases, including cancers, infections, heart dysfunction, and neurodegenerative diseases.[10]

Autophagy in neurodegenerative disorders

A vast number of studies have suggested that neurons rely upon autophagy to maintain their normal function and homeostasis and dysregulation in autophagy is directly linked to neurodegenerative disorders.[12–15] When neuronal cell division becomes dysfunctional, impaired organelles and biological wastes start accumulating. Over the time, neurons lose their ability to clear these wastes and as a result, autolysosomal products such as protein aggregates accumulate abnormally giving rise to degeneration of neurons.[16,17] Enough pieces of evidence have been generated through several years of research to establish the role of autophagy in maintaining neuronal health via delivery of abnormal protein aggregates and toxic molecules to lysosomes for their degradation. It makes autophagy one of the core regulators of senescence and neurodegeneration in the central nervous system.[18,19]

Autophagy is linked to most of the neurodegenerative diseases, which involve the accumulation of misfolded proteins, for example, amyotrophic lateral sclerosis (ALS), polyQ diseases, Parkinson's disease (PD), and Alzheimer's disease (AD). In ALS, several proteins fail to fold properly due to mutations in their genes, e.g., SOD1, TDP-43, FUS, C9orf72, etc.[20–27] It leads to the accumulation of misfolded proteins resulting in progressive degeneration of motor neurons. Autophagy is one of the master regulators of the formation of such pathological aggregates in ALS as overexpression of autophagy-related genes reduces the cellular toxicity induced by accumulation of the misfolded proteins.[28] PolyQ diseases are inherited progressive neurodegenerative diseases that include Huntington's disease, spinal muscular atrophy, ataxia, etc. These diseases are characterized by the expansion of CAG repeats in specific genes leading to the formation of toxic oligomers and aggregates. It has been shown that neuronal autophagic activity affects the disease pathogenesis as autophagy inducers decrease polyQ aggregation.[29,30] AD is a chronic neurodegenerative disease. The neurotoxic *beta-amyloid* (Aβ) plaques and *tau* accumulate in the brain of AD patients. Though the role of autophagy in AD is still controversial, the downregulation of autophagy in AD brains is quite evident. In the early phase of AD, reduced expression of autophagy-associated Beclin 1 was reported in the brain[31]. Also, autophagy inducers reduce the level of toxic Aβ aggregates in AD.[32] PD is a functional failure of dopaminergic neurons leading to muscle rigidity and subsequent neurocognitive impairment. The neurotoxicity in PD is mainly due to high levels of α-synuclein in neurons and its efficient degradation by autophagy is crucial for controlling the disease severity.[33] It has been reported that a defect in autophagy increases α-synuclein accumulation and exacerbates the disease.[34]

RNA-binding proteins in autophagy

As the name suggests RNA-binding proteins (RBPs) bind to RNA in the cell and participate in the formation of ribonucleoprotein complexes.[35] The RBPs contain well-defined structural motifs that can either bind to RNA sequences or some structural motifs within RNA, including RNA recognition motif (RRM), hnRNP K homology domain, or DEAD-box helicase domain[36,37]. The RBPs play a crucial role in RNA metabolism and are considered as key regulators of gene expression in eukaryotes. As a result of recent proteomic advances and the development of new methods to identify RBPs, so many new RBPs have been identified throughout the proteomes of various organisms that helped in establishing their role(s) RNA biology. SGs are the assembly of membrane-less structures that are associated with cellular stress and impaired mRNA translation. These cytoplasmic complexes mainly consist of mRNA and RBPs. The formation of SGs is linked to mutations in autophagy proteins and degenerative disorders of the muscle, bone, and nervous system. It is a general scientific consensus that inhibition of the lysosome-associated autophagy pathway impairs the overall dynamics of SGs resulting in their reduced degradation and, thus, accumulation of SGs. To date, numerous studies have advocated the role of RBPs in autophagy and subsequently in various neurodegenerative diseases. This consideration has been derived from the fact that the misregulation of genes involved in RNA metabolism or autophagy is directly associated with neurodegenerative diseases. Different RBPs are identified and studied in this regard including TDP-43, T-cell intracellular antigen 1 (TIA-1), ataxin-2 (ATXN2), polyadenylate-binding protein (PABP), fused in sarcoma protein (FUS), heterogeneous nuclear ribonucleoprotein A1 (hnRNPA1), heterogeneous nuclear ribonucleoprotein A2/B1 (hnRNPA2/B1), TATA-box–binding protein–associated factor 15 (TAF15), Ras GTPase-activating protein-binding protein 1 (G3BP1), microtubule-associated protein Tau, etc. All these RBPs share some structural and functional similarities. These proteins are associated with RNA metabolism and as a protective response during cellular stress, most of them localize to SGs.[38,39]

TDP-43 and autophagy in ALS

TDP-43 is required for CNS growth from the early embryonic stage to adulthood. It is regarded as one of the main causative factors in diverse neurological disorders, specifically ALS and frontotemporal dementia. A motor neuron disease, ALS causes dysfunction in the nerves that control muscle movements and is generally characterized by the deposition of pathological protein aggregates containing TDP-43.[24,40–42] A wide range of studies have reported that TDP-43 mutations cause ALS.[43] ALS-causing TDP-43 mutations lead to autophagy dysfunction.[44] Additionally, many cell-based studies provided evidence that clearly indicates that mutations of TDP-43 bestow resistance to its degradation or clearance by autophagy.[45] In most of the ALS patients with no known genomic mutations in TDP-43, deposition of wild-type TDP-43 aggregates was observed in their brain and spinal cord.[40] It is believed that either cytoplasmic TDP-43 aggregation causes autophagy dysfunction or autophagy dysfunction causes TDP-43 aggregation in the cytoplasm. Conclusively, the inability of the autophagy system to clear TDP-43 aggregates is quite evident in ALS.

There exists a feedback regulatory loop between autophagy and TDP-43.[46] TDP-43 expression is controlled by autophagy, which affects both forms of TDP-43, wild-type, and aggregated forms. Similarly, TDP-43 also plays a regulatory role in autophagy events. It has been found that the depletion of TDP-43 caused autophagy impairment.[47] Later on, studies elaborated that TDP-43 loss impairs autophagosome and lysosome fusion.[48,49] A recent study has reported enhanced global expression of autophagy—lysosome pathway-associated genes and autophagosomal—lysosomal biogenesis due to loss of TDP-43 in the nucleus.[50] Many of the related studies also showed that TDP-43 clearance was improved through the induction of autophagy.[51–53] TDP-43 localization and its level determine the neurotoxicity and induction of autophagy stimulate TDP-43 clearance that results in reduced TDP-43 levels and enhanced survival in neuronal ALS.[51]

TDP-43 interactome

Freibaum et al. performed global proteomic profiling of TDP-43 interacting proteins and revealed a strong association of TDP-43 with the regulation of RNA metabolism and RNA splicing.[54] They isolated TDP-binding proteins by coimmunoprecipitation and identified them by mass spectrometry. Out of 261 proteins enriched, they found that 126 were exclusively associated with TDP-43. Some of the important TDP-binding proteins are listed in Table 4.1. Mostly linked to RNA metabolism, the list of TDP-43 interacting proteins included heterogeneous nuclear ribonucleoprotein (hnRNPs), RNA helicases, splicing factors, translation regulatory proteins, and proteins involved in mRNA transport and stability. They also observed that disease-associated TDP-43 mutations did not alter this TDP-43 interactome. They further added that TDP-43 exhibits both RNA-dependent and RNA-independent interactions. TDP-43 interactome formed two major clusters: (i) Nuclear/Splicing Cluster consisting of hnRNPs, serine/arginine-rich proteins, small nuclear ribonucleoproteins (snRNPs), an ATP-dependent RNA helicase and nuclear RNA export factors; and (ii) Cytoplasmic/Translation Cluster consisting of translation initiation and elongation factors, and ribosomal subunits.[54] These clusters were connected by an interesting interactor, PABPC1—a specific marker for SGs. The association of TDP-43 with SGs was further supported by the identification of many known proteins of SGs in the TDP-43 interactome (Table 4.2). It was then confirmed by colocalization of TDP-43 with EIF4G

TABLE 4.1 TDP-43 interacting proteins.[54]

Protein name	Symbol	Mol. weight (kDa)	Accession number	Total spectra
40S ribosomal protein S3	RPS3	27	P23396	68
Nucleolin	NCL	77	P19338	57
Polyadenylate-binding protein 1	PABPC1	71	P11940	53
Heat shock cognate 71 kDa protein	HSPA8	71	P11142	51
ATP-dependent RNA helicase A	DHX9	141	Q08211	49
Histone H1.3	HIST1H1D	22	P16402	46
Heat shock 70 kDa protein 1	HSPA1A	70	P08107	44
Heterogeneous nuclear ribonucleoprotein U	HNRNPU (HNRPU)	91	Q00839	43
Interleukin enhancer–binding factor 3	ILF3	95	Q12906	42
Putative ATP-dependent RNA helicase DHX30	DHX30	134	Q7L2E3	41
40S ribosomal protein S9	RPS9	23	P46781	40
60S ribosomal protein L7a	RPL7A	30	P62424	38
Heterogeneous nuclear ribonucleoproteins A2/B1	HNRNPA2B1 (HNRPA2B1)	37	P22626	38
Heterogeneous nuclear ribonucleoproteins C1/C2	HNRNPC (HNRPC)	34	P07910	34

Insulin-like growth factor 2 mRNA-binding protein 1	IGF2BP1	63	Q9NZI8	33
Probable ATP-dependent RNA helicase DDX5	DDX5	69	P17844	32
40S ribosomal protein S18	RPS18	18	P62269	31
40S ribosomal protein SA	RPSA	33	P08865	30
60S ribosomal protein L6	RPL6	33	Q02878	30
Nucleolar RNA helicase 2	DDX21	87	Q9NR30	29

Top 20 interacting proteins based on number of total spectra.

TABLE 4.2 Stress granule proteins coimmunoprecipitated with TDP-43.[54]

S. No.	Name
1	Protein argonaute-2 EIF2C2
2	Caprin-1 CAPRIN1
3	Eukaryotic translation initiation factor 3 (5 subunits) EIF3
4	Eukaryotic translation initiation factor 4 gamma (2 subunits) EIF4G
5	Far upstream element-binding protein 2 KHSRP
6	Ras GTPase-activating protein-binding protein 1 G3BP1
7	ELAV-like protein 1 ELAVL1
8	Polyadenylate-binding protein 1 PABPC1
9	Probable ATP-dependent RNA helicase DDX6 DDX6
10	Plakophilin-1 PKP1
11	Double-stranded RNA-binding protein Staufen homolog 1 STAU1
12	Ubiquitin carboxyl-terminal hydrolase 10 USP10
13	Eukaryotic initiation factor 4A (2 subunits) EIF4A
14	Nuclease-sensitive element-binding protein 1 YBX1
15	40S Ribosome (20 subunits) 40S

and G3BP1, two major proteins associated with SGs. Interestingly, RNA binding is required for TDP-43 localization in SGs. Based on all these findings, it was concluded that TDP-43 plays a translational regulatory role.

Blokhuis et al. performed a comparative interactomics analysis of different ALS-associated proteins.[55] The list of proteins included wild-type and ALS-associated mutant versions of ATXN2, C9orf72, FUS, OPTN, TDP-43, and UBQLN2. Several previously known TDP-43 interactors were identified in this study, such as hnRNPA2B1 and hnRNPA1, hnRNPH, hnRNPK, hnRNPQ, interleukin enhancer—binding factor 3, Matrin3, Pabpc, and RNA helicase DDX5. Several of the TDP-43-binding proteins identified, such as hnRNPA1, hnRNPA2B1, Matrin3, and FMRP, were associated with neurodegenerative disorders. The interaction maps of different ALS-associated proteins converged to some common pathways, namely RNA metabolism, autophagy, and NF-κB signaling. The extensive analysis found that FMRP is a common interactor of ATXN2, FUS, and TDP-43. As FMRP reverses pathogenic effects caused by mutant FUS and TDP-43, it can be explored as a common therapeutic target in ALS pathogenesis. An exploration of protein—protein functional network database STRING (Version 11.0) provided a protein

TABLE 4.3 STRING db analysis: 74 Nodes of TDP-43 protein network.

ALS2, ALYREF, ATXN2, C9orf72, CDC5L, CHMP2B, CHTOP, CPSF6, CTNNBL1, DDX39B, DHX9, ELAVL1, ERH, EWSR1, EXOSC10, FUS, GRN, HNRNPA0, HNRNPA1, HNRNPA2B1, HNRNPA3, HNRNPC, HNRNPF, HNRNPH1, HNRNPH3, HNRNPK, HNRNPL, HNRNPM, HNRNPR, HNRNPU, HTT, ILF2, ILF3, KHDRBS1, KHSRP, MAPT, MATR3, MBNL1, NDRG3, OPTN, PABPN1, PARK2, PCBP1, PTBP1, RBM14, RBMX, SFPQ, SNRNP70, SNRPD2, SNRPD3, SNRPE, SNRPF, SNRPG, SOD1, SQSTM1, SRRT, SRSF1, SRSF3, SRSF7, SSB, SYNCRIP, TAF15, **TARDBP**, TIAL1, TMEM106B, TRA2B, U2AF2, U2AF35, UBE2I, UBQLN1, UBQLN2, VAPB, VCP, XRN2.

This gene (TARDBP) encodes TDP-43. It was used as an input for network analysis and thus, highlighted by bold text.

network of 74 nodes for TDP-43 at a high confidence level (Table 4.3). Analysis of this protein network (Fig. 4.2) reveals that most of these proteins are associated with mRNA processing (biological processes, GO:0006397 and GO:0016071) and RNA splicing (biological processes, GO:0000375, GO:0008380, and GO:0000398). Also, more than half of the interactors were RNA-binding which supports the fact that interaction between TDP-43 and its interactors can be both RNA dependent or independent.

Most of the ALS patients feature abnormalities in TDP-43 and loss of nuclear TDP-43 impairs DNA repair.[56] A very recent study identified changes induced by DNA damage in TDP-43 and FUS interactomes.[57] It was observed that changes in TDP-43 interactome were fewer and the most affected interactions were shared by both TDP-43 and FUS. This interactomics study concluded that TDP-43 was essential for genomic stability and DNA damage repair.

Therapeutic applications: targeting autophagy in the treatment of neurological disorders

Autophagy is considered one of the most important biological events as deregulation of associated pathways gives rise to a range of human diseases. Numerous recent scientific studies have highlighted that malfunctions generated by alterations in autophagy cause many human diseases including cancers, inflammatory diseases, cardiovascular and metabolic dysfunctions, and neurodegenerative diseases. Due to this, autophagy has immense clinical significance. Autophagy is closely linked to the development of neurological disorders and, thus, the process and its regulatory biological molecules are explored extensively for searching new drug targets for improving pharmacological regulation of neuroprotection.[58] Recently, Ramesh et al. have reviewed the role of autophagy in different diseases. This review gives a detailed account of several diseases associated with alterations in the autophagy-related genes.[59] Many of these autophagy genes were associated

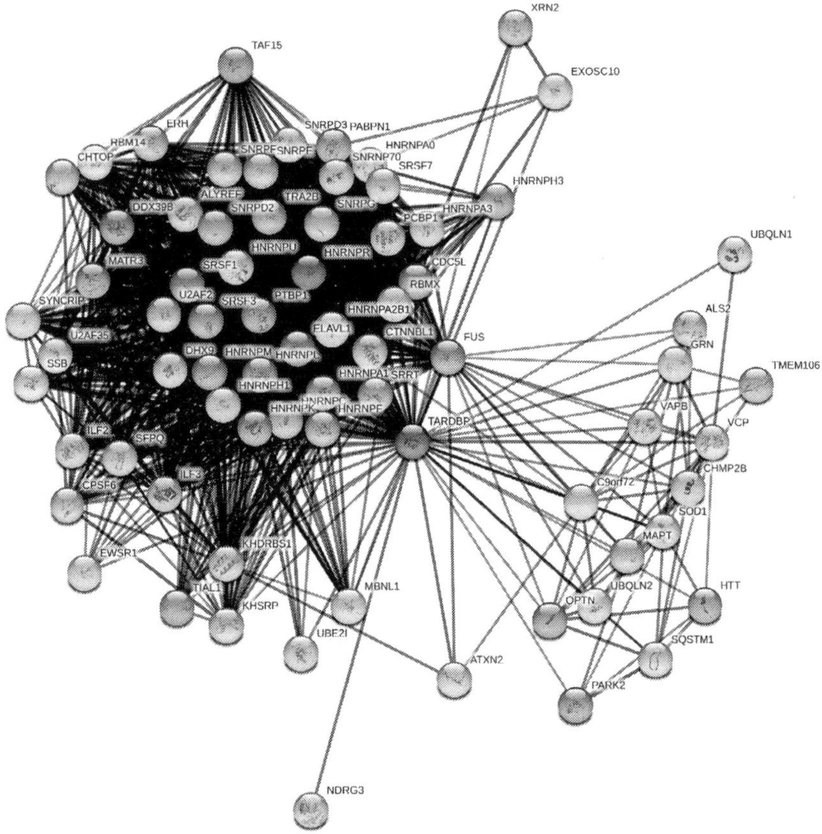

FIGURE 4.2 TDP-43 interactome. *Source: STRING database.*

with different neurological disorders, such as Ambra1, Beclin1, cathepsin, cln3, etc. There are several viewpoints related to the therapeutic approaches that vary for different diseases, targeting the regulation of the autophagy process through various inducers and inhibitors. As autophagy impairment is linked to many diseases, induction of autophagy using effective inducers becomes the foremost basic therapeutic approach. For example, autophagy is usually induced in cancer treatment in addition to the enhancement of its tumor-suppressing traits. Many of the anticancer drugs induce cell death in a way similar to autophagy, e.g., rapamycin, tamoxifen, histone deacetylase inhibitors, ionizing radiation, and vitamin D analogues.[60] Most of the human neurodegenerative disorders are characterized by the deposition of aggregated proteins that accumulate in neurons and it is well known that these can be removed by the process of autophagy.[61] Thus, upregulation of autophagy is a powerful tool and probably a promising therapeutic solution for neurodegenerative diseases as it has the potential to protect neurons from the toxicity of

TABLE 4.4 Potential drugs that target autophagy-related molecular events for the treatment of neurodegenerative diseases.

Disease	Drug
Alzheimer's disease	Rapamycin, Arctigenin, Latrepirdine, Resveratrol, Nilotinib, Trehalose, Temsirolimus, Berberine
Parkinson's disease	Curcumin, Resveratrol, Trehalose, Beclin-1 expression mimetics, Nilotinib, Piperine, Sestrin2, Calcitriol
Huntington's disease	Rapamycin, Rilmenidine, Histone deacetylase inhibitor, CTEP, Liraglutide, Neferine
Amyotrophic lateral sclerosis	Rapamycin, Trehalose, Berberine, S6K1 inhibitor, n-butylidenephthalide (BP)

protein accumulation.[62] Two widely targeted mechanisms applied for inducing autophagy are inhibition of the mammalian target of rapamycin (mTOR) and activation of AMP-activated protein kinase (AMPK). mTOR is a Ser/Thr kinase that regulates cell growth and various biological processes. mTOR-dependent signal transduction pathway controls autophagy through growth factors and amino acids. Activation of mTOR inhibits autophagy via inhibitory phosphorylation of ULK1. It has been reported that mTOR inhibition by amino acid depletion or by the mTOR inhibitors (such as rapamycin and resveratrol) induces autophagy.[63,64] Impairment of cellular ATP production activates autophagy via activation of AMPK, which is a key cellular energy sensor protein. AMPK gets activated when the cell reaches a low energy state. In cell starvation conditions, AMPK suppresses mTOR to induce autophagy via activation of autophagy-initiating kinase ULK1.[65] AMPK activators such as resveratrol, quercetin, metformin, polyphyllin I, and berberine are known to induce autophagy.[66–68] Several autophagy-targeting drugs are widely accepted as a potential treatment option for various neurodegenerative diseases. Recently, Meng et al. have reviewed the role of autophagy in neurological disorders and compiled a list of autophagy-related potential drugs for the treatment of neurodegenerative diseases, *see* Table 4.4.[69] Conclusively, autophagy is closely associated with neurodegenerative disorders like ALS, and induction of autophagy via different processes can serve as a potential treatment target for these diseases.

Conclusion and future prospectives

There exists a feedback regulatory loop between TDP-43 and autophagy, where autophagy controls the expression of wild and aggregated forms of TDP-43 and TDP-43 regulates the events of autophagy. Based on available scientific evidence, a balanced synergy between TDP-43 expression and autophagy reduces neurotoxicity and is capable of enhancing the survival rate in patients with neurodegenerative diseases. Thus, autophagy and regulation of autophagy by TDP-43 can further be explored and targeted for developing potential therapeutic mediations for treating neurodegenerative disorders.

Conflict of interest

The authors declare no conflict of interest.

Acknowledgments

The authors thank Indian Council of Medical Research (ICMR), New Delhi, for financial support.

References

1. Buratti E, Baralle FE. The multiple roles of TDP-43 in pre-mRNA processing and gene expression regulation. *RNA Biol* 2010;**7**(4):420–9.
2. Colombrita C, et al. TDP-43 is recruited to stress granules in conditions of oxidative insult. *J Neurochem* 2009;**111**(4):1051–61.
3. Tollervey JR, et al. Characterizing the RNA targets and position-dependent splicing regulation by TDP-43. *Nat Neurosci* 2011;**14**(4):452–8.
4. Ravanan P, Srikumar IF, Talwar P. Autophagy: the spotlight for cellular stress responses. *Life Sci* 2017;**188**:53–67.
5. Tasset I, Cuervo AM. Role of chaperone-mediated autophagy in metabolism. *FEBS J* 2016;**283**(13):2403–13.
6. Feng Y, et al. The machinery of macroautophagy. *Cell Res* 2014;**24**(1):24–41.
7. Sahu R, et al. Microautophagy of cytosolic proteins by late endosomes. *Dev Cell* 2011;**20**(1):131–9.
8. Cuervo AM, Wong E. Chaperone-mediated autophagy: roles in disease and aging. *Cell Res* 2014;**24**(1):92–104.
9. Kaushik S, Cuervo AM. Chaperone-mediated autophagy: a unique way to enter the lysosome world. *Trends Cell Biol* 2012;**22**(8):407–17.
10. Fernandez AF, Lopez-Otin C. The functional and pathologic relevance of autophagy proteases. *J Clin Invest* 2015;**125**(1):33–41.
11. Mizushima N, Komatsu M. Autophagy: renovation of cells and tissues. *Cell* 2011;**147**(4):728–41.
12. Budini M, et al. Autophagy and its impact on neurodegenerative diseases: new roles for TDP-43 and C9orf72. *Front Mol Neurosci* 2017;**10**:170.
13. Cirulli ET, et al. Exome sequencing in amyotrophic lateral sclerosis identifies risk genes and pathways. *Science* 2015;**347**(6229):1436–41.
14. Shimizu H, et al. Sporadic ALS with compound heterozygous mutations in the SQSTM1 gene. *Acta Neuropathol* 2013;**126**(3):453–9.
15. Teyssou E, et al. Mutations in SQSTM1 encoding p62 in amyotrophic lateral sclerosis: genetics and neuropathology. *Acta Neuropathol* 2013;**125**(4):511–22.

16. Hara T, et al. Suppression of basal autophagy in neural cells causes neurodegenerative disease in mice. *Nature* 2006;**441**(7095):885–9.
17. Komatsu M, et al. Loss of autophagy in the central nervous system causes neurodegeneration in mice. *Nature* 2006;**441**(7095):880–4.
18. Plaza-Zabala A, Sierra-Torre V, Sierra A. Autophagy and microglia: novel partners in neurodegeneration and aging. *Int J Mol Sci* 2017;**18**(3).
19. Wang MM, et al. The relationship between autophagy and brain plasticity in neurological diseases. *Front Cell Neurosci* 2019;**13**:228.
20. Prakash A, et al. Comparative analysis of thermal unfolding simulations of RNA recognition motifs (RRMs) of TAR DNA-binding protein 43 (TDP-43). *J Biomol Struct Dyn* 2019;**37**(1):178–94.
21. Prakash A, et al. Solvent sensitivity of protein aggregation in Cu, Zn superoxide dismutase: a molecular dynamics simulation study. *J Biomol Struct Dyn* 2018;**36**(10):2605–17.
22. Kumar V, et al. TFE-induced local unfolding and fibrillation of SOD1: bridging the experiment and simulation studies. *Biochem J* 2018;**475**(10):1701–19.
23. Kumar V, Hasan GM, Hassan MI. Unraveling the role of RNA mediated toxicity of C9orf72 repeats in C9-FTD/ALS. *Front Neurosci* 2017;**11**:711.
24. Kumar V, et al. Protein aggregation and neurodegenerative diseases: from theory to therapy. *Eur J Med Chem* 2016;**124**:1105–20.
25. Kumar V, et al. Structural insight into C9orf72 hexanucleotide repeat expansions: towards new therapeutic targets in FTD-ALS. *Neurochem Int* 2016;**100**:11–20.
26. Kumar V, et al. Delineating the effect of mutations on the conformational dynamics of N-terminal domain of TDP-43. *Biophys Chem* 2019;**250**:106174.
27. Kumar V, Prakash A, Lynn AM. Alterations in local stability and dynamics of A4V SOD1 in the presence of trifluoroethanol. *Biopolymers* 2018;**109**(3):e23102.
28. Chen Y, et al. The altered autophagy mediated by TFEB in animal and cell models of amyotrophic lateral sclerosis. *Am J Trans Res* 2015;**7**(9):1574–87.
29. Chen ZZ, et al. Trehalose attenuates the gait ataxia and gliosis of spinocerebellar ataxia type 17 mice. *Neurochem Res* 2015;**40**(4):800–10.
30. Jiang W, et al. Therapeutic effect of berberine on Huntington's disease transgenic mouse model. *PLoS One* 2015;**10**(7):e0134142.
31. Pickford F, et al. The autophagy-related protein beclin 1 shows reduced expression in early Alzheimer disease and regulates amyloid beta accumulation in mice. *J Clin Invest* 2008;**118**(6):2190–9.
32. Steele JW, et al. Latrepirdine improves cognition and arrests progression of neuropathology in an Alzheimer's mouse model. *Mol Psychiatry* 2013;**18**(8):889–97.
33. Rott R, et al. alpha-Synuclein fate is determined by USP9X-regulated monoubiquitination. *Proc Natl Acad Sci USA* 2011;**108**(46):18666–71.
34. Fujikake N, Shin M, Shimizu S. Association between autophagy and neurodegenerative diseases. *Front Neurosci* 2018;**12**:255.
35. Dreyfuss G, Kim VN, Kataoka N. Messenger-RNA-binding proteins and the messages they carry. *Nat Rev Mol Cell Biol* 2002;**3**(3):195–205.
36. Hentze MW, et al. A brave new world of RNA-binding proteins. *Nat Rev Mol Cell Biol* 2018;**19**(5):327–41.
37. Kumar V, et al. Exploring the aggregation-prone regions from structural domains of human TDP-43. *Biochim Biophys Acta Proteins Proteom* 2019;**1867**(3):286–96.
38. Buchan JR, Parker R. Eukaryotic stress granules: the ins and outs of translation. *Mol Cell* 2009;**36**(6):932–41.

39. Monahan Z, Shewmaker F, Pandey UB. Stress granules at the intersection of autophagy and ALS. *Brain Res* 2016;**1649**(Pt B):189−200.
40. Neumann M, et al. Ubiquitinated TDP-43 in frontotemporal lobar degeneration and amyotrophic lateral sclerosis. *Science* 2006;**314**(5796):130−3.
41. Sami N, et al. Protein aggregation, misfolding and consequential human neurodegenerative diseases. *Int J Neurosci* 2017;**127**(11):1047−57.
42. Kumar V, et al. Therapeutic progress in amyotrophic lateral sclerosis-beginning to learning. *Eur J Med Chem* 2016;**121**:903−17.
43. Lagier-Tourenne C, Cleveland DW. Rethinking ALS: the FUS about TDP-43. *Cell* 2009;**136**(6):1001−4.
44. Ramesh N, Pandey UB. Autophagy dysregulation in ALS: when protein aggregates get out of hand. *Front Mol Neurosci* 2017;**10**:263.
45. Ling SC, et al. ALS-associated mutations in TDP-43 increase its stability and promote TDP-43 complexes with FUS/TLS. *Proc Natl Acad Sci USA* 2010;**107**(30):13318−23.
46. Wang X, et al. Activation of ER stress and autophagy induced by TDP-43 A315T as pathogenic mechanism and the corresponding histological changes in skin as potential biomarker for ALS with the mutation. *Int J Biol Sci* 2015;**11**(10):1140−9.
47. Bose JK, Huang CC, Shen CK. Regulation of autophagy by neuropathological protein TDP-43. *J Biol Chem* 2011;**286**(52):44441−8.
48. Xia Q, et al. TDP-43 loss of function increases TFEB activity and blocks autophagosome-lysosome fusion. *EMBO J* 2016;**35**(2):121−42.
49. Ying Z, et al. TARDBP/TDP-43 regulates autophagy in both MTORC1-dependent and MTORC1-independent manners. *Autophagy* 2016;**12**(4):707−8.
50. Sun L, et al. TAR DNA binding protein-43 loss of function induced by phosphorylation at S409/410 blocks autophagic flux and participates in secondary brain injury after intracerebral hemorrhage. *Front Cell Neurosci* 2018;**12**:79.
51. Barmada SJ, et al. Autophagy induction enhances TDP43 turnover and survival in neuronal ALS models. *Nat Chem Biol* 2014;**10**(8):677−85.
52. Scotter EL, et al. Differential roles of the ubiquitin proteasome system and autophagy in the clearance of soluble and aggregated TDP-43 species. *J Cell Sci* 2014;**127**(Pt 6):1263−78.
53. Wang X, et al. Degradation of TDP-43 and its pathogenic form by autophagy and the ubiquitin-proteasome system. *Neurosci Lett* 2010;**469**(1):112−6.
54. Freibaum BD, et al. Global analysis of TDP-43 interacting proteins reveals strong association with RNA splicing and translation machinery. *J Proteome Res* 2010;**9**(2):1104−20.
55. Blokhuis AM, et al. Comparative interactomics analysis of different ALS-associated proteins identifies converging molecular pathways. *Acta Neuropathol* 2016;**132**(2):175−96.
56. Mitra J, et al. Motor neuron disease-associated loss of nuclear TDP-43 is linked to DNA double-strand break repair defects. *Proc Natl Acad Sci USA* 2019;**116**(10):4696−705.
57. Kawaguchi T, et al. Changes to the TDP-43 and FUS interactomes induced by DNA damage. *J Proteome Res* 2020;**19**(1):360−70.
58. Russo R, et al. Autophagy dysregulation and the fate of retinal ganglion cells in glaucomatous optic neuropathy. *Prog Brain Res* 2015;**220**:87−105.
59. Ramesh J, et al. Autophagy intertwines with different diseases-recent strategies for therapeutic approaches. *Diseases* 2019;**7**(1).
60. Shimizu S, et al. Role of Bcl-2 family proteins in a non-apoptotic programmed cell death dependent on autophagy genes. *Nat Cell Biol* 2004;**6**(12):1221−8.
61. Frake RA, et al. Autophagy and neurodegeneration. *J Clin Invest* 2015;**125**(1):65−74.

62. Karabiyik C, Lee MJ, Rubinsztein DC. Autophagy impairment in Parkinson's disease. *Essays Biochem* 2017;**61**(6):711–20.
63. Meijer AJ, Codogno P. Autophagy: regulation and role in disease. *Crit Rev Clin Lab Sci* 2009;**46**(4):210–40.
64. Park D, et al. Resveratrol induces autophagy by directly inhibiting mTOR through ATP competition. *Sci Rep* 2016;**6**:21772.
65. Jang M, et al. AMPK contributes to autophagosome maturation and lysosomal fusion. *Sci Rep* 2018;**8**(1):12637.
66. Kim J, et al. AMPK activators: mechanisms of action and physiological activities. *Exp Mol Med* 2016;**48**:e224.
67. Wang Y, et al. Metformin induces autophagy and G0/G1 phase cell cycle arrest in myeloma by targeting the AMPK/mTORC1 and mTORC2 pathways. *J Exp Clin Cancer Res* 2018;**37**(1):63.
68. Wu Y, et al. Polyphyllin I activates AMPK to suppress the growth of non-small-cell lung cancer via induction of autophagy. *Arch Biochem Biophys* 2020:108285.
69. Meng T, et al. Recent progress in the role of autophagy in neurological diseases. *Cell Stress* 2019;**3**(5):141–61.

Chapter 5

Prion-like behavior of TDP-43 aggregates and its implication to disease

Ankit Srivastava[1], Nikita Admane[2], Shiv Pratap Singh Yadav[3], Himanshi Kukrety[4]

[1]*LPVD, Rocky Mountain Laboratories, NIAID, NIH, Hamilton, MT, United States;* [2]*Department of Biological Sciences, BITS Pilani - K K Birla Goa Campus, Goa, India;* [3]*Division of Nephrology, Indiana University School of Medicine (IUSM), Indianapolis, IN, United States;* [4]*Department of Biotechnology, Jawaharlal Nehru University, New Delhi, India*

Introduction

TAR DNA-binding protein 43 (TDP-43) is a ubiquitously expressed RNA/DNA-binding nuclear protein mainly involved in regulating mRNA stability, trafficking, alternative splicing, and miRNA production.[1,2] It shuttles between the nucleus and cytoplasm for its regulatory functions and reportedly forms cytoplasmic inclusions associated with amyotrophic lateral sclerosis (ALS) and frontotemporal lobar degeneration (FTLD-TDP).[3–5] Accumulation of TDP-43 inclusions forms a shared pathological feature in 95% ALS and 50% FTLD cases making them a potential biomarker for these neurodegenerative maladies.[6] Several identified TDP-43 mutations form a prominent pathophysiological link for the onset of familial forms of ALS.[7] Emerging evidence has linked these mutations with increased aggregation, increased half-life, altered protein–protein interactions, and enhanced cytoplasmic localization of TDP-43.[8] Several clinical reports have also associated posttranslational modifications (PTMs) in TDP-43 with alterations in its structure, function, and localization properties.[9,10]

Dissemination of neuronal degeneration in motor neurons of brain and spinal cord during ALS progression bears conspicuous resemblance with prion diseases, including the sporadic Creutzfeldt–Jakob disease and Gerstmann–Straussler–Scheinker disease.[11,12]Although evidence of protease-resistant prion protein (PrP^{res}) and spongiform degeneration typical to prion diseases is not yet reported in ALS or FTD, the onset of clinical features and its spread to neighboring sites are akin to prion-like propagation.[13] In case of prion

TDP-43 and Neurodegeneration. https://doi.org/10.1016/B978-0-12-820066-7.00004-7

diseases, cell-to-cell propagation of misfolded PrP^{res} species culminates in pathological spread throughout the CNS and systemic organs. These misfolded species have been categorized into different subtypes, referred to as prion strains that entail variations in regional distribution and clinical features.[14]

Although it is not yet clearly understood how prions cause infectivity and toxicity, the most widely accepted hypothesis is the seeding/nucleation model. The model suggests that misfolded oligomeric prion conformers act as nuclei/seed to recruit normal, monomeric prions (cellular prion protein; PrP^{C}) into an aggregated polymer through a process that is much faster than the initial nuclei/seed formation.[15,16] Recent reports have demonstrated that TDP-43 aggregates could also participate in similar seeding reactions in cell culture.[17] TDP-43 seeded from diseased brains showed protease resistance and induced cellular toxicity, proteasome dysfunction, and cell-to-cell infectivity.[18] Phosphorylated TDP-43 (pTDP-43) in postmortem brain and spinal cord of ALS patients suggests a dual spreading mechanism of TDP-43 pathology via axonal trajectories and synaptic connections along with noncontiguous spread.[19,20]

Several structural reports have projected the C-terminus of TDP-43 as the prion-like domain (PLD) having glutamine/asparagine (Q/N)-rich region that might potentially regulate aggregation and amyloid formation.[21,22] This is supported by identification of majority of ALS mutations in the C-terminus of TDP-43, which may alter its native interactions and intracellular localization properties.[8,23] Besides, motor neurons accumulating TDP-43 and other ALS-linked proteins including SOD1 and FUS are now considered hotspots for prion-like seeding and propagation in ALS and FTD.[24] Indeed, ALS pathological spreading indicates prion-like spreading mechanism that propagates from the CNS to muscles via the spinal cord. Against this background, the prion-like behavior of TDP-43 aggregates might be a crucial link between pathological onset and disease propagation in both ALS and FTD. In this chapter, we comprehensively discuss pieces of evidence and contexts of prion-like behavior of TDP-43 aggregates as they relate to disease and etiology.

Aggregation and amyloidogenic conversion of TDP-43

TDP-43 comprises of 414 amino acid residues having two fully functional RNA recognition regions (or motifs RRM1 and RRM2) subsequently carrying a GQ-rich C-terminal domain (CTD) (Fig. 5.1). The RNA recognition regions show distinct nucleic acid recognition and binding properties along with arbitrating protein—protein interactions.[25] TDP-43 sequence also contains a nuclear localization sequence and a nuclear export sequence which mediates nucleocytoplasmic translocations of it.[26] The glycine-rich CTD harbors the widely held ALS-associated mutation hotspots. Different structural domains of TDP-43 and especially the CTD have been implicated in its aggregation and amyloidogenic conversion.

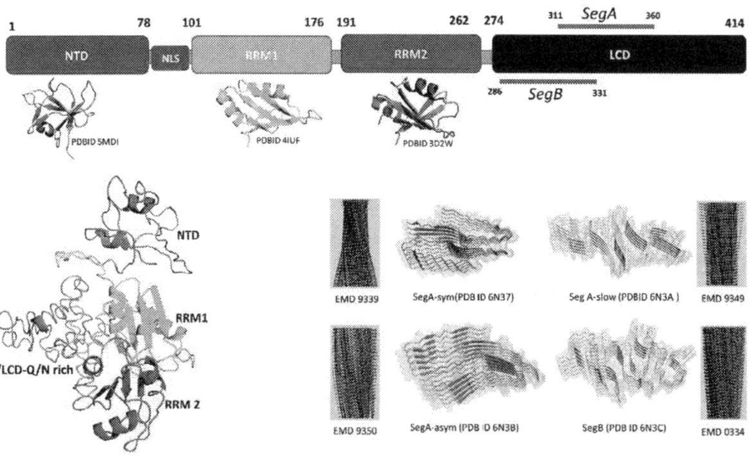

FIGURE 5.1 The schematic representation of TDP-43 domain arrangement showing N-terminus, RRM domains, and C-terminus region and their putative lengths as shown in various studies. Left panel represents full-length homology model of TDP-43 with structured N-terminal domain (NTD) that includes RRM1 and RRM2 regions followed by an unstructured C-terminal domain (CTD). Right panel shows cryo-EM structures of polymorphic amyloidogenic cores of TDP-43 known as SegA-sym/asym/slow and SegB along with their orthogonal views of polymeric forms.

Structural regions of TDP-43 and their role in amyloidogenic conversion

Individual domains of TDP-43 and their dynamics have been characterized through a variety of structural methods including NMR, X-ray crystallography, Cryo-Electron microscopy, small-angle X-ray scattering, and more recently by electron crystallography. The available structural reports provide crucial insights into physiological and pathological roles N-terminal domain (NTD) (residues 1–102), RRM1 (residues 106–177), RRM2 (residues 192–259), and the CTD (residues 274–414).[9] Physiological functions of TDP-43 are majorly mediated by NTD that contains a nuclear localization signal sequence (82–98 residues) and the RRM2 that includes a nuclear export signal sequence (239–250).[27,28] Besides, the intrinsically disordered CTD remains central to pathological manifestations of TDP-43 that includes formation of cytotoxic fragments and disease-associated amyloid aggregates.[29–31]

CTD consists of a Q/N-rich prion-like aggregation-prone region and incorporates all ALS/FTD-associated mutations. Perhaps, structural studies of full-length TDP-43 were historically limited due to strong aggregation propensity and disorder in CTD. Aggregated TDP-43 forms detergent insoluble inclusions like polyQ-proteins such as Huntingtin (Htt).[32] It is proposed that the Q/N-rich stretch constituted within the CTD might trigger the formation of fibrillar β-sheet rich inclusions by TDP-43. Intriguingly, the Q/N-rich stretch is now considered as a PLD identified in several other RNA-binding proteins such as FUS, SS18L1, HNRNPA1, TAF15, and EWSR1 associated with known ALS-FTLD pathology.[33–35]

Overall, the CTD appears to adopt transient and dynamic secondary structures ranging from α-helix to β-strand conformations (**Left panel**, Fig. 5.1). The structural study of the CTD is difficult because of its flexible nature and presence of the low complexity domain (LCD) which is believed to be involved in both in reversible stress granules and in irreversible pathogenic amyloid.[36] Recent Cryo-EM structures of fibrils formed by pathogenic core stretches SegA and SegB in LCD indicated four polymorphs (**Right panel**, Fig. 5.1). The SegA segment forms three polymorphs with dagger-shaped folds and the SegB segment results in a polymorph with R-shaped folds.[37] Evidently, the dagger-shaped polymorphs constituted irreversible fibril structures, in contrast to the SegB polymorph that could participate in both reversible and irreversible fibrillation. Besides, the LCD can also undergo liquid−liquid phase separation (LLPS) to form dynamic protein droplets that under persistent stress conditions like aging and mutations form irreversible cytoplasmic inclusions.[38] The C-terminal truncated TDP-43 protein fragments have been obtained within insoluble aggregates derived from the brain tissue of affected individual.[39,40] Thus, the CTD or the LCD/QN-rich domain of TDP-43 remains crucial for understanding its pathological behavior as well as its prion-like transmission across the nervous system.

Amyloidogenic conversion of TDP-43

NMR spectroscopy of NTD revealed that physiological concentration of ATP specifically binds to this region to enhance its oligomerization.[41] Intriguingly, addition of certain solution ingredients such as NaCl attenuates this functional oligomerization of TDP-43. The amyloidogenic transition of TDP-43 is shown to involve oxidation of cysteine containing domains of TDP-43 (NTD, RRM-1, RRM-2). It was found that both RRM-1 and RRM-2 were capable of oxidation and oligomer formation, but only RRM-1 oxidation led to aggregation of TDP-43.[42] Eventually, sequential oxidation Cys^{173} and Cys^{175} residues of RRM-1 destabilizes hydrophobic interactions and leads to insoluble RRM-1 dimer formation. Both NTD and RRM-2 were found to be facilitators in enhancing speed and severity of aggregation of RRM-1 domain. TDP-43 aggregation happens through a biphasic process (oligomerization followed by aggregation) demonstrated in recombinant TDP-43 (rTDP-43) by SDD-AGE.[43] Here in the initial oligomer formation was found to gradually transition into high molecular weight aggregates. An increase in aggregation and LLPS formation was also found prevalent with addition of polyethylene glycol. The relevance of these in vitro observations confirmed a previous study on TDP-43 aggregates isolated from FTLD patients.[44]

Further, the behavior of C-terminal fragments (CTFs) of TDP-43 in vitro and their fibril formation has been well characterized by several studies.[46–48] In a study comprising several peptides from TDP-43 sequence, two small TDP-43 peptides (246–258 and 311–323) exhibited significantly high Thioflavin T fluorescence during aggregation, indicating their strong amyloid-like characteristics.[45] Strikingly, these peptides also exhibited similar amyloid morphologies that were distinct from the worm-like fibrils formed by recombinant CTD (Fig. 5.2). A study on another CTF (residues 193–414) of TDP-43 showed that the anions in aggregation reaction follow reverse Hofmeister series order. Kosmotropic anions enhanced aggregation by decreasing the lag phase, whereas the chaotropic anions impede aggregation rates which in turn differently impacted the morphology of end-stage aggregates.[49]

It was shown that the tethered-RRM domain of TDP-43 (TDP-43^{tRRM}) exist in a pH-dependent and reversible thermodynamic equilibrium with a protonated, nanosized, 40-metric A-form conformer.[50] Interestingly, low-pH stress induces formation of A form having loosely packed core constituting intrinsically disordered and aggregation-prone regions in proximity. The metastable A form gradually transitions into a highly stable amyloid-like fibrillar form (β form). A follow-up study showed that a metastable low pH form of TDP-43 (RRM-1 domain) or L-form contains an intact native-like secondary structure even after disruption of its tertiary structure due to low pH stress. This metastable form fully misfolded on addition of KCl and further oligomerized into β form later forming worm-like amyloid fibrils.[51]

Role of posttranslational modifications in TDP-43 aggregation

TDP-43 is a target for various PTMs that can change its structure, localization, function, and its aggregation propensity. Ubiquitination and phosphorylation are common PTMs in TDP-43.[9] Lys 79 along with many other potential lysine residues are prone to ubiquitination which promote TDP-43 mislocalization and aggregation.[52,53] While residues ser 409 and ser 410 are well documented

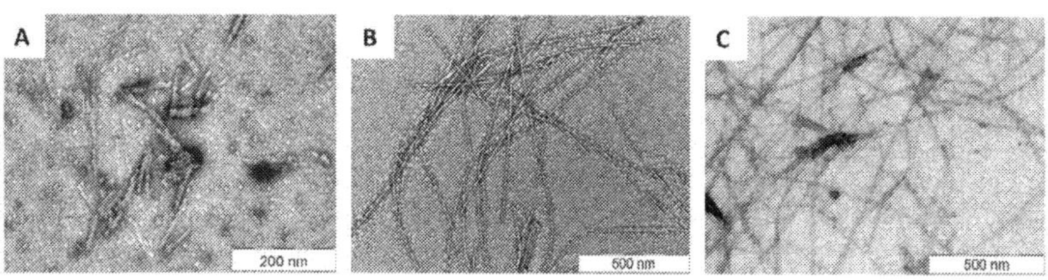

FIGURE 5.2 Electron micrographs depicting differences in amyloid morphology of fibrils formed by the CTD and two core-peptide sequences of TDP-43, (A) CTD, (B) 246–258 peptide in RRM2 region, and (C) 311–323 peptide in LCD region. *The image is modified and presented with permission from the authors of reference Saini A, Chauhan VS. Delineation of the core aggregation sequences of TDP-43 C-terminal fragment.* Chembiochem *2011;**12**(16): 2495–2501.*

targets for phosphorylation, few other studies suggest that all serine and threonine in CTD are prone to phosphorylation.[54,55] This modification may either increase propensity of TDP-43 to aggregate or to be hydrolyzed into CTFs. Some other PTMs such as oxidation, acetylation, and SUMOylation are also reported.[56,57] Acetylation on Lysine 145 in TDP-43 is associated with a loss of its RNA-binding ability. On the other hand, SUMOylation occurs at Lysine 136 but its role in toxicity or aggregation propensity is yet to be established. In addition, disulfide formation between Cys39 and Cys50 is found critical for dimerization of NTD, oxidation at these sites thought to increase aggregation propensity.[58]

TDP-43 mutation spectrum and its relationship with ALS pathology

Pathological mutations in TDP-43 have been found associated with both familial and sporadic forms of ALS, establishing a prominent pathophysiological link between TDP-43 and ALS.[7,59] Majority of the ALS-associated mutations are identified in CTD of TDP-43 that also acts as a binding domain for heterogeneous nuclear ribonucleoprotein (hnRNP) and remains majorly phosphorylated in the insoluble cytoplasmic inclusions present in the neurons and glia of sporadic ALS patients.[60] Studies on ALS-affected patient populations have established more than 50 mutations linked with this fatal disease pathology.[61,62] Emerging evidence indicate that TDP-43 mutations are directly involved in ALS-associated cellular events such as increased aggregation of TDP-43, increased half-life, altered protein–protein interactions, and enhanced cytoplasmic localization.[63]

Prion-like properties of TDP-43 aggregates

Prion diseases and ALS show persuasive similarities which are majorly attributed to the typical propagation of their disease pathology, accumulation of aberrant misfolded protein aggregates within neurons, multiple clinical phenotypes, and elusive transmission features which are difficult to abrogate. These resemblances strongly specify that ALS and prion diseases share some common pathological mechanisms and propagation features.[64] Furthermore, the "prion-like" propagations of misfolded aberrant proteins (like TDP-43, tau, synuclein) have been recognized as the major pathogenic drivers of ALS, FTLD, Alzheimer's disease, and Parkinson disease. Recent evidence suggests that the signature cortical neuronal cytoplasmic inclusions of TDP-43 in different molecular subtypes of FTLD share identical spatial patterns in the frontal and temporal cortex.[65] These evidences corroborate a common "prion-like" propagation of TDP-43 alongside the anatomical pathways across the nervous system. Overall, the prion-like aggregation of TDP-43, leading to formation of neurotoxic inclusion and loss of neuronal population, may provide mechanistic insights into the intricate pathomechanisms.

Prion-like seeded aggregation of TDP-43

The aberrant protein assemblies accumulating in various neurodegenerative diseases behave as infectious seeding nuclei of aggregation, further recruiting their native isoforms into the growing anomalous seed.[66] This self-perpetuating nature of the highly ordered infectious seeds generates large aggregates which ultimately fragment into new seeds. Fragmentation of the aggregates and propagation of seeds subsequently occur in a template-dependent manner within the cellular milieu. Eventually, these seeds are released into extracellular spaces, where cell-to-cell transmission of the seeds results in reiteration of the propagation cycle and amplification of the disease phenotype.[67]

In cellular stress conditions, the PLD of TDP-43, instigate aggregation and nucleation within the highly dynamic stress granules. Conformational antibodies specific for the cytoplasmic mislocalized aggregates of TDP-43 have detected misfolded TDP-43 in the spinal cord inclusions of patients affected with ALS.[68,69] The prion-like characteristics of TDP-43 have been further corroborated by the recognition of a glutamine/asparagine (Q/N)-rich domain in the C-terminal, resembling prion protein. This was further corroborated by induction of TDP-43 into insoluble inclusions by polyglutamine proteins in yeast culture.[21] Misfolded conformations of protein recruit its natively folded conformers to form an infectious seed. This leads to aggregated self-templating conformational isomers of proteins, which were initially discovered in prion diseases.[14]

Interestingly, the seeding behavior of recombinant wild-type, and mutant TDP-43 aggregates in cell models has now been well documented. These fibrillar aggregates showed the potential of seeding fibrillization of native TDP-43, consequently producing intracellularly seeded aggregates containing protease-resistant cores and were Sarkosyl insoluble.[17]

More evidence of seeding was established in a model involving transfection of human ALS-affected pathological extracts of brain-derived TDP-43 into cells.[18] The transfected cells formed phosphorylated, ubiquitinated, insoluble cytoplasmic inclusions of TDP-43. The induction of pathologically similar TDP-43 inclusions strongly indicated a successful seeding reaction leading to protease and heat-resistant aggregates. It was also established in this model that these inclusions proliferated between cells, which further helped in serial propagation of the seeds. Furthermore, impairment of the proteasome and enhanced cellular toxicity associated with aggregated TDP-43 deposits, signifying the efficiency and disease-relevance of this model for TDP-43 neuropathies. Another study demonstrated that intracerebral injections of pathogenic FTLD-TDP seeds into transgenic mice expressing cytoplasmic human TDP-43 and even nontransgenic mice induced de novo TDP-43 pathology.[70]

TDP-43 oligomers mediate self-propagating aggregation

Symptomatic progression of ALS disease pathology is postulated to be correlated with the release of TDP-43 oligomeric aggregates from the affected cells and efficient nucleation and oligomerization of native TDP-43 of recipient cells. Intriguing evidence further establish horizontal transmission of TDP-43 oligomeric aggregates between cell somata possibly via microvesicles or exosomes, prompting oligomerization of TDP-43 subsequently leading to higher neurotoxicity in recipient cells as compared to free TDP-43.[71,72] Studies involving microfluidic neuronal culture have established neuron-to-neuron propagation of TDP-43 oligomers involving both anterograde and retrograde transsynaptic transmission.[73] In fact, TDP-43 oligomer seeding activity has been successfully established by corresponding infectious material derived from both ALS-infected brain lysates and cell culture extracts. Thus, substantial evidence indicates prion-like seeding of TDP-43 and preferential microvesicular uptake of oligomeric aggregates via soma-to-soma "horizontal" and bidirectional "vertical" transsynaptic intercellular transmission.[71]

Seeded aggregation of TDP-43 peptide fragments

Further investigations for decoding the molecular mechanisms of TDP-43 aggregation, studies have been instigated to find out the amino acid stretches responsible for aggregation. Studies using deletion mutants of TDP-43 expressed in human neuroblastoma cells demonstrated that residues 274–313 in glycine-rich region of CTD are predominantly critical for aggregation.[74] Synthetic peptides of the region 274–313 and 314–353 of TDP-43 are aggregation prone and form amyloid-like fibrils, which were capable of inducing seed-dependent aggregation of TDP-43 in cell models. These infected cells were later found to contain distinct trypsin-resistant bands and phosphorylated CTFs of TDP-43. These and other similar reports suggested the importance of CTFs in triggering conformational transition of natively folded TDP-43 into amyloid-like fibrils.[45] Apparently, CTFs may be acting like prion conformational strains that give rise to distinct clinical pathology associated with different TDP-43 neuropathies. Eventually, the identified CTF sequences might help in developing in vivo models for studying TDP-43 neuropathies.

Intracellular aggregation and localization of TDP-43 prions

The pathomechanisms of TDP-43 aggregation majorly involves formation of intracellular aggregates and inclusion like other major neurodegenerative disorders such as AD, PD, and prion diseases. Role of intracellular pathogenesis of TDP-43 was established by studies involving addition of ALS or FTD brain fractions into cells that result in TDP-43-positive inclusions.[75]

The intracellular TDP-43 aggregates have been found to be associated with proteasomal dysfunction and neuronal death. Proteasome dysfunction is mainly accredited to congestion of the ubiquitin–proteasome system which might be overloaded by ubiquitinated proteins, majorly involving TDP-43 aggregates.[76] Neurotoxicity and cell death primarily occur due to suppression of proteasomal activity. Similar seed-dependent intracellular aggregation has also been reported in other aggregation prone proteins like poly(Q) repeats proteins and α-synuclein, where aggregates cause proteasome dysfunction and cellular impairment.[77,78] Such findings also support the notion that insoluble TDP-43 aggregates present in patient brains may possess prion-like propagative properties.

Stress granules and intracellular aggregation of TDP-43

The propagation of pathogenic protein assemblies of TDP-43 is mainly arbitrated by key mediators like dysfunctional stress granule assembly and RNA processing errors. Evidences demonstrating the involvement of stress granules rich in TDP-43 and other RNA-binding proteins indicate augmented oxidative stress recapitulating the pathological progression in ALS.[79,80] Notably, there is a strong correlation between cytoplasmic TDP-43 aggregates and stress granules as these cytoplasmic aggregates tend to colocalize with stress granule markers. During cellular stress conditions, stress granules represent a kind of cytoplasmic inclusions that suppress the translation of various neurodegeneration-associated proteins (like Ataxin-2 and SMN) and repress a particular subset of RNAs.[81] Consequently, probing the interplay between ALS-associated TDP-43 mutations, stress granules and TDP-43 aggregation might provide crucial insights into intracellular aggregation of TDP-43.

For understanding the role of SGs in ALS-associated TDP-43 intracellular aggregation, two models have been proposed namely "independent model" and "precursor model" (Fig. 5.3). The "independent model" suggests that intracellular aggregation of TDP-43 remains independent of SG formation. The more favored "precursor model" stipulates that SGs are precursors of TDP-43 aggregation.[82] Here, the SGs remain crucial for the formation of TDP-43 aggregate "seed" and persistent stress conditions, environmental factors, or mutations may influence the concentration-dependent TDP-43 aggregation. Interestingly, larger stress granules are developed in case of TDP-43 possessing ALS-associated mutations, suggesting that concentration-dependent aggregation is achieved more readily. Apparently, this model is also in accordance with the "prion-like propagation model" that emphasizes concentration-dependent aggregation of neuropathic proteins.

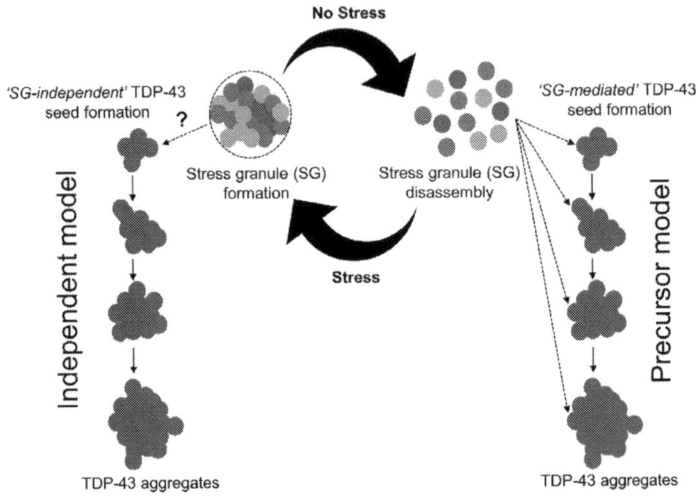

FIGURE 5.3 The two proposed models for understanding the role of stress granules in ALS-associated TDP-43 aggregation process. Left panel depicts the "independent model" that involves formation of TDP-43 aggregates due to an unknown factor independent of the formation and dissociation of stress granules (SG). The right panel shows the "precursor model" where the SG formation helps in generating TDP-43 seeds that further accumulate and aggregate due to persistent stress, mutations, or other environmental factors.

Cytoplasmic inclusions of TDP-43 in ALS

In neurodegenerative disorders, mislocalization of neuronal and glial TDP-43 into the cytoplasm outrages its aggregation as stress granules and insoluble inclusion bodies. Although present scenario warrants more insights to establish whether the toxicity is attributed to inclusions or soluble species, pathological correlation of TDP-43 with atrophy of brain and dementia corroborates its role in the downstream mechanisms of neurodegeneration in ALS.[49] The impairment of nucleocytoplasmic shuttle of TDP-43 consequently instigates its cytoplasmic localization and associated toxicity.[83] Nucleocytoplasmic shuttling of TDP-43 is governed mainly by two signal sequences i.e. nuclear localization and nuclear export signal sequences. Therefore, deletion of either of the two sequences leads to impairment of functions in TDP-43. While the deletion of Nuclear export signal in TDP-43 reportedly enhances its tendency to form nuclear aggregates, deletion of nuclear localization signal leads to accumulation of cytoplasmic aggregates accompanied by sequestration of native TDP-43. Moreover, this further alters chromatin assembly and histone processing due to alteration of transcripts.[84]

The cytoplasmic aggregation of TDP-43 averts its return to the nucleus that further leads to nuclear clearance. Specific biochemical signatures of cytoplasmic aggregates of TDP-43 include hyperphosphorylations, polyubiquitinations, and caspase-dependent proteolytic cleavage of CTD, which in turn aggregates in the cytoplasm in consort with its full-length counterparts.[85] Majority of the point mutations linked to ALS are found in this intrinsically aggregative prion-like CTD and CTFs of TDP-43, which suggestively clues to a causative link between TDP-43 cytoplasmic inclusions and disease.[86] The detailed pathogenic events triggered by TDP-43 are yet to be deciphered and more evidences are required to clearly define whether the toxicity is due to loss of its normal function, obnoxious cytoplasmic inclusions, or a combination of both.[87]

It has been revealed that the inclusions or aggregates formed by pathologically derived CTFs of TDP-43, namely TDP-35 (~35 kDa) and TDP-25 (~35 kDa), recruit its full-length counterpart TDP-43 into triggering neuronal cytoplasmic inclusions. This sequestration happens due to binding of TDP-35 with RNAs that help in enriching the neuronal cytoplasmic inclusions. Besides, TDP-35 also plays a dominant role in sequestering TDP-43 into cytoplasmic inclusions deviating it from nuclear translocations, and RNA binding essentially controls this process.[88] Intriguingly, the factors driving nuclear clearance of TDP-43 still need to be deciphered appropriately. Perhaps, the sustenance of phase-separated TDP-43 is self-determining and not dependent on the conventional stress granules. Apparently, this phase separation inhibits nuclear transport of TDP-43 and because of which it deposits in the cytoplasm. Later recruitment of importin-α and Nup62 by the TDP-43 LLPS triggers mislocalization of other nuclear proteins including Ran, RanGap1, and Nup107. This further aggravates inhibition of nucleocytoplasmic transport and depletion of nuclear TDP-43 eventually leading to toxicity and neuronal cell death.[89]

Profilin-dependent seeding of TDP-43

Mutations in the profilin 1 (PFN1) (an actin monomer—binding protein) gene have been linked with familial ALS.[90] Neuropathological studies indicate coaggregation of TDP-43 with mutated PFN1 which accumulate in the brains of patients. Expression of mutated PFN1 resulted in the formation of pathological prion-like, ubiquitin-positive cytoplasmic aggregates which further sequestered endogenous TDP-43. Administration of proteasome or lysosome inhibitor facilitated in the accumulation of TDP-43 aggregates. The C-terminal region of TDP-43 proves indispensable for its interaction with PFN1 and coexpression of PFN1 mutants and TDP-43 enhances the production of detergent-insoluble and phosphorylated TDP-43.[91] Moreover, detergent-insoluble fractions isolated from cells expressing ALS-linked PFN1 mutants act as infectious seeds. These evidences suggest that the expression of PFN1 mutants encourages conformational transition of normal TDP-43 into an abnormal form. Thus, aggregation of TDP-43 might be also triggered by sequestration of TDP-43 into PFN1 aggregates and further aggravated by self-perpetuating TDP-43 aggregation.

Interestingly, this seed-dependent accretion of aberrant TDP-43 aggregates increases with time, further increasing the concentration of PFN1 aggregates in the detergent-insoluble fraction. This was further corroborated by identification of PFN1 accumulations in brain of sporadic ALS patients showing deposits of phosphorylated TDP-43.[92] Thus, TDP-43 sequestration into PFN1 aggregates appropriates template-guided and seed-dependent prion-like pathological conversion and fibrillization in TDP-43.

Intercellular transmission and infectivity of TDP-43 prions

The evidence of in vitro seeding capabilities of TDP-43 aggregates has intrigued researchers into studying their in vivo seeding efficacy and mechanisms. Seeding and templated aggregation of oligomeric intermediates and their transmissibility has been already demonstrated in major neurodegenerative disorders.[93,94] Indeed, recent reports have linked TDP-43 oligomerization to toxicity and have also shown the presence of toxic TDP-43 oligomers in both FTLD-TDP mice models and FTLD patients.[18,95] Moreover, TDP-43 oligomers are capable of cross-seeding Alzheimer's amyloid-β to form amyloid oligomers, demonstrating interconvertibility between the amyloid species.[43]Therefore, methods of abrogating the seeding and formation of toxic oligomers of TDP-43 might be beneficial to counter its neurotoxicity and pathological progression. Altogether, progression of TDP-43 neuropathies essentially involve spreading of abnormal TDP-43 aggregate seeds that get transferred to different anatomical regions thereby infecting surrounding neuronal tissues and worsening the neurodegenerative pathologies in diseases like ALS and FTLD.[96]

In vivo seeding and spreading of TDP-43 aggregates

Studies have demonstrated that injection of patient-derived samples enriched with TDP-43 aggregates developed pathologically pertinent TDP-43 inclusions in the recipient mice.[70,97] Interestingly, these reports also established that the endogenous mouse TDP-43 responds and aggregates in presence of active, infectious seeds. These studies involving transgenic mice models have advocated the importance of N-terminal regions of TDP-43 in its self-recognition of endogenous isoforms and prion-like propagation.[49,98]

Another study on mice models expressing human TDP-43 also suggested that controlling this regulatable protein expression and targeting the cytoplasmic, intracellular TDP-43 inclusions can cripple the neurodegeneration process. A dramatic decrease in the deposition of phosphorylated, aberrant insoluble cytoplasmic TDP-43 was observed when the transgene expression was controlled or supressed, which even reversed the loss of endogenous nuclear TDP-43 in the mice model prohibiting the subsequent motor neuron loss.[99] These findings further corroborated the importance of seeding, affirming that only the accumulation of intracellular pathological deposits of TDP-43 at a certain point are insufficient for the propagation of disease pathology. Perhaps, cellular transmissions might not sustain without the pathological recruitment from a continuous cytoplasmic pool of TDP-43. Also, neurodegenerative pathology in ALS appears focally and then spreads spatiotemporally across both upper and lower motor neuron systems of the human body.[100] Thus, it remains crucial to identify the molecular factors that are involved in pathological spread of ALS and related neuropathies for developing newer therapeutics that could interfere or deaccelerate ALS progression.

Transcellular transmission of pathological TDP-43 aggregates

The propagation events involving intricate mechanisms that trigger the transport of obnoxious TDP-43 aggregates across neurons require further probing. The misfolded protein seeds might get released from the affected cell and efficiently internalized by the adjacent cell, where encryption of other endogenous monomers prompts their conformational conversion into pathogenic isoforms thereby amplifying the disease.[101] The mechanisms supporting transmission and propagation of seeds include exocytosis, transsynaptic transmission, vesicle-mediated release (exosomes or presynaptic vesicles), or passive release by cellular degeneration.[102] The adjacent receiving neurons internalize the aberrant protein aggregates via receptor-mediated endocytosis, micropinocytosis, vesicle membrane fusion, or direct penetration. The "tunneling nanotubes" could also help in direct transfer of the infectious seeds. Thus, the pathways and mechanisms of transport includes both active and passive processes of uptake and release.[103,104]

Nevertheless, the properties dictating efficient transmission of disease-associated TDP-43 aggregates are not yet fully understood. Intriguing evidences have suggested the horizontal transmission of TDP-43 in dimeric/oligomeric forms via the exosome/microsome pathway in ALS.[71] In a way, the TDP-43 dimers/oligomers signify a distinct physiological state, unlike the aggregation assemblies found in FTD patients and transgenic mice which represent small pore-like structures.[95] Hence, future studies should be focused on obtaining more insights into uptake and seeding mechanisms of TDP-43 aggregates from donor to a healthy recipient cell.

Moreover, it has been shown that impaired endocytosis often contributes to the intracellular cytotoxicity of TDP-43. Cell models and yeast culture studies have reported that TDP-43 obstructs the endocytosis pathway at abnormally high concentrations by colocalizing with the endocytosis-associated proteins.[105] Such colocalization has also been reported in the frontal cortex tissues of ALS-affected patients. Indeed, these findings clearly established that endocytosis dysfunction correlates with TDP-43 toxicity, while enhancing endocytosis was found to reverse the motor neuron dysfunction and toxic phenotypes.[106] Studies have also emphasized that the endo-lysosomal pathways and endocytosis are evidently distressed by TDP-43 interference. The endo-lysosomal pathway safeguards the cells from detrimental effects of TDP-43 by contributing in its clearance independent of autophagy. Contrastingly, enhancement in autophagy worsens TDP-43 cytotoxicity, regardless of its involvement in degradation of the aberrant protein[107]. This interference in the activity of endo-lysosomal system poses two malicious upshots for the cell. Firstly, by making use of this pathway, TDP-43 interferes in its own clearance, thereby promoting its accumulation. Secondly, it contributes to aberrant autophagy because of impaired vacuolar proteolytic activity provoked due to defective endosomal-vacuolar trafficking.[108] Convincingly, ameliorating the activity of endo-lysosomal pathway delivers reversal of toxicity and enhances cell survival in TDP-43 neuropathies.[109]

Conclusion

Last two decades have established that pathological spread of proteopathic seeds in neurodegenerative disorders involve prion-like mechanisms. In case of ALS and related TDP-43 neuropathies, biophysical, biochemical, as well as cell-based and in vivo evidence strongly indicate prion-like seeding and propagation. Henceforth, it will be interesting to test established findings from prion biology and their correlations in ALS and related neuropathies. Additionally, implementing seed amplification assays like protein misfolding cyclic amplification (PMCA) and real-time quaking induced conversion (RT-QuIC) that were initially used to in vitro amplify and detect pathological prion aggregates could also help in corroborating prion-like behavior of TDP-43 aggregates.[110,111] In the long run, prion-like properties of TDP-43 aggregates might also assist in developing better in vitro assays and animal models for screening therapeutics against ALS and related neuropathies.

Acknowledgments

The authors acknowledge infrastructural support from the Rocky Mountain laboratories, NIH, and Indiana University School of Medicine, USA, along with BITS Pilani—K K Birla Goa Campus and Jawaharlal Nehru University, India.

References

1. Guo L, Shorter J. Biology and pathobiology of TDP-43 and emergent therapeutic strategies. *Cold Spring Harb Perspect Med* 2017;**7**(9).
2. Chen-Plotkin AS, Lee VM, Trojanowski JQ. TAR DNA-binding protein 43 in neurodegenerative disease. *Nat Rev Neurol* 2010;**6**(4):211–20.
3. Suk TR, Rousseaux MWC. The role of TDP-43 mislocalization in amyotrophic lateral sclerosis. *Mol Neurodegener* 2020;**15**(1):45.

4. Kumar V, et al. Protein aggregation and neurodegenerative diseases: from theory to therapy. *Eur J Med Chem* 2016;**124**:1105–20.
5. Kumar V, et al. Therapeutic progress in amyotrophic lateral sclerosis-beginning to learning. *Eur J Med Chem* 2016;**121**:903–17.
6. Majumder V, et al. TDP-43 as a potential biomarker for amyotrophic lateral sclerosis: a systematic review and meta-analysis. *BMC Neurol* 2018;**18**(1):90.
7. Sreedharan J, et al. TDP-43 mutations in familial and sporadic amyotrophic lateral sclerosis. *Science* 2008;**319**(5870):1668–72.
8. Buratti E. Functional significance of TDP-43 mutations in disease. *Adv Genet* 2015;**91**:1–53.
9. Francois-Moutal L, et al. Structural insights into TDP-43 and effects of post-translational modifications. *Front Mol Neurosci* 2019;**12**:301.
10. Kametani F, et al. Mass spectrometric analysis of accumulated TDP-43 in amyotrophic lateral sclerosis brains. *Sci Rep* 2016;**6**:23281.
11. Scheckel C, Aguzzi A. Prions, prionoids and protein misfolding disorders. *Nat Rev Genet* 2018;**19**(7):405–18.
12. Lee S, Kim HJ. Prion-like mechanism in amyotrophic lateral sclerosis: are protein aggregates the key? *Exp Neurobiol* 2015;**24**(1):1–7.
13. Polymenidou M, Cleveland DW. The seeds of neurodegeneration: prion-like spreading in ALS. *Cell* 2011;**147**(3):498–508.
14. Collinge J, Clarke AR. A general model of prion strains and their pathogenicity. *Science* 2007;**318**(5852):930–6.
15. Soto C. Prion hypothesis: the end of the controversy? *Trends Biochem Sci* 2011;**36**(3):151–8.
16. Marin-Moreno A, et al. Transmission and Replication of prions. *Prog Mol Biol Transl Sci* 2017;**150**:181–201.
17. Furukawa Y, et al. A seeding reaction recapitulates intracellular formation of Sarkosyl-insoluble transactivation response element (TAR) DNA-binding protein-43 inclusions. *J Biol Chem* 2011;**286**(21):18664–72.
18. Nonaka T, et al. Prion-like properties of pathological TDP-43 aggregates from diseased brains. *Cell Rep* 2013;**4**(1):124–34.
19. Brettschneider J, et al. Stages of pTDP-43 pathology in amyotrophic lateral sclerosis. *Ann Neurol* 2013;**74**(1):20–38.
20. Braak H, et al. Amyotrophic lateral sclerosis–a model of corticofugal axonal spread. *Nat Rev Neurol* 2013;**9**(12):708–14.
21. Fuentealba RA, et al. Interaction with polyglutamine aggregates reveals a Q/N-rich domain in TDP-43. *J Biol Chem* 2010;**285**(34):26304–14.
22. Udan M, Baloh RH. Implications of the prion-related Q/N domains in TDP-43 and FUS. *Prion* 2011;**5**(1):1–5.
23. Pesiridis GS, Lee VM, Trojanowski JQ. Mutations in TDP-43 link glycine-rich domain functions to amyotrophic lateral sclerosis. *Hum Mol Genet* 2009;**18**(R2):R156–62.
24. Ciryam P, et al. Spinal motor neuron protein supersaturation patterns are associated with inclusion body formation in ALS. *Proc Natl Acad Sci USA* 2017;**114**(20):E3935–43.
25. Buratti E, Baralle FE. Characterization and functional implications of the RNA binding properties of nuclear factor TDP-43, a novel splicing regulator of CFTR exon 9. *J Biol Chem* 2001;**276**(39):36337–43.

26. Ayala YM, et al. Structural determinants of the cellular localization and shuttling of TDP-43. *J Cell Sci* 2008;**121**(Pt 22):3778–85.
27. Qin H, et al. TDP-43 N terminus encodes a novel ubiquitin-like fold and its unfolded form in equilibrium that can be shifted by binding to ssDNA. *Proc Natl Acad Sci USA* 2014;**111**(52):18619–24.
28. Jiang LL, et al. The N-terminal dimerization is required for TDP-43 splicing activity. *Sci Rep* 2017;**7**(1):6196.
29. Mompean M, et al. Electrostatic repulsion governs TDP-43 C-terminal domain aggregation. *PLoS Biol* 2016;**14**(4):e1002447.
30. Yang C, et al. The C-terminal TDP-43 fragments have a high aggregation propensity and harm neurons by a dominant-negative mechanism. *PLoS One* 2010;**5**(12):e15878.
31. Kumar V, et al. Exploring the aggregation-prone regions from structural domains of human TDP-43. *Biochim Biophys Acta Proteins Proteom* 2019;**1867**(3):286–96.
32. Baloh RH. TDP-43: the relationship between protein aggregation and neurodegeneration in amyotrophic lateral sclerosis and frontotemporal lobar degeneration. *FEBS J* 2011;**278**(19):3539–49.
33. Vance C, et al. Mutations in FUS, an RNA processing protein, cause familial amyotrophic lateral sclerosis type 6. *Science* 2009;**323**(5918):1208–11.
34. Chesi A, et al. Exome sequencing to identify de novo mutations in sporadic ALS trios. *Nat Neurosci* 2013;**16**(7):851–5.
35. Kim HJ, et al. Mutations in prion-like domains in hnRNPA2B1 and hnRNPA1 cause multisystem proteinopathy and ALS. *Nature* 2013;**495**(7442):467–73.
36. Guenther EL, et al. Atomic structures of TDP-43 LCD segments and insights into reversible or pathogenic aggregation. *Nat Struct Mol Biol* 2018;**25**(6):463–71.
37. Cao Q, et al. Cryo-EM structures of four polymorphic TDP-43 amyloid cores. *Nat Struct Mol Biol* 2019;**26**(7):619–27.
38. Babinchak WM, et al. The role of liquid-liquid phase separation in aggregation of the TDP-43 low-complexity domain. *J Biol Chem* 2019;**294**(16):6306–17.
39. Cohen TJ, Lee VM, Trojanowski JQ. TDP-43 functions and pathogenic mechanisms implicated in TDP-43 proteinopathies. *Trends Mol Med* 2011;**17**(11):659–67.
40. Vanden Broeck L, Callaerts P, Dermaut B. TDP-43-mediated neurodegeneration: towards a loss-of-function hypothesis? *Trends Mol Med* 2014;**20**(2):66–71.
41. Wang L, et al. A novel mechanism for ATP to enhance the functional oligomerization of TDP-43 by specific binding. *Biochem Biophys Res Commun* 2019;**514**(3):809–14.
42. Chang CK, et al. Molecular mechanism of oxidation-induced TDP-43 RRM1 aggregation and loss of function. *FEBS Lett* 2013;**587**(6):575–82.
43. French RL, et al. Detection of TAR DNA-binding protein 43 (TDP-43) oligomers as initial intermediate species during aggregate formation. *J Biol Chem* 2019;**294**(17):6696–709.
44. Bosque PJ, Boyer PJ, Mishra P. A 43-kDa TDP-43 species is present in aggregates associated with frontotemporal lobar degeneration. *PLoS One* 2013;**8**(5):e62301.
45. Saini A, Chauhan VS. Delineation of the core aggregation sequences of TDP-43 C-terminal fragment. *Chembiochem* 2011;**12**(16):2495–501.
46. Chen AK, et al. Induction of amyloid fibrils by the C-terminal fragments of TDP-43 in amyotrophic lateral sclerosis. *J Am Chem Soc* 2010;**132**(4):1186–7.
47. Shenoy J, et al. Structural dissection of amyloid aggregates of TDP-43 and its C-terminal fragments TDP-35 and TDP-16. *FEBS J* 2020;**287**(12):2449–67.
48. Berning BA, Walker AK. The pathobiology of TDP-43 C-terminal fragments in ALS and FTLD. *Front Neurosci* 2019;**13**:335.

49. Prasad A, et al. Molecular mechanisms of TDP-43 misfolding and pathology in amyotrophic lateral sclerosis. *Front Mol Neurosci* 2019;**12**:25.
50. Pillai M, Jha SK. The folding and aggregation energy landscapes of tethered RRM domains of human TDP-43 are coupled via a metastable molten globule-like oligomer. *Biochemistry* 2019;**58**(6):608–20.
51. Pillai M, Jha SK. Early metastable Assembly during the stress-induced formation of worm-like amyloid fibrils of nucleic acid binding domains of TDP-43. *Biochemistry* 2020;**59**(3):315–28.
52. Hans F, et al. Identification and characterization of ubiquitinylation sites in TAR DNA-binding protein of 43 kDa (TDP-43). *J Biol Chem* 2018;**293**(41):16083–99.
53. Nakayama Y, et al. Linear polyubiquitin chain modification of TDP-43-positive neuronal cytoplasmic inclusions in amyotrophic lateral sclerosis. *J Neuropathol Exp Neurol* 2020;**79**(3):256–65.
54. Hasegawa M, et al. Phosphorylated TDP-43 in frontotemporal lobar degeneration and amyotrophic lateral sclerosis. *Ann Neurol* 2008;**64**(1):60–70.
55. Neumann M, et al. Phosphorylation of S409/410 of TDP-43 is a consistent feature in all sporadic and familial forms of TDP-43 proteinopathies. *Acta Neuropathol* 2009;**117**(2):137–49.
56. Dangoumau A, et al. Protein SUMOylation, an emerging pathway in amyotrophic lateral sclerosis. *Int J Neurosci* 2013;**123**(6):366–74.
57. Cohen TJ, et al. An acetylation switch controls TDP-43 function and aggregation propensity. *Nat Commun* 2015;**6**:5845.
58. Wang A, et al. A single N-terminal phosphomimic disrupts TDP-43 polymerization, phase separation, and RNA splicing. *EMBO J* 2018;**37**(5).
59. Kumar V, et al. Delineating the effect of mutations on the conformational dynamics of N-terminal domain of TDP-43. *Biophys Chem* 2019;**250**:106174.
60. D'Ambrogio A, et al. Functional mapping of the interaction between TDP-43 and hnRNP A2 in vivo. *Nucleic Acids Res* 2009;**37**(12):4116–26.
61. Gros-Louis F, Gaspar C, Rouleau GA. Genetics of familial and sporadic amyotrophic lateral sclerosis. *Biochim Biophys Acta* 2006;**1762**(11–12):956–72.
62. Volk AE, et al. Current knowledge and recent insights into the genetic basis of amyotrophic lateral sclerosis. *Med Genet* 2018;**30**(2):252–8.
63. Austin JA, et al. Disease causing mutants of TDP-43 nucleic acid binding domains are resistant to aggregation and have increased stability and half-life. *Proc Natl Acad Sci USA* 2014;**111**(11):4309–14.
64. Smethurst P, Sidle KC, Hardy J. Review: prion-like mechanisms of transactive response DNA binding protein of 43 kDa (TDP-43) in amyotrophic lateral sclerosis (ALS). *Neuropathol Appl Neurobiol* 2015;**41**(5):578–97.
65. Armstrong RA, Cairns NJ. Spatial patterns of TDP-43 neuronal cytoplasmic inclusions (NCI) in fifteen cases of frontotemporal lobar degeneration with TDP-43 proteinopathy (FTLD-TDP). *Neurol Sci* 2011;**32**(4):653–9.
66. Brundin P, Melki R, Kopito R. Prion-like transmission of protein aggregates in neurodegenerative diseases. *Nat Rev Mol Cell Biol* 2010;**11**(4):301–7.
67. Maniecka Z, Polymenidou M. From nucleation to widespread propagation: a prion-like concept for ALS. *Virus Res* 2015;**207**:94–105.
68. Saberi S, et al. Neuropathology of amyotrophic lateral sclerosis and its variants. *Neurol Clin* 2015;**33**(4):855–76.

69. Shodai A, et al. Conserved acidic amino acid residues in a second RNA recognition motif regulate assembly and function of TDP-43. *PLoS One* 2012;**7**(12):e52776.
70. Porta S, et al. Patient-derived frontotemporal lobar degeneration brain extracts induce formation and spreading of TDP-43 pathology in vivo. *Nat Commun* 2018;**9**(1):4220.
71. Feiler MS, et al. TDP-43 is intercellularly transmitted across axon terminals. *J Cell Biol* 2015;**211**(4):897−911.
72. Iguchi Y, et al. Exosome secretion is a key pathway for clearance of pathological TDP-43. *Brain* 2016;**139**(Pt 12):3187−201.
73. Zhu L, et al. An ALS-mutant TDP-43 neurotoxic peptide adopts an anti-parallel beta-structure and induces TDP-43 redistribution. *Hum Mol Genet* 2014;**23**(25):6863−77.
74. Shimonaka S, et al. Templated aggregation of TAR DNA-binding protein of 43 kDa (TDP-43) by seeding with TDP-43 peptide fibrils. *J Biol Chem* 2016;**291**(17):8896−907.
75. Ishii T, et al. Formation and spreading of TDP-43 aggregates in cultured neuronal and glial cells demonstrated by time-lapse imaging. *PLoS One* 2017;**12**(6):e0179375.
76. Picher-Martel V, et al. Neuronal expression of UBQLN2(P497H) exacerbates TDP-43 pathology in TDP-43(G348C) mice through interaction with ubiquitin. *Mol Neurobiol* 2019;**56**(7):4680−96.
77. Verhoef LG, et al. Aggregate formation inhibits proteasomal degradation of polyglutamine proteins. *Hum Mol Genet* 2002;**11**(22):2689−700.
78. Zondler L, et al. Proteasome impairment by alpha-synuclein. *PLoS One* 2017;**12**(9):e0184040.
79. Besnard-Guerin C. Cytoplasmic localization of amyotrophic lateral sclerosis-related TDP-43 proteins modulates stress granule formation. *Eur J Neurosci* 2020;**52**(8):3995−4008.
80. Li YR, et al. Stress granules as crucibles of ALS pathogenesis. *J Cell Biol* 2013;**201**(3):361−72.
81. Vanderweyde T, et al. Role of stress granules and RNA-binding proteins in neurodegeneration: a mini-review. *Gerontology* 2013;**59**(6):524−33.
82. Dewey CM, et al. TDP-43 aggregation in neurodegeneration: are stress granules the key? *Brain Res* 2012;**1462**:16−25.
83. Kim HJ, Taylor JP. Lost in transportation: nucleocytoplasmic transport defects in ALS and other neurodegenerative diseases. *Neuron* 2017;**96**(2):285−97.
84. Winton MJ, et al. Disturbance of nuclear and cytoplasmic TAR DNA-binding protein (TDP-43) induces disease-like redistribution, sequestration, and aggregate formation. *J Biol Chem* 2008;**283**(19):13302−9.
85. Neumann M, et al. Ubiquitinated TDP-43 in frontotemporal lobar degeneration and amyotrophic lateral sclerosis. *Science* 2006;**314**(5796):130−3.
86. Johnson BS, et al. TDP-43 is intrinsically aggregation-prone, and amyotrophic lateral sclerosis-linked mutations accelerate aggregation and increase toxicity. *J Biol Chem* 2009;**284**(30):20329−39.
87. Ling SC, et al. ALS-associated mutations in TDP-43 increase its stability and promote TDP-43 complexes with FUS/TLS. *Proc Natl Acad Sci USA* 2010;**107**(30):13318−23.
88. Che MX, et al. TDP-35 sequesters TDP-43 into cytoplasmic inclusions through binding with RNA. *FEBS Lett* 2015;**589**(15):1920−8.
89. Gasset-Rosa F, et al. Cytoplasmic TDP-43 de-mixing independent of stress granules drives inhibition of nuclear import, loss of nuclear TDP-43, and cell death. *Neuron* 2019;**102**(2):339−357 e7.
90. Wu CH, et al. Mutations in the profilin 1 gene cause familial amyotrophic lateral sclerosis. *Nature* 2012;**488**(7412):499−503.

91. Tanaka Y, Hasegawa M. Profilin 1 mutants form aggregates that induce accumulation of prion-like TDP-43. *Prion* 2016;**10**(4):283–9.
92. Tanaka Y, et al. Gain-of-function profilin 1 mutations linked to familial amyotrophic lateral sclerosis cause seed-dependent intracellular TDP-43 aggregation. *Hum Mol Genet* 2016;**25**(7):1420–33.
93. Simoneau S, et al. In vitro and in vivo neurotoxicity of prion protein oligomers. *PLoS Pathog* 2007;**3**(8):e125.
94. Ward SM, et al. Tau oligomers and tau toxicity in neurodegenerative disease. *Biochem Soc Trans* 2012;**40**(4):667–71.
95. Fang YS, et al. Full-length TDP-43 forms toxic amyloid oligomers that are present in frontotemporal lobar dementia-TDP patients. *Nat Commun* 2014;**5**:4824.
96. Arai T, et al. Phosphorylated and cleaved TDP-43 in ALS, FTLD and other neurodegenerative disorders and in cellular models of TDP-43 proteinopathy. *Neuropathology* 2010;**30**(2):170–81.
97. Mitchell JC, et al. Wild type human TDP-43 potentiates ALS-linked mutant TDP-43 driven progressive motor and cortical neuron degeneration with pathological features of ALS. *Acta Neuropathol Commun* 2015;**3**:36.
98. Kawakami I, Arai T, Hasegawa M. The basis of clinicopathological heterogeneity in TDP-43 proteinopathy. *Acta Neuropathol* 2019;**138**(5):751–70.
99. Walker AK, et al. Functional recovery in new mouse models of ALS/FTLD after clearance of pathological cytoplasmic TDP-43. *Acta Neuropathol* 2015;**130**(5):643–60.
100. Ravits J, Paul P, Jorg C. Focality of upper and lower motor neuron degeneration at the clinical onset of ALS. *Neurology* 2007;**68**(19):1571–5.
101. Walker LC, et al. Mechanisms of protein seeding in neurodegenerative diseases. *JAMA Neurol* 2013;**70**(3):304–10.
102. Ferrara D, et al. Role of extracellular vesicles in amyotrophic lateral sclerosis. *Front Neurosci* 2018;**12**:574.
103. Hock EM, Polymenidou M. Prion-like propagation as a pathogenic principle in frontotemporal dementia. *J Neurochem* 2016;**138**(Suppl. 1):163–83.
104. Sackmann C, Sackmann V, Hallbeck M. TDP-43 is efficiently transferred between neuron-like cells in a manner enhanced by preservation of its N terminus but independent of extracellular vesicles. *Front Neurosci* 2020;**14**:540.
105. Liu G, et al. Cdc48/VCP and endocytosis regulate TDP-43 and FUS toxicity and turnover. *Mol Cell Biol* 2020;**40**(4).
106. Liu G, et al. Endocytosis regulates TDP-43 toxicity and turnover. *Nat Commun* 2017;**8**(1):2092.
107. Hergesheimer RC, et al. The debated toxic role of aggregated TDP-43 in amyotrophic lateral sclerosis: a resolution in sight? *Brain* 2019;**142**(5):1176–94.
108. Leibiger C, et al. TDP-43 controls lysosomal pathways thereby determining its own clearance and cytotoxicity. *Hum Mol Genet* 2018;**27**(9):1593–607.
109. Leibiger C, et al. Endolysosomal pathway activity protects cells from neurotoxic TDP-43. *Microb Cell* 2018;**5**(4):212–4.
110. Barria MA, Gonzalez-Romero D, Soto C. Cyclic amplification of prion protein misfolding. *Methods Mol Biol* 2012;**849**:199–212.
111. Atarashi R, et al. Ultrasensitive human prion detection in cerebrospinal fluid by real-time quaking-induced conversion. *Nat Med* 2011;**17**(2):175–8.

Chapter 6

Search for functions of intrinsically disordered prion-like domains for FET proteins involved in amyotrophic lateral sclerosis and frontotemporal dementia

Oxana V. Galzitskaya[1,2]

[1]*Institute of Protein Research of Russian Academy of Sciences, Head of Laboratory, Pushchino, Russia;* [2]*Institute of Theoretical and Experimental Biophysics of Russian Academy of Sciences, Pushchino, Russia*

Introduction

Prions are infective proteins able to accept different conformations including amyloid fibrils that can serve as a matrix and "infect" other proteins both within and between the cells and between organisms.[1,2] An increase in the number of prion diseases in recent years caused a broad interest in the biomedical studies on the fundamental mechanisms of amyloid formation. It is currently believed that the main cause of these diseases is the transition of "healthy" protein or peptide molecules from the native conformation into the stable "pathological" form. The molecule in the "pathological" conformation becomes capable to specific aggregation by forming amyloid fibrils. These fibrils can grow indefinitely. Aggregation is associated with such diseases as Alzheimer's, Parkinson's, and Huntington's diseases, cystic fibrosis, diabetes, types of hereditary cataracts.

However, it was later found that yeast prions help them adapt to different environmental conditions.[3] In mammals, prion proteins from one side can cause various diseases, sometimes even leading to death, and from the other side they perform important functions: activate cell-based immunity (MAVS protein), form stress granules (SGs) in unfavorable conditions for the cell, provide long-term memory (CPEB protein), etc.[4]

TDP-43 and Neurodegeneration. https://doi.org/10.1016/B978-0-12-820066-7.00003-5

Many proteins contain prion-like domains, which endow the protein with properties close to those of prions including the ability to self-assembly. Prion-like domains having low complexity (LC) amino acid composition and are aggregation-prone with only one exception, which is nonheritable prion-like complexes.[5] Many of these proteins contain RNA-binding domains. It turns out that the same proteins are now emerging in connection with several neurodegenerative diseases.[6] For example, TAR DNA-binding protein-43 (TDP-43) causes frontotemporal lobar degeneration and amyotrophic lateral sclerosis (ALS), mutations in protein fused in sarcoma (FUS), and TATA-binding protein-associated factor 2N (TAF15) connected with certain familial forms of ALS. Cytotoxic granule—associated RNA-binding protein (TIA1) is connected with SGs formation and has been widely used as SGs marker.[7] In addition, TIA-1 provides a direct interaction between tau and SGs.[8] It is known that RNA associated with SGs might further promote tau aggregation and cause tauopathies.[9] The amino acid composition of prion-like domains of FUS, TDP-43, and TIA1 is close to prion domains of yeast proteins such as Sup35 and Ure2.[10] At the same time, the glycine-rich domain of TIA-1 shares homology with the yeast prion protein, Sup35. In contrast, there is no homology between Sup-35 and TDP-43.[4,10] It should be noted that prion-like domains in these proteins are intrinsically disordered regions.[11] Further, we will designate these domains as prion-like IDRs. Probably, these proteins should be the general players in the misfolding process, proteinopathies.

Prion-like domains endowed RNA-binding proteins with the ability to transfer from the unfolded three-dimensional structure to intermediate states by folding, when the protein has a tendency to various conformational transitions, including the formation of amyloid fibrils[12–14] (Fig. 6.1). Prion-like domains can assemble into structured amyloidogenic oligomers that are then transformed into pathological aggregates of a nonamyloid type or into amyloid fibrils.[15–19]

Considering 29 RNA-binding proteins with prion-like domain, we have revealed that presence of imperfect repeats and homo-repeats is the important characteristics of these proteins.[20,21] The larger the number of repeats and the length of homo-repeats, the higher propensity to aggregate and to be toxic in the aggregated state.[21,22] It is interesting to note that if to delete 100 residues out of the 165 from the prion-like domain of FUS, then it can aggregate,[23] but lost the ability to form gel and was toxic. Based on these results we can suggest that imperfect repeats in the prion-like domain are necessary for quick transition of RNA-binding proteins in a gel-like phase of SGs. The biological problem of how proteins contribute to the formation of reversible subcellular domains like SGs is sufficiently interesting and not entirely clear.

It should be mentioned that from many points of view FET proteins are a perfect model for studying the processes involved in both conformational transitions by the prion mechanism and the formation of protein aggregates.

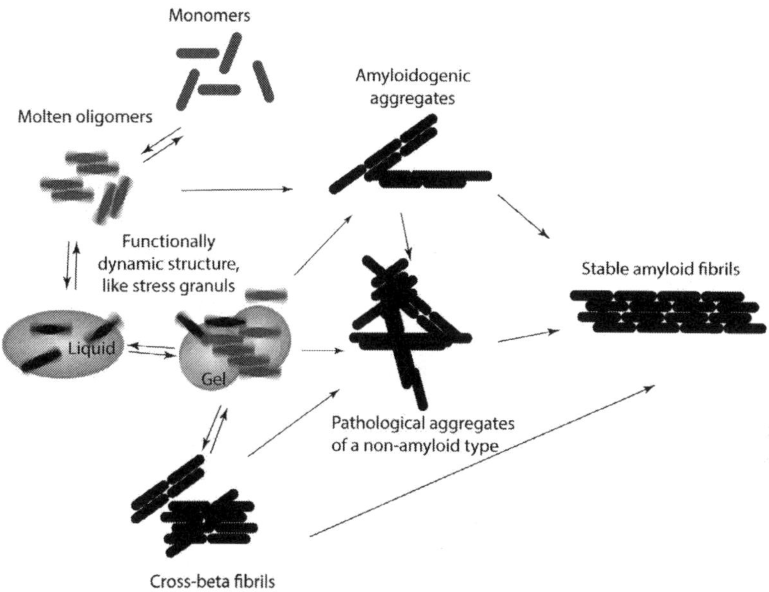

FIGURE 6.1 Schematic representation of possible reversible and irreversible transitions for proteins with prion-like domains. *Modified and adapted from Li YR, King OD, Shorter J, Gitler AD. Stress granules as crucibles of ALS pathogenesis.* J Cell Biol *2013;**201**:361–372.*

One suggestion was made that the FET protein family has evolved to facilitate a more complex combination of transcription with RNA processing that occurs in multicellular organisms.

In this chapter, we are interested in the possible functions of prion-like domains in the FET proteins and search for important partners for interactions. Using bioinformatics analysis including evolutionary and phylogenetic approaches, we have found differences in the prion-like domains, which can determine specific functions of these proteins.

Results and discussion

FET family of proteins with prion-like domains

FET family of proteins with prion-like domains includes three proteins: proteins fused in sarcoma (FUS/TLS), Ewing's sarcoma (EWS), and TATA-binding protein-associated factor 2N (TAF15).[25] FET proteins have several important functions within the cell: they are involved in modulation of mRNA transcription, splicing, mRNA transport, regulation of gene expression, and participate in DNA repair.[26–28] All three proteins interact with several transcription factors and with the transcription preinitiation complex, formed by the RNA Polymerase II and the TFIID complex.[29–32] FET proteins also have nonredundant functions in RNA metabolism.[33] Moreover, all three proteins take part in the formation of SGs.

There are several domains in the structure of FET proteins: N-terminal prion-like domain or LC domain with imperfect repeats, RNA-binding domain (RRM), zinc-finger motif, several RGG domains, and nuclear location signal (NLS) (Fig. 6.2). If to align three proteins using the ClustalO program, the identity is about 20% with 163 identical positions. The functions of some domains and motifs are known. For example, NLS is recognized by the nuclear receptor and is responsible for protein transport from the cytoplasm into the nucleus and back.[34–38] RGG domains affect RNA binding.[39] RNA-recognition and zinc finger motifs are involved in nucleic acid binding.[29] Prion-like domains of these proteins are critical for aggregation. It has been shown that FUS (1–359) is sufficient for aggregation in neuroblastoma cell models.[40] The RGG3 domain does not affect aggregation, and mutations in 502–526 only lead to the accumulation of protein in the cytoplasm and its insertion into SGs. In yeast, FUS accumulated only in the cytoplasm.[41] FET proteins involved not only in ALS,[42] Huntington's diseases, spinocerebral ataxy, and dentatorubral pallidoluysian atrophy[43] but also in the formation of human myxoid liposarcoma.[44–46]

Determination of domains participating in different processes and functions remain to be clarified. Now we know that the LC and RGG1 domains are responsible for such process as assemble into fibers, activate transcription, bind C-terminal domain (CTD) of RNA Pol II, and translocated in sarcomas.[47] RRM domain binds RNA, RGG2, ZnF, and RGG3 domains also bind RNA, assemble into fibers, bind CTD of RNA Pol II. Prion-like domain is weakly conserved in FET proteins for multicellular organisms.[47]

To understand a possible role in regulating aggregation ability of FET proteins, in future it is interesting to consider wet experiments using the deletion and addition of repeats to test repeats effect on protein aggregation formation.

Disordered regions and repeats in FUS/TLS protein

According to the prediction of the IsUnstruct program, the prion-like domain, RGG domains, and NLS part are disordered in FUS protein (Fig. 6.3).[48] FUS human protein enriches with amino acid residues such as glycine, tyrosine, serine, and arginine in comparison with the human proteome (Fig. 6.4). The fraction of glutamine, asparagine, tyrosine, and glycine is 50% of the total

FIGURE 6.2 Domain structure of FET proteins. Low complexity domain includes SYGQ-repeats. RGG1 domain is not very noticeable in TAF15.

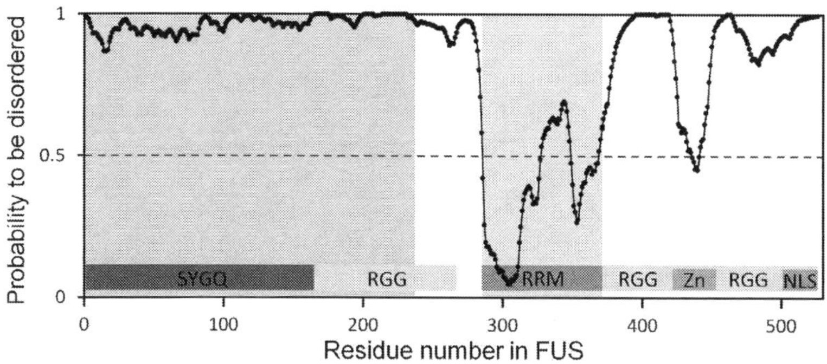

FIGURE 6.3 Probability profiles of amino acid residues in human FUS protein (UniProt code is P35637) to be ordered or disordered calculated by the IsUnstruct program. Green color corresponds to the prion-like domain.

protein. The N-terminal prion-like domain of FUS has 27 different variants of GYG, GYS, SYG, and SYS triplets or 15 imperfect repeated motifs (QPGQGYSQQSS). It has been demonstrated that namely tyrosine residues are responsible for the formation of hydrogel.[14] The C-terminus of the FUS protein includes four imperfect repeats as DDRRGGRGGY. FUS protein includes five disordered patterns from the library of disordered motifs: GSHM, GGGGSGG, GGGGG, GGSGGGGSGGG, and RGGGGSG.[48–50]

Disordered regions and repeats in TAF15

According to the prediction of the IsUnstruct program, the prion-like domain, RGG domains, and NLS part are disordered in TAF15 protein (Fig. 6.5). For TAF15 protein RGG repeated motifs have been considered, but at a more thorough consideration one can find a more attractive repetition at the C-terminus: DRGGGYGG, DRSSGGGYSG, and DRGSRGGYGG. The number of repeats is counted up 23.[51] Moreover, what is more important that this motif contains a tyrosine residue flanked by glycine residues. The N-terminus (1–208) enriches with such amino acid residues as glutamine, glycine, serine, and tyrosine, and one can find such repeats as SYGQSGGEQQ. The fraction of glutamine, asparagine, tyrosine, and glycine is also 50% of the total protein. TAF15 is enriched also such residues as asparagine acid and arginine in comparison with the human proteome (Fig. 6.4).

Disordered regions and repeats in EWS (Ewing's sarcoma) protein

The N-terminal part and most part of the prion-like domain, RGG domains, and NLS part are disordered in EWS protein (Fig. 6.6). The EWS protein includes imperfect repeats such as SAYTAQPTQGYAQTTQAYGCQQ at the N-terminus and DRGRGGPGG repeats at the C-terminus. We can observe 31 repeats in the prion-like domain (the N-terminal part) and six repeats

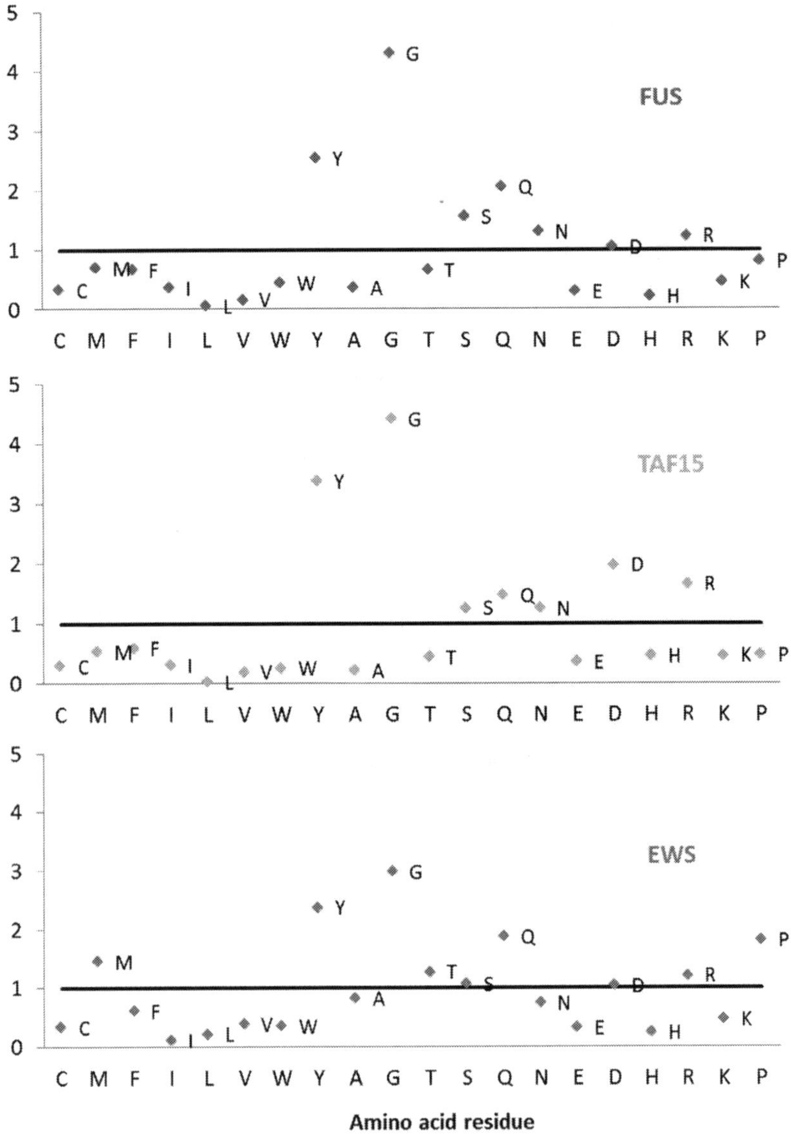

FIGURE 6.4 Comparison of the amino acid composition of the selected human FET proteins with the human proteome.

(DRGRGGPGG) at the C-terminal part (Fig. 6.6). According to the prediction of prion-like domains for 29 candidates of RNA-binding proteins, FUS and TAF15 occupy the first and second places,[52] the third place belongs to EWS protein[29]. The fraction of glutamine, asparagine, tyrosine, and glycine is 38%

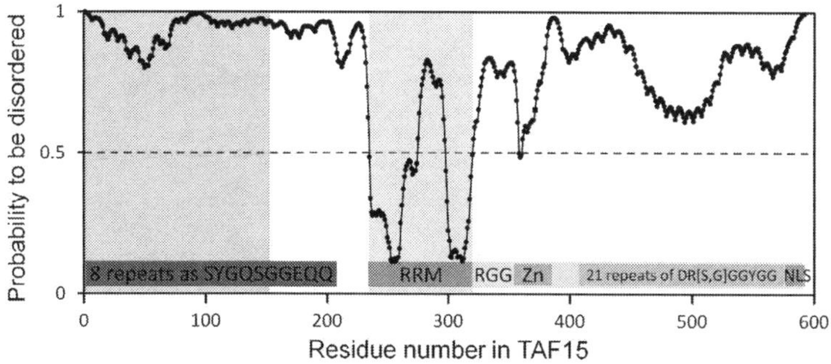

FIGURE 6.5 Probability profiles of amino acid residues in human TAF15 protein (UniProt code is Q92804) to be ordered or disordered calculated by the IsUnstruct program. Green color corresponds to the prion-like domain.

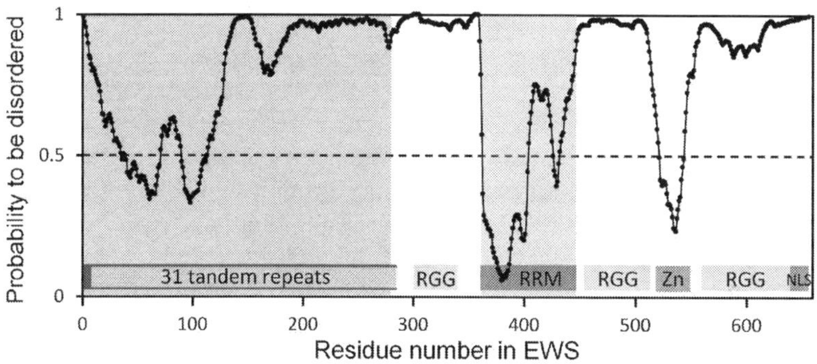

FIGURE 6.6 Probability profiles of amino acid residues in human EWS protein (UniProt code is Q01844) to be ordered or disordered calculated by the IsUnstruct program. Green color corresponds to the prion-like domain.

of the total protein. EWS enriches proline residue in comparison with FUS and TAF15 (Fig. 6.4). Moreover, there are absent homo-repeats with length larger than three residues (AAA, GGG, TTT, SSS, PPP).

Formation of stress granules

The pioneer of SGs (N. Kedersha) was shown that it is a prion-like domain is responsible for their formation: if to remove this domain, the SGs are not formed in vivo.[4] At the moment, specific representatives of SGs are several proteins deserving particular attention because of their involvement in many cell processes. One of such RNA-binding proteins within SGs is RNA-binding protein FUS, but due to its low content in the cytoplasm its concentration in SGs is also low. Mutations in FUS can affect its subcellular localization, which in turn alters its nuclear–cytoplasmic distribution, resulting in changes in the relative abundance within SGs. However, mutations associated with lateral amyotrophic sclerosis, which are located mainly in the nuclear localization signal and cause relocation of the protein from the nucleus, result in its accumulation in the cytoplasm and more efficient protein sequestration in SGs.[34–38] Again, this warrants further study as the molecular mechanisms are not yet fully understood.

Such SGs with a high content of FUS were called pathological or constitutive SGs.[24,53] The hypothesis was also proposed on direct conversion of FUS-containing SGs into insoluble, sometimes ubiquitinated protein aggregates characteristic of FUS proteinopathy.[23] It is assumed that due to self-assembly, prion-like domains of RNA-binding proteins can facilitate the formation of SGs.[53,54]

It should be noted that FUS protein has many glycine homo-repeats in the prion-like domain. Recently we have demonstrated that the RNA-binding proteins with prion-like domain have many homo-repeats including such proteins as hnRNPA0, hnRNPA1, hnRNPA2, and hnRNPA3.[21] For some of these proteins has been shown that they can bind to hydrogel.[14]

Partners for interactions with FET proteins

Disordered proteins have usually many partners for interactions.[55] We used the STRING database to find how many partners present for FET proteins.[56] The number of partners is different for these three proteins: 34 for TAF15 (Fig. 6.7), 112 for EWS, and 275 for FUS. According to the BioGrid database the numbers of partners are different: 143 for TAF15, 661 for EWS, and 420 for FUS.

One of the prominent partners of FUS is RNA-polymerase II—essential enzyme, which is responsible for synthesis of most mRNA, as well as snRNA and microRNA.[57,58] Through direct interaction with CTD of RNA-polymerase II FUS regulates phosphorylation and aggregation of polymerase.[59,60] Moreover, CTD is also a LC region with repetitive pattern (YSPTSPS)n which quite unexpected evolutionary trajectory throughout the nature. There is evidence that it has arisen separately in different organisms several times during the course of evolution and converged later. Thus, it differs in length and level of degeneracy of the repeats.[61] According to the STRING database, the probability of interaction with PolII for all three proteins is 0.9 for FUS, 0.5 for TAF15, and 0.7 for EWS. Considering above, the diversity and evolution of FET family proteins is of great interest.

LOGO patterns

Logo patterns based on multiple sequence alignments from up to 40 eukaryotic taxons reveal clear repetitive structure of prion-like domains with tyrosines serving as markup points (Fig. 6.8). All three proteins share the same common basic pattern "SYGQQS" though with different degree of variability. A unique feature of FUS is occurrence of multiple glycine-rich tracks of variable length

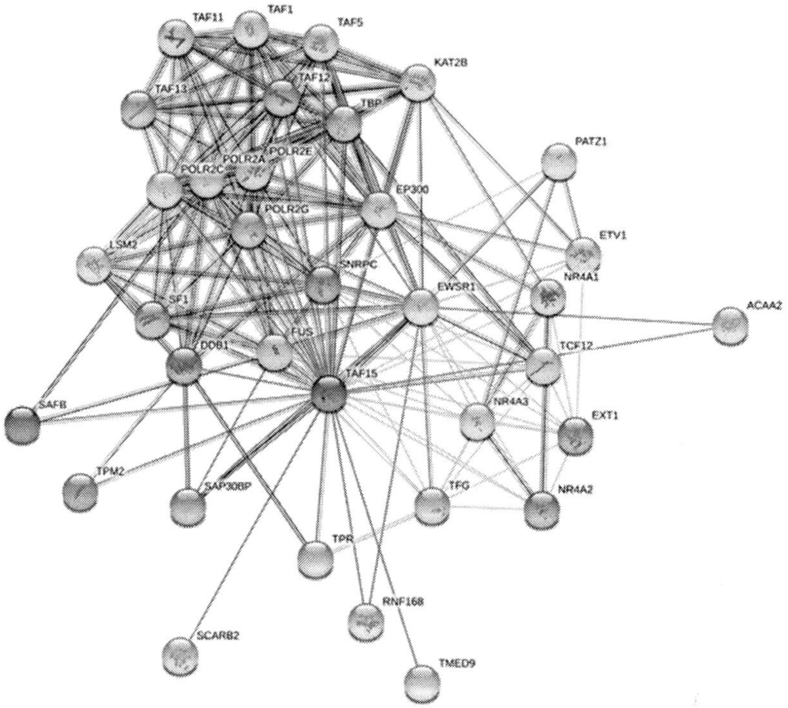

FIGURE 6.7 Schematic representation of partners for TAF15 has been obtained from the STRING database. Probability of interactions equal to 0.4 is considered as the boundary condition for entry in the list of partners. The more transparent of edge in the graph, the less probability of interaction.

which may work as a linker between prion-like and RNA-binding motifs and are required for its functioning.[62] Another interesting difference between FUS, EWS, and TAF15 is the abundance of QQ repeats in the first two, though increased frequency of asparagines. This small, but notable difference might affect aggregation properties of the proteins, due to the fact that both glutamine and asparagine are capable of spontaneous cross-linking, though with their own peculiarities.[63,64]

Evolution of proteins of FET family

Considering data about FUS interaction with CTD domain of RNA-polymerase II, it is worthwhile to compare evolutionary relations between all three sets of proteins. Phylogeny reconstruction with Maximum Parsimony algorithm demonstrates high degree of separation between sets with bootstrap support approaching 100%. It should be noted that there is a remarkable difference between trees reconstructed on full-length sequences (Fig. 6.9A) and prion-like domains (Fig. 6.9B). In full-length case FUS is closely related

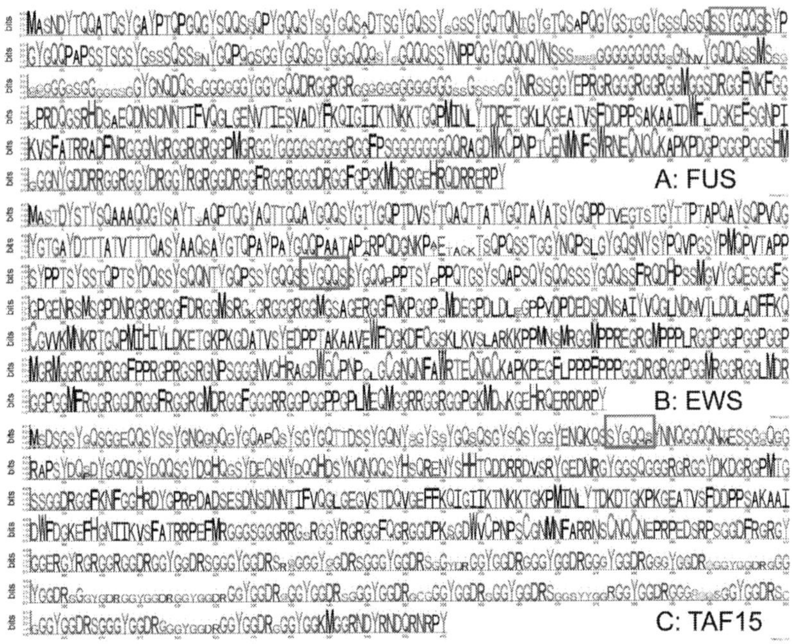

FIGURE 6.8 LOGO patterns of (A) FUS, (B) EWS, and (C) TAF15 proteins. Red frames mark examples of repetitive patterns.

to TAF15, while in the prion-like domain the TAF15 tree is closer to the EWS one. Such phylogenic relationships are of great interest, because TAF15 also is thought to interact with RNA-polymerase II CTD by its prion-like domain[65], thus one might expect that prion domains of FUS and TAF15 should be closely related, although other possible proteins or nucleic partners might exist.

The differences in evolutionary pressures on these proteins have defined their specific functions. Until now, the functions of the RNA binding of FET proteins are not completely elucidated. Gene expression responses to FUS, EWS, and TAF15 reduction and SG sequestration analyses identifies FET-protein nonredundant functions that FUS and TAF15 locate to cellular SGs to a larger extend than EWS, moreover, FUS and TAF15 are more functionally related to each other.[33]

To further investigate evolutionary relationships of FUS, EWS, and TAF15, we calculated *dN/dS* rations [6.1], which describes strength of purifying selection acting on the proteins:

$$dN/dS = \frac{d_n}{d_s}, \tag{6.1}$$

where d_n means nonsynonymous substitutions per nonsynonymous site and d_s means synonymous substitutions per synonymous site.

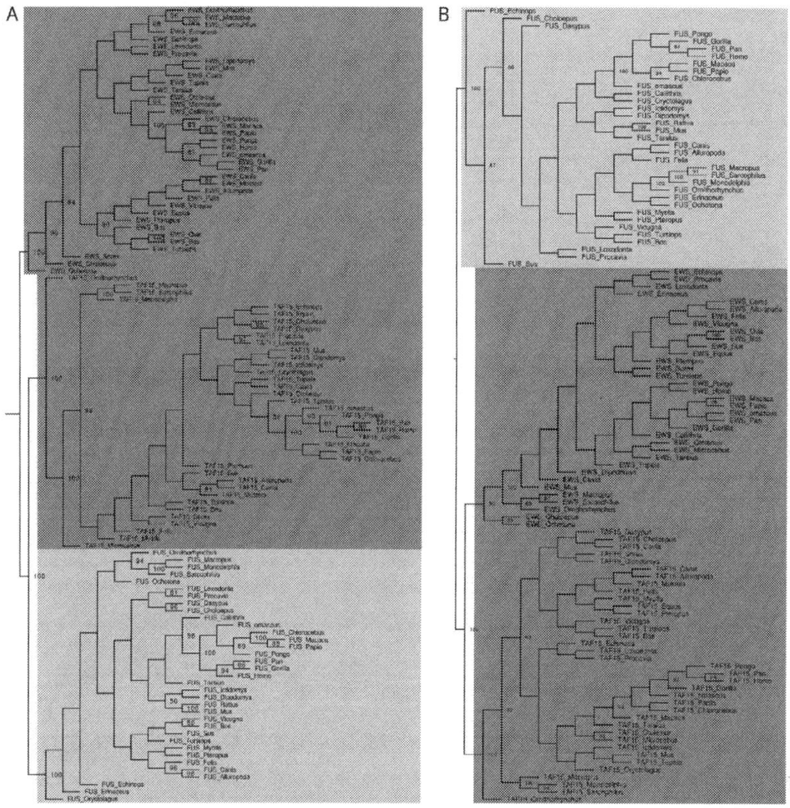

FIGURE 6.9 Phylogeny reconstruction of (A) full-length and (B) prion-like domains of FUS, EWS, and TAF15 proteins. Numbers represent bootstrap values (only values greater than 80 are shown).

Such low values which are close to zero mean that all three proteins are under high selection pressure. However, EWS ratio is approximately 1.5 times lower than that for FUS and TAF15. Closer look (Fig. 6.10) reveals that three proteins have very different evolutionary trajectories. Albeit FUS is under high pressure of purifying selection, it is subjected to large numbers of both synonymous and non-synonymous mutations with a moderate number of insertion or deletions (indels) in the prion-like domain. Unlike FUS, EWS has a much greater number of substitutions of indels, while having lowest number of other mutations. It is also worth to mention that all three proteins have a distinct plateau on the indels graphs corresponding to the RNA-binding motif, that means preserving of total size of this domain. Another expected observation is that all indels are a multiple of three nucleobases, meaning that we do not observe any frame shifts or other complex evolutionary events.

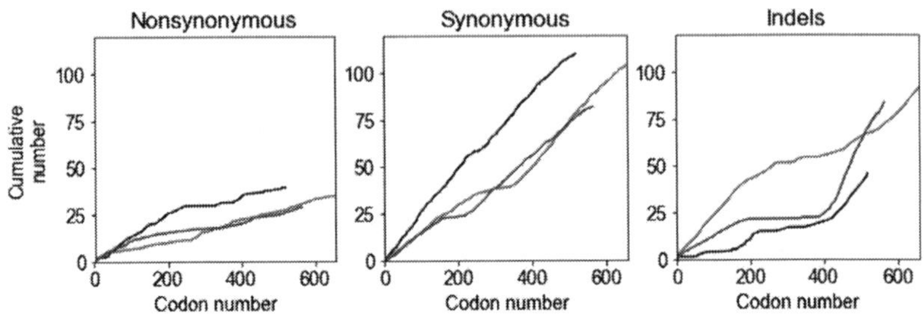

FIGURE 6.10 Cumulative number of substitutions along sequence. FUS in black, EWS in red, and TAF15 in blue.

Materials and methods

Logo

Sequences for Logo alignments were retrieved from UniProt. In total, we obtained 104 amino acid sequences from 84 unique taxa for FUS, 108 sequences from 42 unique taxa for EWS, and 33 sequences from 24 unique taxa. Alignments were performed with Muscle with default settings.[66] Logo was created with WebLogo desktop version.[67]

Phylogeny reconstruction

Sequence retrieved at the previous step was combined and aligned with Muscle. Two independent alignments were created: for full-length sequences and for prion-like domain only (boundaries of domains were determined based on exons). Phylogeny reconstruction was performed with Maximum Parsimony algorithm as implemented in MPBoot package with 1000 bootstrap replicas and rerooted into the middle point.[68] Visualization of phylogenetic trees was done in FigTree software package (https://scholar.google.com/scholar?hl=en&as_sdt=0,5&cluster=10570788131709508105).

Evolutionary analysis

We used prealigned and filtered sequence nucleotide sequences obtained from OrthoMaM database.[69] We calculated dN/dS ratios, synonymous, nonsynonymous, and indel ratios with SNAP web-service.[70] Resulting ratios are in Table 6.1.

STRING database

The list of partners for FET proteins has been obtained from the STRING database using the information from the databases, data about homologies, possible coexpression, experimental confirmation about interactions, etc.[71]

TABLE 6.1 Amino acid replacement rates (dN/dS ratios).

Protein	Meaning
FUS	0.052
EWS	0.037
TAF15	0.058

To obtain the list of all partners for FUS interactions the user should use this link: https://string-db.org/api/tsv/interaction_partners?identifiers=FUS&species=9606&required_score=400. For construction of figure we changed the confidence cutoff to 0.4. Transparency of edges connecting vertices in the graph designates probability of interaction. The more transparent of edge, the less probability of interaction.

Conclusions

Using bioinformatics analysis including evolutionary and phylogenetic approaches we have found differences in the prion-like domains, which can determine specific functions of these proteins. Altogether presented data can be interpreted such as FUS, EWS, and TAF15 are an example of duplication and neofunctionalization with an ongoing evolutionary process. In the case of FUS it's driven by necessity to adapt to changes in CTD domain of RNA-polymerase II. The obtained results are in agreement with the recent results that FET-protein reduction had no influence on SG formation suggesting that FET proteins are not vital for SG formation.[33] Our results may help shed novel insights into our understanding of the differences in function between the FET family members.

Acknowledgments

O.G. is grateful to Arthur O. Zalevsky, Aleksandr A. Cheblokov, and Vladimir N. Solovev for alignments and phylogenetic reconstruction and valuable discussion on the evolution of FET proteins, Elena Leonova for preparing Fig. 6.1, Anna Glyakina for preparing Fig. 6.4, and Professor Pavel Ivanov (Brigham and Women's Hospital, Harvard Medical School, Harvard Initiative for RNA Medicine, Boston) for reading the manuscript and valuable comments. I am grateful to the Institute of Bioinformatics and EPAM Company for organization of the BioHack2018, which was held in St. Petersburg. Some part of the work was done in this BioHack2018.

References

1. Aguzzi A, Rajendran L. The transcellular spread of cytosolic amyloids, prions, and prionoids. *Neuron* 2009;**64**:783–90.

2. Prusiner SB, Scott MR, DeArmond SJ, Cohen FE. Prion protein biology. *Cell* 1998;**93**:337–48.
3. Halfmann R, Alberti S, Lindquist S. Prions, protein homeostasis, and phenotypic diversity. *Trends Cell Biol* 2010;**20**:125–33.
4. Gilks N, Kedersha N, Ayodele M, Shen L, Stoecklin G, Dember LM, Anderson P. Stress granule assembly is mediated by prion-like aggregation of TIA-1. *Mol Biol Cell* 2004;**15**:5383–98.
5. Alberti S, Halfmann R, King O, Kapila A, Lindquist S. A systematic survey identifies prions and illuminates sequence features of prionogenic proteins. *Cell* 2009;**137**:146–58.
6. Walsh DM, Hartley DM, Kusumoto Y, Fezoui Y, Condron MM, Lomakin A, Benedek GB, Selkoe DJ, Teplow DB. Amyloid beta-protein fibrillogenesis. Structure and biological activity of protofibrillar intermediates. *J Biol Chem* 1999;**274**:25945–52.
7. Wolozin B. Regulated protein aggregation: stress granules and neurodegeneration. *Mol Neurodegener* 2012;**7**:56.
8. Vanderweyde T, Yu H, Varnum M, Liu-Yesucevitz L, Citro A, Ikezu T, Duff K, Wolozin B. Contrasting pathology of the stress granule proteins TIA-1 and G3BP in tauopathies. *J Neurosci* 2012;**32**:8270–83.
9. Kampers T, Friedhoff P, Biernat J, Mandelkow E-M, Mandelkow E. RNA stimulates aggregation of microtubule-associated protein tau into Alzheimer-like paired helical filaments. *FEBS Lett* 1996;**399**:344–9.
10. Serio TR, Lindquist SL. [PSI+], SUP35, and chaperones. *Adv Protein Chem* 2001;**57**:335–66.
11. Malinovska L, Kroschwald S, Alberti S. Protein disorder, prion propensities, and self-organizing macromolecular collectives. *Biochim Biophys Acta* 2013;**1834**:918–31.
12. Halfmann R, Alberti S, Krishnan R, Lyle N, O'Donnell CW, King OD, Berger B, Pappu RV, Lindquist S. Opposing effects of glutamine and asparagine govern prion formation by intrinsically disordered proteins. *Mol Cell* 2011;**43**:72–84.
13. Han TW, Kato M, Xie S, Wu LC, Mirzaei H, Pei J, Chen M, Xie Y, Allen J, Xiao G, et al. Cell-free formation of RNA granules: bound RNAs identify features and components of cellular assemblies. *Cell* 2012;**149**:768–79.
14. Kato M, Han TW, Xie S, Shi K, Du X, Wu LC, Mirzaei H, Goldsmith EJ, Longgood J, Pei J, et al. Cell-free formation of RNA granules: low complexity sequence domains form dynamic fibers within hydrogels. *Cell* 2012;**149**:753–67.
15. Selivanova OM, Glyakina AV, Gorbunova EY, Mustaeva LG, Suvorina MY, Grigorashvili EI, Nikulin AD, Dovidchenko NV, Rekstina VV, Kalebina TS, et al. Structural model of amyloid fibrils for amyloidogenic peptide from Bgl2p-glucantransferase of *S. cerevisiae* cell wall and its modifying analog. New morphology of amyloid fibrils. *Biochim Biophys Acta* 2016;**1864**:1489–99.
16. Selivanova OM, Suvorina MY, Surin AK, Dovidchenko NV, Galzitskaya OV. Insulin and lispro insulin: what is common and different in their behavior? *Curr Protein Pept Sci* 2017;**18**:57–64.
17. Galzitskaya OV, Selivanova OM. Rosetta stone for amyloid fibrils: the key role of ring-like oligomers in amyloidogenesis. *J Alzheimers Dis* 2017;**59**:785–95.
18. Selivanova OM, Surin AK, Ryzhykau YL, Glyakina AV, Suvorina MY, Kuklin AI, Rogachevsky VV, Galzitskaya OV. To be fibrils or to be nanofilms? Oligomers are building blocks for fibril and nanofilm formation of fragments of Aβ peptide. *Langmuir ACS J Surf Colloids* 2018;**34**:2332–43.

19. Galzitskaya OV, Surin AK, Glyakina AV, Rogachevsky VV, Selivanova OM. Should the treatment of amyloidosis be personified? Molecular mechanism of amyloid formation by Aβ peptide and its fragments. *J Alzheimers Dis Rep* 2018;**2**:181–99.
20. Galzitskaya OV. Repeats are one of the main characteristics of RNA-binding proteins with prion-like domains. *Mol Biosyst* 2015;**11**:2210–8.
21. Galzitskaya OV, Lobanov MY. Proteome-scale understanding of relationship between homo-repeat enrichments and protein aggregation properties. *PLoS One* 2018;**13**:e0206941.
22. Barton S, Jacak R, Khare SD, Ding F, Dokholyan NV. The length dependence of the polyQ-mediated protein aggregation. *J Biol Chem* 2007;**282**:25487–92.
23. Sun Z, Diaz Z, Fang X, Hart MP, Chesi A, Shorter J, Gitler AD. Molecular determinants and genetic modifiers of aggregation and toxicity for the ALS disease protein FUS/TLS. *PLoS Biol* 2011;**9**:e1000614.
24. Li YR, King OD, Shorter J, Gitler AD. Stress granules as crucibles of ALS pathogenesis. *J Cell Biol* 2013;**201**:361–72.
25. Anderson P, Kedersha N. Stress granules: the Tao of RNA triage. *Trends Biochem Sci* 2008;**33**:141–50.
26. Fujii R, Okabe S, Urushido T, Inoue K, Yoshimura A, Tachibana T, Nishikawa T, Hicks GG, Takumi T. The RNA binding protein TLS is translocated to dendritic spines by mGluR5 activation and regulates spine morphology. *Curr Biol* 2005;**15**:587–93.
27. Ishigaki S, Masuda A, Fujioka Y, Iguchi Y, Katsuno M, Shibata A, Urano F, Sobue G, Ohno K. Position-dependent FUS-RNA interactions regulate alternative splicing events and transcriptions. *Sci Rep* 2012;**2**:529.
28. Rogelj B, Easton LE, Bogu GK, Stanton LW, Rot G, Curk T, Zupan B, Sugimoto Y, Modic M, Haberman N, et al. Widespread binding of FUS along nascent RNA regulates alternative splicing in the brain. *Sci Rep* 2012;**2**:603.
29. Tan AY, Manley JL. The TET family of proteins: functions and roles in disease. *J Mol Cell Biol* 2009;**1**:82–92.
30. Paronetto MP. Ewing sarcoma protein: a key player in human cancer. *Int J Cell Biol* 2013;**2013**:642853.
31. Bertolotti A, Lutz Y, Heard DJ, Chambon P, Tora L. hTAF(II)68, a novel RNA/ssDNA-binding protein with homology to the pro-oncoproteins TLS/FUS and EWS is associated with both TFIID and RNA polymerase II. *EMBO J* 1996;**15**:5022–31.
32. Bertolotti A, Melot T, Acker J, Vigneron M, Delattre O, Tora L. EWS, but not EWS-FLI-1, is associated with both TFIID and RNA polymerase II: interactions between two members of the TET family, EWS and hTAFII68, and subunits of TFIID and RNA polymerase II complexes. *Mol Cell Biol* 1998;**18**:1489–97.
33. Blechingberg J, Luo Y, Bolund L, Damgaard CK, Nielsen AL. Gene expression responses to FUS, EWS, and TAF15 reduction and stress granule sequestration analyses identifies FET-protein non-redundant functions. *PLoS One* 2012;**7**:e46251.
34. Dormann D, Rodde R, Edbauer D, Bentmann E, Fischer I, Hruscha A, Than ME, Mackenzie IRA, Capell A, Schmid B, et al. ALS-associated fused in sarcoma (FUS) mutations disrupt Transportin-mediated nuclear import. *EMBO J* 2010;**29**:2841–57.
35. Gal J, Zhang J, Kwinter DM, Zhai J, Jia H, Jia J, Zhu H. Nuclear localization sequence of FUS and induction of stress granules by ALS mutants. *Neurobiol Aging* 2011;**32**:27–40.
36. Ito D, Seki M, Tsunoda Y, Uchiyama H, Suzuki N. Nuclear transport impairment of amyotrophic lateral sclerosis-linked mutations in FUS/TLS. *Ann Neurol* 2011;**69**:152–62.

37. Kino Y, Washizu C, Aquilanti E, Okuno M, Kurosawa M, Yamada M, Doi H, Nukina N. Intracellular localization and splicing regulation of FUS/TLS are variably affected by amyotrophic lateral sclerosis-linked mutations. *Nucleic Acids Res* 2011;**39**:2781–98.
38. Niu C, Zhang J, Gao F, Yang L, Jia M, Zhu H, Gong W. FUS-NLS/Transportin 1 complex structure provides insights into the nuclear targeting mechanism of FUS and the implications in ALS. *PLoS One* 2012;**7**:e47056.
39. Ozdilek BA, Thompson VF, Ahmed NS, White CI, Batey RT, Schwartz JC. Intrinsically disordered RGG/RG domains mediate degenerate specificity in RNA binding. *Nucleic Acids Res* 2017;**45**:7984–96.
40. Shelkovnikova TA, Robinson HK, Connor-Robson N, Buchman VL. Recruitment into stress granules prevents irreversible aggregation of FUS protein mislocalized to the cytoplasm. *Cell Cycle Georget Tex* 2013;**12**:3194–202.
41. Ju S, Tardiff DF, Han H, Divya K, Zhong Q, Maquat LE, Bosco DA, Hayward LJ, Brown RH, Lindquist S, et al. A yeast model of FUS/TLS-dependent cytotoxicity. *PLoS Biol* 2011;**9**:e1001052.
42. Couthouis J, Hart MP, Shorter J, DeJesus-Hernandez M, Erion R, Oristano R, Liu AX, Ramos D, Jethava N, Hosangadi D, et al. A yeast functional screen predicts new candidate ALS disease genes. *Proc Natl Acad Sci USA* 2011;**108**:20881–90.
43. Doi H, Koyano S, Suzuki Y, Nukina N, Kuroiwa Y. The RNA-binding protein FUS/TLS is a common aggregate-interacting protein in polyglutamine diseases. *Neurosci Res* 2010;**66**:131–3.
44. Crozat A, Aman P, Mandahl N, Ron D. Fusion of CHOP to a novel RNA-binding protein in human myxoid liposarcoma. *Nature* 1993;**363**:640–4.
45. Kwiatkowski Jr TJ, Bosco DA, Leclerc AL, Tamrazian E, Vanderburg CR, Russ C, Davis A, Gilchrist J, Kasarskis EJ, Munsat T, et al. Mutations in the FUS/TLS gene on chromosome 16 cause familial amyotrophic lateral sclerosis. *Science* 2009;**323**:1205–8.
46. Vance C, Rogelj B, Hortobágyi T, De Vos KJ, Nishimura AL, Sreedharan J, Hu X, Smith B, Ruddy D, Wright P, et al. Mutations in FUS, an RNA processing protein, cause familial amyotrophic lateral sclerosis type 6. *Science* 2009;**323**:1208–11.
47. Schwartz JC, Cech TR, Parker RR. Biochemical properties and biological functions of FET proteins. *Annu Rev Biochem* 2015;**84**:355–79.
48. Lobanov MY, Sokolovskiy IV, Galzitskaya OV. IsUnstruct: prediction of the residue status to be ordered or disordered in the protein chain by a method based on the Ising model. *J Biomol Struct Dyn* 2013;**31**:1034–43.
49. Lobanov MY, Galzitskaya OV. The Ising model for prediction of disordered residues from protein sequence alone. *Phys Biol* 2011;**8**:035004.
50. Lobanov MY, Sokolovskiy IV, Galzitskaya OV. HRaP: database of occurrence of Homo-Repeats and patterns in proteomes. *Nucleic Acids Res* 2014;**42**:D273–8.
51. Galzitskaya OV. Reversible and irreversible aggregation of proteins from the FET family: influence of repeats in protein chain on its aggregation capacity. *Curr Protein Pept Sci* 2016;**17**:319–31.
52. King OD, Gitler AD, Shorter J. The tip of the iceberg: RNA-binding proteins with prion-like domains in neurodegenerative disease. *Brain Res* 2012;**1462**:61–80.
53. Liu-Yesucevitz L, Bassell GJ, Gitler AD, Hart AC, Klann E, Richter JD, Warren ST, Wolozin B. Local RNA translation at the synapse and in disease. *J Neurosci* 2011;**31**:16086–93.

54. Bosco DA, Lemay N, Ko HK, Zhou H, Burke C, Kwiatkowski TJ, Sapp P, McKenna-Yasek D, Brown RH, Hayward LJ. Mutant FUS proteins that cause amyotrophic lateral sclerosis incorporate into stress granules. *Hum Mol Genet* 2010;**19**:4160–75.
55. Kim PM, Sboner A, Xia Y, Gerstein M. The role of disorder in interaction networks: a structural analysis. *Mol Syst Biol* 2008;**4**.
56. Jensen LJ, Kuhn M, Stark M, Chaffron S, Creevey C, Muller J, Doerks T, Julien P, Roth A, Simonovic M, et al. STRING 8–a global view on proteins and their functional interactions in 630 organisms. *Nucleic Acids Res* 2009;**37**:D412–6.
57. Kornberg RD. Eukaryotic transcriptional control. *Trends Cell Biol* 1999;**9**:M46–9.
58. Sims RJ, Mandal SS, Reinberg D. Recent highlights of RNA-polymerase-II-mediated transcription. *Curr Opin Cell Biol* 2004;**16**:263–71.
59. Schwartz JC, Ebmeier CC, Podell ER, Heimiller J, Taatjes DJ, Cech TR. FUS binds the CTD of RNA polymerase II and regulates its phosphorylation at Ser2. *Genes Dev* 2012;**26**:2690–5.
60. Burke KA, Janke AM, Rhine CL, Fawzi NL. Residue-by-residue view of in vitro FUS granules that bind the C-terminal domain of RNA polymerase II. *Mol Cell* 2015;**60**:231–41.
61. Chapman RD, Heidemann M, Hintermair C, Eick D. Molecular evolution of the RNA polymerase II CTD. *Trends Genet* 2008;**24**:289–96.
62. Shang Y, Huang EJ. Mechanisms of FUS mutations in familial amyotrophic lateral sclerosis. *Brain Res* 2016;**1647**:65–78.
63. Heck T, Faccio G, Richter M, Thöny-Meyer L. Enzyme-catalyzed protein crosslinking. *Appl Microbiol Biotechnol* 2013;**97**:461–75.
64. Friedrich MG, Wang Z, Schey KL, Truscott RJW. Spontaneous cross-linking of proteins at aspartate and asparagine residues is mediated via a succinimide intermediate. *Biochem J* 2018;**475**:3189–200.
65. Janke AM, Seo DH, Rahmanian V, Conicella AE, Mathews KL, Burke KA, Mittal J, Fawzi NL. Lysines in the RNA polymerase II C-terminal domain contribute to TAF15 fibril recruitment. *Biochemistry* 2018;**57**:2549–63.
66. Edgar RC. MUSCLE: multiple sequence alignment with high accuracy and high throughput. *Nucleic Acids Res* 2004;**32**:1792–7.
67. Crooks GE, Hon G, Chandonia J-M, Brenner SE. WebLogo: a sequence logo generator. *Genome Res* 2004;**14**:1188–90.
68. Hoang DT, Vinh LS, Flouri T, Stamatakis A, von Haeseler A, Minh BQ. MPBoot: fast phylogenetic maximum parsimony tree inference and bootstrap approximation. *BMC Evol Biol* 2018;**18**:11.
69. Ranwez V, Delsuc F, Ranwez S, Belkhir K, Tilak M-K, Douzery EJOMM. A database of orthologous genomic markers for placental mammal phylogenetics. *BMC Evol Biol* 2007;**7**:241.
70. Korber B. In: Rodrigo AG, Learn GH, editors. *HIV signature and sequence variation analysis. Computational analysis of HIV molecular sequences, chapter 4*. Dordrecht, Netherlands: Kluwer Academic Publishers; 2000. p. 55–72.
71. Szklarczyk D, Franceschini A, Wyder S, Forslund K, Heller D, Huerta-Cepas J, Simonovic M, Roth A, Santos A, Tsafou KP, et al. STRING v10: protein-protein interaction networks, integrated over the tree of life. *Nucleic Acids Res* 2015;**43**:D447–52.

Chapter 7

A systems biology approach to understand the role of TDP-43 in amyotrophic lateral sclerosis

Mohd Maksuf Ul Haque[1], Md Sheeraz Anwar[1], Md Zubbair Malik[2], R.K. Brojen Singh[2], Nidhi Verma[3], Saurabh Kumar Sharma[4]

[1]*Department of Computer Science, Jamia Millia Islamia, New Delhi, India;* [2]*School of Computational and Integrative Sciences, Jawaharlal Nehru University, New Delhi, India;* [3]*Medical School, University of Texas Health Science Centre, Department of Microbiology and Molecular Genetics, Houston, TX, United States;* [4]*School of Computer & Systems Sciences, Jawaharlal Nehru University, New Delhi, India*

Introduction

Unstable aggregates of transactivating response region (TAR) DNA-binding protein of 43 kDa (TDP-43) have been found in patients suffering from the inherited form of amyotrophic lateral sclerosis (ALS) and frontal temporal lobar degeneration[1–4] and have also been associated with patients of some neurodegenerative diseases such as Alzheimer's, Huntington's, and Parkinson's diseases.[5–7] The interaction of the fraction of TDP-43 bound with fused in sarcoma/translocation in liposarcoma (FUS/TLS) has turned out to be considerably advanced by TDP-43 mutants. The abnormal balance of ALS-linked mutant TDP-43 and its increased binding to normal FUS/TLS implicates its pathogenic role in ALS.[8] TDP-43 is an RNA-binding protein that is engaged in various cytological processes in particular RNA metabolism (transcription, translation, miRNA processing, transport of mRNA within the nucleus),[8–10] apoptosis,[11] cytokinesis,[12] embryo development,[13] as well as stress response.[14] TDP-43 contains two stable folded RNA recognition motifs (RRMs) that play a key role in these functions.[15] TDP-43 contains a disrupted, glycine-rich C-terminal domain with every single major pathologically linked ALS mutations[16] having a high tendency to self-assemble,[17,18] aggregate,[3] and phase separate.[3,19] The second RNA recognition motif (RRM2) has found to be playing a major role in aggregation of TDP-43 and the resulting toxicity.[20,21] TDP-43 also comprises an N-terminal domain (NTD) whose structure and structural integrity forms the basis for the vital functions of TDP-43. The NTD

TDP-43 and Neurodegeneration. https://doi.org/10.1016/B978-0-12-820066-7.00006-0

consists of five β-strands (β2-β1-β5-β3-β4) which are packed against an α-helix. Within the loop between β4 and β5, an additional β-hairpin is present that is representative of the structural domain of NTD being similar to the Ubiquitin and DIX domains.[46]

Round about 97% of both sporadic and familial ALS patients depicts the aggregation of TAR-binding protein (TDP-43) in cytoplasmic region and associated with TDP-43 loss in nuclear leads to loss of normal function of TDP-43 in nucleus and results in formation of inclusion foremost to cellular. Overexpression of wild TDP-43 can cause fragmentation of protein in 35 and 25 kDa fragments seen in human ALS cases.[47] Additionally, unusual distribution and accumulation of TDP-43 in ALS, various modifications in post-translational are related with pathologic TDP-43, such as ubiquitination, proteolytic cleavage, and phosphorylation.[22] TDP-43 is a member of the heterogeneous nuclear ribonucleoprotein (hnRNP) family of proteins that is highly conserved and ubiquitously expressed.[23] TDP-43 contains two motifs for RNA recognition and mainly binds RNA via the first of these. Interaction with several other hnRNP proteins, specifically hnRNPs A311, A2/B1, A1 and C1/C2,[24] has been shown to mediate the glycine-rich C-terminus of TDP-43, although the full extent of TDP-43 interactions has not been previously defined. TDP-43 has been shown to shuttle between the nucleus and the cytoplasm as a primarily nuclear protein.[25] Interestingly, as a physiological reaction to neuronal damage, TDP-43 redistributes to cytosolic granules, and nuclear localization is restored after recovery.[26,27]

We applied network theoretical approaches to explore the important of TDP-43. Our key hypothesis is that a significant role in neurodegenerative diseases is likely to be played by most core gene in the gene-interaction network. In the current study, in the context of a network, we mapped and analyzed the interactions of TDP-43. The proposed study is based on network analysis approaches to anticipate some relevant genes, miRNAs, transcription factors to directly bind to TDP-43, which can be regarded (or can be tested) as reliable candidates for further (in vivo/in vitro) studies. The approach for this is based on previous studies[28,29] that have proven the merit of network analysis in finding key regulators and biomarkers.

Expression analysis

We used the GTEx portal[30] to obtain the expression of TDP-43 in different areas of the brain in form of a violin plot and organized the data based on the median. Also, a spatiotemporal expression heatmap is generated based on the calculated expression levels of TDP-43 in RNAseq data from Brainspan,[31] through BEST—a web server for brain expression spatiotemporal pattern analysis.[32]

Construction of protein–protein interaction network

The genes involved in ALS were collected from an online database Amyotrophic Lateral Sclerosis Online Database "ALSoD."[33] A protein–protein interaction network was formed by obtaining interaction data from the STRING[34] database and rendering it with Cytoscape[35] for our gene list. Further MCODE[36,37] was used to find communities or modules in the network. Another network was prepared using the STRING database where we used TDP-43 as the sole input and then obtained the network of all of the proteins it interacts with. NetworkAnalyst[38] was used to find relevant data on TDP-43, using the protein name, the relevant networks found were Gene-disease Associations and TF-miRNA Coregulatory Network. DGIdb[39] was used to find drugs interacting with the TDP-43, the drug–gene interaction network was then rendered in Cytoscape. Finally, TDP-43 role in major pathways was obtained through KEGG.[40]

Enrichment analysis

The enrichment analysis was done on Cytoscape's STRING plug-in[41]. Firstly, we found GO process terms for the main network, after which we used the network-based around TDP-43 and obtained the GO processes related to it. Then we looked up the common GO terms from the two enrichment analyses and rendered them using ggplot[42] in R.

Results

Expression analysis

The expression data obtained from the GTEx portal and BEST tool are shown in Fig. 7.1A and B, respectively. Both the results show high and continuous expression of TDP-43 in most of the regions of the brain especially the cerebellum. In the violin plot in Fig. 7.1A, the TPM (transcripts per million) is the highest, and in Fig. 7.1B, we can see that TDP-43 is expressed throughout the life cycle in the cerebellum and the frontal, parietal, temporal, and occipital cortexes, the expression in amygdala and hippocampus is also prominent.

Protein–protein interaction network

A list of 126 genes that have been experimentally found to play significant roles in ALS and TDP-43 (TDP-43) is one of them, showing that it plays important role in the neurodegenerative disease. The protein–protein interaction network formed from these ALS genes is shown in Fig. 7.2A. The network is made up of 123 nodes and 528 edges. By analyzing the properties of the network, we can easily see that TDP-43 is one of the most connected genes (hub gene) and tops the list in other properties too, it has fourth highest degree. The top 10 genes from the network and their properties are given in Table 7.1.

We obtained four modules of TDP-43 direct interaction to other protein, the first module had TDP-43 and it was the second most important genes based on analyzed properties of the network. Further the three modules in which

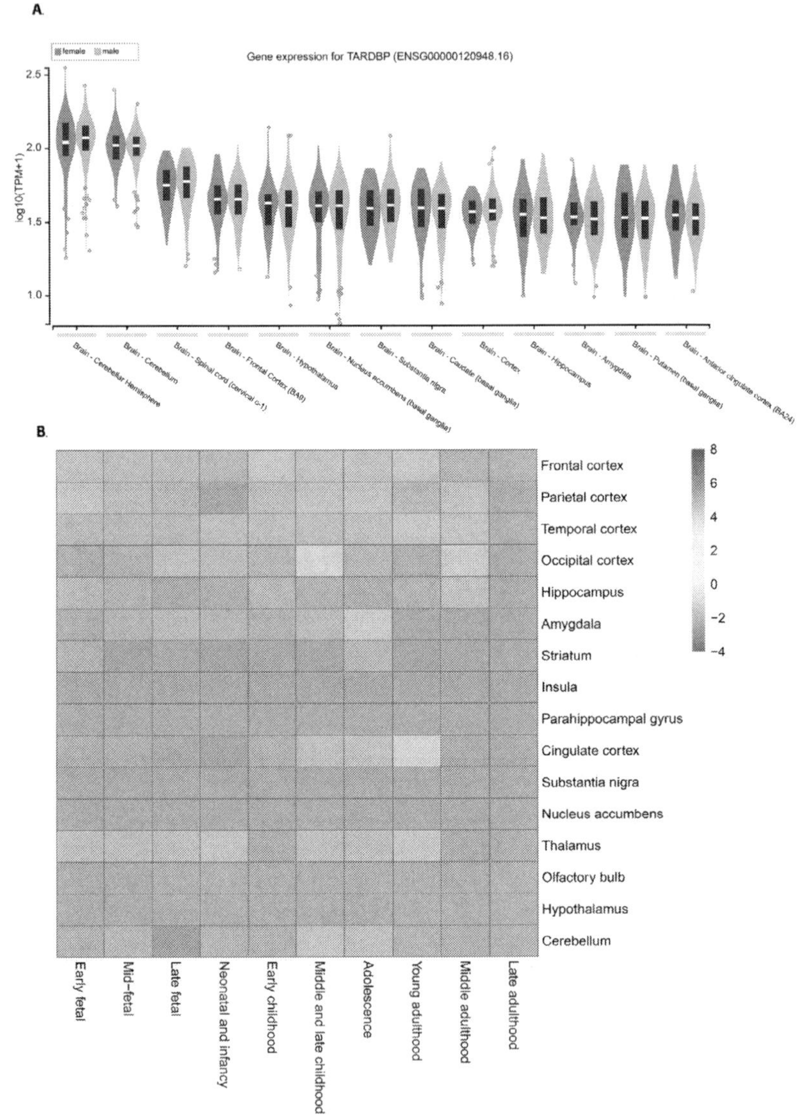

FIGURE 7.1 Expression level of TDP-43: (A) Violin plot shows TDP-43 gene expression in multiple normal tissues from GTEx portal in different regions of brain. (B) BEST results of the TDP-43 genes. Spatiotemporal expression heatmap of TDP-43 genes from RNA-Seq Brain Span dataset analysis from BEST tool.

gene were directly connected to TDP-43 (Fig. 7.2B). SOD1, ALSIN, SETX, SPG11, FUS, VAPB, TARDBP, OPTN, ATXN2, and C9orf72 were discovered to be involved in the pathogenesis of ALS. Gene enrichment of TDP-43 and these associated were significant in pathogenesis of ALS. The associated genes

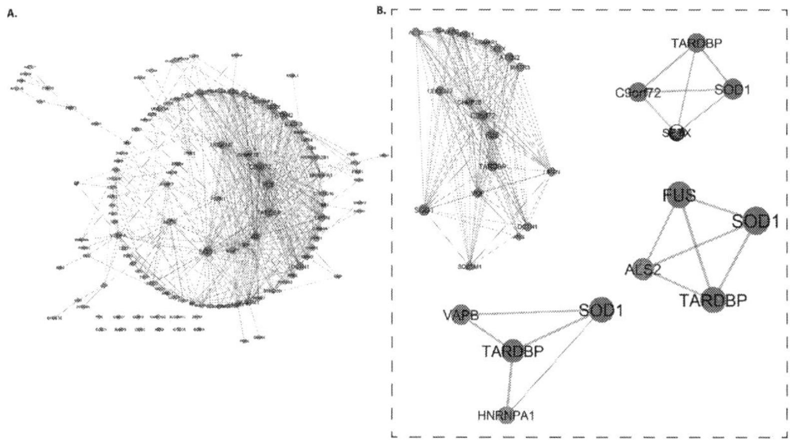

FIGURE 7.2 Protein–protein interaction (PPI) network. (A) PPI regulatory network of highly associated gene to TDP-43 in ALS, constructed using Cytoscape on the basis of human interactome (Nodes with higher degree values (hubs) are depicted with large shape). (B) Modules of PPI regulatory network of TDP-43.

to TDP-43 were mainly enriched in the gene ontology terms related to motor neuron atrophy, paralyzed, motor neuron disease, fatigable weakness of swallowing muscles, laryngospasm, agitation, muscular fasciculation, spasticity, dysarthria, abnormal lower motor neuron morphology, and ALS with dementia (Fig. 7.3A).

Enrichment analysis

We obtained 728 terms from enrichment analysis of the ALS genes network and filtered terms of GO Process, GO Function, KEGG Pathways, and Reactome Pathways, hence obtained 568 terms. 82 terms had TDP-43's involvement and there were several terms related to neuron and nervous system development and differentiation, e.g., ALS pathway, development of central nervous system, central nervous system, neuronal maturation, neuromuscular junction development, glial cell activation, glial cell development, etc. We found 101 terms network and filtered terms of GO Process, GO Function, KEGG Pathways, and Reactome Pathways (Table 7.2). We obtained 67 terms including major biological processes like mRNA Splicing—Major Pathway, RNA splicing, mRNA processing, mRNA splicing, via spliceosome, Spliceosome, regulation of mRNA metabolic process, RNA binding, nucleic acid binding, regulation of RNA splicing, mRNA 3′-UTR binding, mRNA binding, downregulation of mRNA metabolic process, and regulation of gene expression (Fig. 7.3B).

On analyzing the network centered around TDP-43 shown in Fig. 7.4A we found 8 nodes/genes to be its immediate neghbours, and they had 55 edges/

TABLE 7.1 List of top 10 highest degree genes and its topological properties (Degree, clustering-coefficient, neighborhood connectivity) and centrality measurement average shortest, pathlength, betweenness, closeness.

Display name	Degree	Average shortest pathlength	Betweenness centrality	Closeness centrality	Clustering coefficient	Neighborhood connectivity
SOD1	41	1.95	0.15	0.51	0.34	18.17
FUS	36	2.14	0.05	0.46	0.43	19.66
C9orf72	36	2.08	0.06	0.48	0.44	19.94
TARDBP	35	2.10	0.05	0.47	0.46	20.08
VCP	31	2.30	0.02	0.43	0.50	20.96
UBQLN2	30	2.31	0.02	0.43	0.55	21.70
ATXN2	30	2.32	0.01	0.42	0.54	21.56
SETX	27	2.39	0.03	0.41	0.54	21.29
DCTN1	27	2.36	0.01	0.42	0.53	21.70
OPTN	26	2.40	0.01	0.41	0.64	23.00

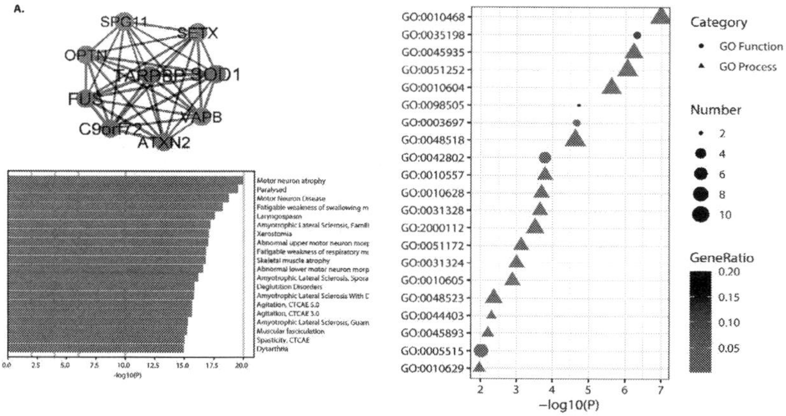

FIGURE 7.3 (A) PPI network of TDP-43 and direct interacted gene and gene enrichment. (B) Dot plot of TDP-43 and associated genes enriched biological process and molecular function.

TABLE 7.2 Gene ontology of TDP-43 and associated gene. The significant enrichment of biological process.

GO term	*P*-value	Description	#	Total
GO:0005515	.0097	Protein binding	8	6607
GO:0042802	.00016	Identical protein binding	6	1754
GO:0048518	2.36E-05	Positive regulation of biological process	10	5459
GO:0048523	.0043	Negative regulation of cellular process	7	4454
GO:0031324	.001	Negative regulation of cellular metabolic process	6	2463
GO:0010604	2.44E-06	Positive regulation of macromolecule metabolic process	9	3081
GO:0010605	.0013	Negative regulation of macromolecule metabolic process	6	2558
GO:0010557	.00016	Positive regulation of macromolecule biosynthetic process	6	1758
GO:0045935	5.86E-07	Positive regulation of nucleobase-containing compound metabolic process	8	1770
GO:0031328	.00022	Positive regulation of cellular biosynthetic process	6	1846
GO:0051172	.00074	Negative regulation of nitrogen compound metabolic process	6	2307

Continued

TABLE 7.2 Gene ontology of TDP-43 and associated gene. The significant enrichment of biological process.—cont'd

GO term	*P*-value	Description	#	Total
GO:0010468	1.04E-07	Regulation of gene expression	11	4533
GO:2000112	.0003	Regulation of cellular macromolecule biosynthetic process	8	4050
GO:0035198	4.78E-07	miRNA binding	3	26
GO:0045893	.0063	Positive regulation of transcription, DNA-templated	4	1435
GO:0010628	.0002	Positive regulation of gene expression	6	1826
GO:0098505	1.89E-05	G-rich strand telomeric DNA binding	2	10
GO:0003697	.000022	Single-stranded DNA binding	3	99
GO:0051252	8.79E-07	Regulation of RNA metabolic process	10	3890
GO:0044403	.005	Symbiont process	3	650
GO:0010629	.0108	Negative regulation of gene expression	4	1670

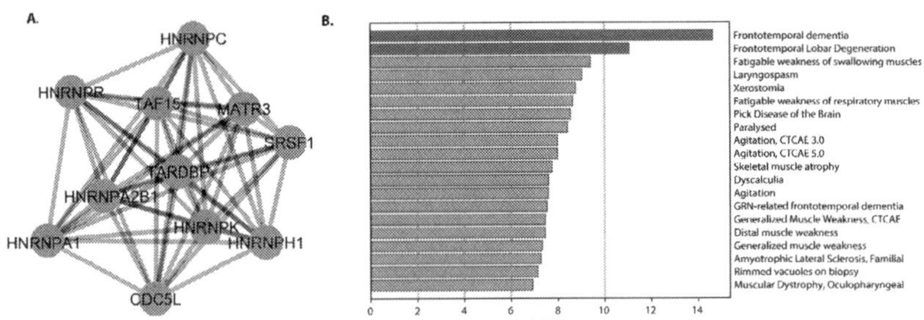

FIGURE 7.4 PPI network of first neighbors of TDP-43, including 8 nodes and 55 edges and Pathway enrichment values for the genes involved in TDP-43 associated network.

interactions amongst themselves. Gene list enrichments are identified in the DisGeNET by using Metascape: a Gene annotation and Analysis Resource (https://metascape.org/gp/index.html#/main/step1). All genes in the genome have been used as the enrichment background. Terms with a *P*-value < .01, a minimum count of 3, and an enrichment factor >1.5 (the enrichment factor is the ratio between the observed counts and the counts expected by chance) are collected and grouped into clusters based on their membership similarities. The top few enriched clusters (one term per cluster) are shown in Fig. 7.4B. The algorithm used here is the same as that is used for pathway and process enrichment analysis.

Analysis of TDP-43 using network analyst

Gene–disease associations

The gene–disease associations that were plotted in Fig. 7.5 by using data from NetworkAnalyst clearly show that TDP-43 has a major involvement in neurodegenerative diseases and muscular diseases like Amyotrophic Lateral Sclerosis, Stereotypic Movement Disorder, Xerostomia, Apathy, Mood swings, muscle degeneration, Neurogenic Muscular Atrophy, Social disinhibition, Paralysed, Neurodegenerative Disorders, Skeletal muscle atrophy, Generalized muscle weakness, Frontotemporal Lobar Degeneration, Respiratory Failure, Rapidly progressive, Neurogenic muscle atrophy, especially in the lower limbs, Rapidly progressive disorder, etc. In total it is involved in 36 diseases. As you can see in Fig. 7.5 the green-colored nodes show muscle-related diseases and the yellow-colored nodes are brain/neuron related diseases. All the related diseases are listed in Table 7.3.

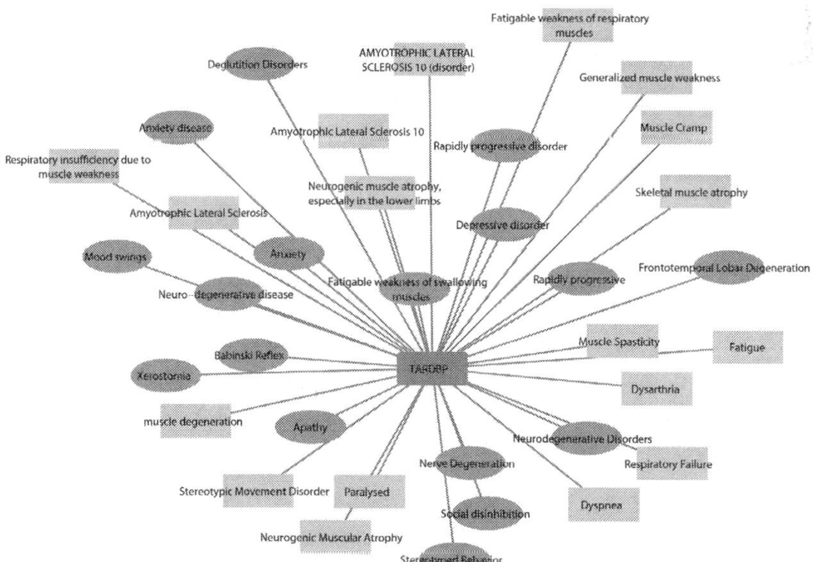

FIGURE 7.5 Gene–disease interaction network. TDP-43 associated with the diseases. The figure represents the interaction of TDP-43 with their related brain disorder.

TABLE 7.3 List of disease associated with TDP-43.

Amyotrophic Lateral Sclerosis	Generalized muscle weakness	Neurogenic muscle atrophy, especially in the lower limbs
Anxiety	Respiratory failure	Stereotypic movement disorder
Rapidly progressive	Xerostomia	Neurogenic muscular atrophy
Depressive disorder	Apathy	Frontotemporal lobar degeneration
Dysarthria	Mood swings	Rapidly progressive disorder
Dyspnea	Muscle degeneration	Amyotrophic lateral sclerosis 10 (disorder)
Fatigue	Deglutition disorders	Amyotrophic lateral sclerosis 10
Muscle cramp	Social disinhibition	Respiratory insufficiency due to muscle weakness
Muscle spasticity	Paralyzed	Neurodegenerative disease
Nerve degeneration	Anxiety disease	Neurodegenerative disorders
Pain	Skeletal muscle atrophy	Fatigable weakness of respiratory muscles
Babinski reflex	Stereotyped behavior	Fatigable weakness of swallowing muscles

TF-miRNA coregulatory network

Fig. 7.6 shows the involvement of multiple transcription factors (70) and five miRNAs in the coregulatory network of TDP-43.

Drug–gene interactions

A total of 112 drug interactions were found for TDP-43 showing that it is a highly targeted gene. We selected 30 interactions out of those with high interaction scores and made a drug–gene interaction network of TDP-43 which is shown in Fig. 7.7. The top 30 interactions and their interaction score are given in Table 7.4.

KEGG pathway

TDP-43 has a role in four KEGG pathways, two of them are disease-related pathways, namely map05014 (ALS) and map05022 (pathways of neurodegeneration - multiple diseases). The other two are normal biological pathways, RNA transport, and mRNA surveillance pathway. In the pathway for ALS, it is shown as a disease-causing gene marked in red color and appears

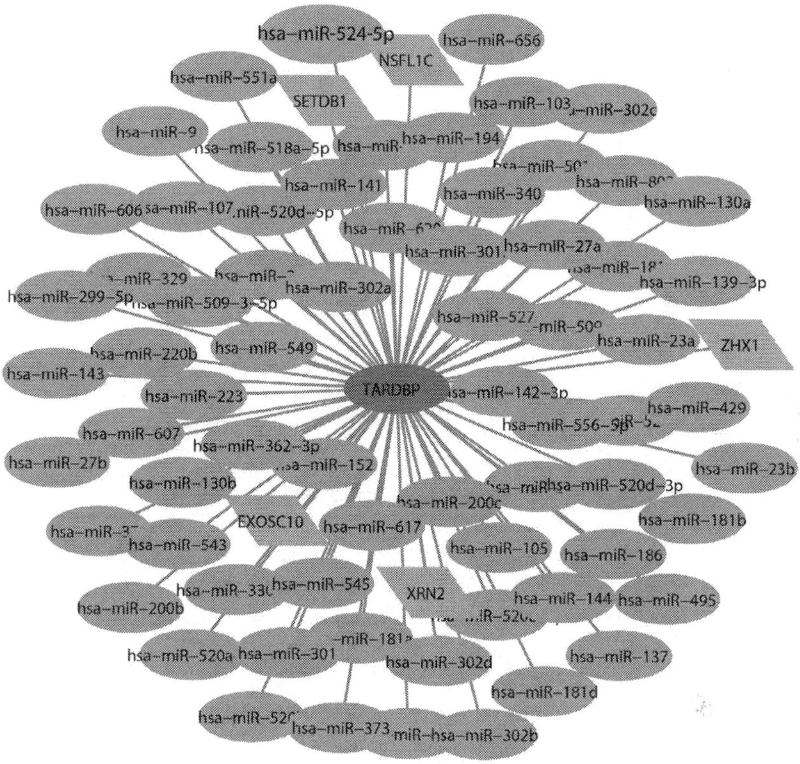

FIGURE 7.6 MicroRNA- (miRNA-) target regulatory networks of TDP-43. Circular shape represents miRNAs, yellow diamonds represent transaction factor targeted to TDP-43.

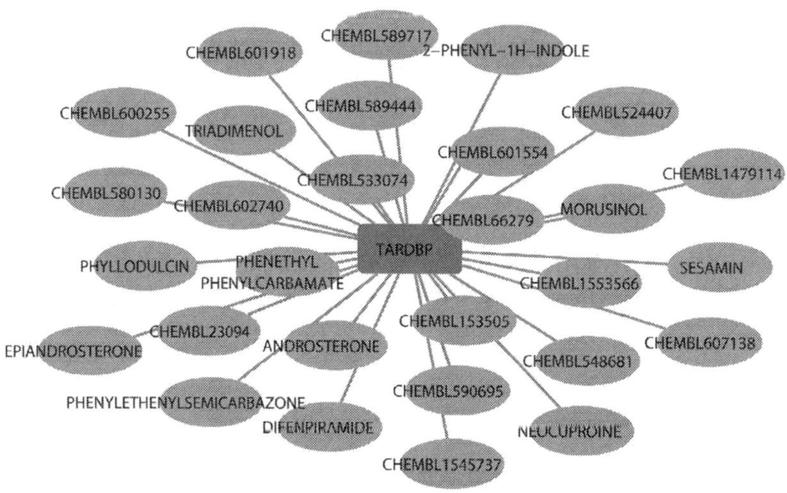

FIGURE 7.7 The network of a drug–gene interaction. The drug prediction databases DGIdb was used to predict drugs that interact with the TDP-43 gene.

TABLE 7.4 List of drugs associated with TDP-43.

Drug	Interaction score	Drug	Interaction score
CHEMBL589717	0.51	CHEMBL601554	0.51
CHEMBL533074	0.51	PHENETHYL PHENYLCARBAMATE	0.51
CHEMBL548681	0.51	PHENYLETHENYLSEMI-CARBAZONE	0.51
CHEMBL601918	0.51	CHEMBL580130	0.51
CHEMBL589444	0.51	ANDROSTERONE	0.51
CHEMBL524407	0.51	2-PHENYL-1H-INDOLE	0.51
CHEMBL23094	0.51	CHEMBL1479114	0.25
CHEMBL607138	0.51	CHEMBL153505	0.25
SESAMIN	0.51	EPIANDROSTERONE	0.25
CHEMBL1553566	0.51	CHEMBL1545737	0.25
CHEMBL600255	0.51	MORUSINOL	0.25
CHEMBL590695	0.51	NEOCUPROINE	0.25
PHYLLODULCIN	0.51	CHEMBL602740	0.25
DIFENPIRAMIDE	0.51	CHEMBL66279	0.25
CHEMBL600107	0.51	TRIADIMENOL	0.25

multiple times in the whole pathway. The map05014 pathway is shown in Fig. 7.8. According to the pathway's description on KEGG, in 90% of patients, ALS is sporadic, with no clear genetic linkage. On the other hand, the remaining 10% of cases show familial inheritance, with mutations in SOD1, TDP43 (TDP-43), FUS, or C9orf72 genes being the most frequent causes,[43] again we can see the role of TDP-43.

Discussion and conclusion

Upon observing the results from expression analysis, we can see that TDP-43 is expressed in almost all parts of the brain but the TPM (transcripts per million) is highest in the cerebellum from where the voluntary muscle movement is coordinated and it plays a vital role in cognition too.[44,45] Hence, any mutation in the gene is bound to cause some kind of movement impairment and cognitive dysfunctions. This is made clearer in the spatiotemporal heatmap where we see that TDP-43 is expressed throughout the lifecycle in the

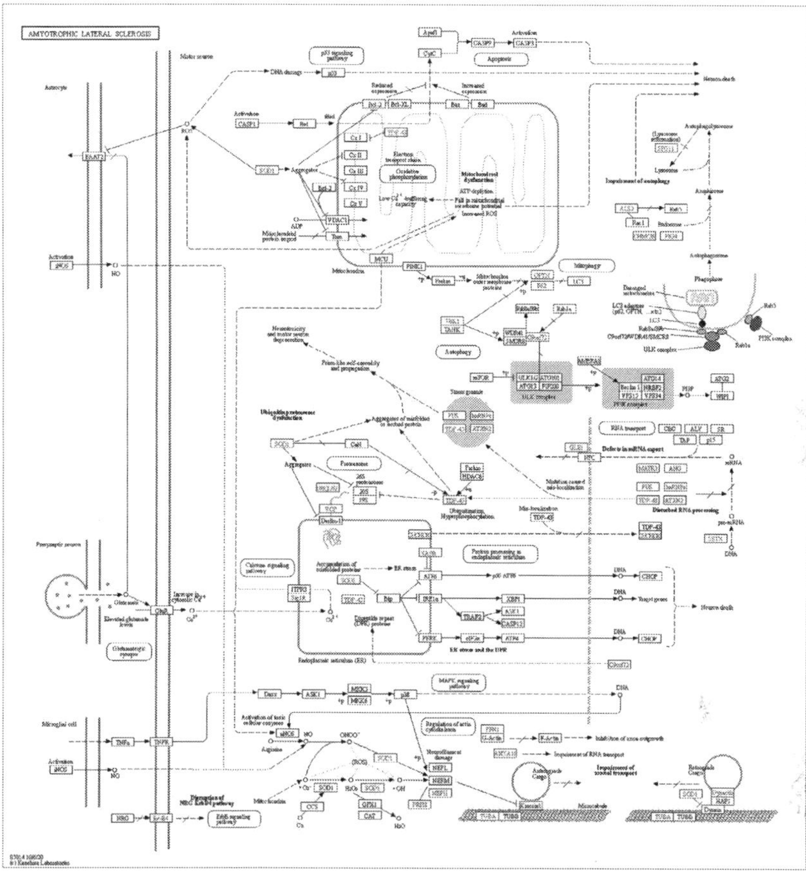

FIGURE 7.8 KEGG pathway of Amyotrophic lateral sclerosis—Reference pathway (map05014).

cerebellum, the frontal, parietal, temporal, and occipital cortexes and other parts showing that it plays a vital role from fetal stages to adulthood.

From the start, it gets evident that TDP-43 plays an important role in ALS and other neurodegenerative diseases as it shows up on ALSoD. Further, while preparing the protein−protein interaction network we saw that TDP-43 is crucial to ALS disease, it had the fourth-highest degree among 126 proteins, and it was also an important part of models in the network. The connection between the disease and the protein becomes more evident when we do enrichment analysis where weekly find processes and functions that are crucial for neuro and muscular cells to function properly. Visa that the proteins in the network are involved in the development of neuromuscular junctions, glial cell activation, and also central nervous system development, etc. Upon analyzing the network-based solely on TDP-43 we find its involvement in important biological processes like mRNA splicing and processing, RNA-binding nucleic acid binding, etc. This does not imply that TDP-43 is not a part of neuro and muscular processes, rather it plays a role in connecting the proteins that are directly involved in those processes as we saw in our main network which was based on genes from ALSoD. TDP-43 is a hub connecting to other nodes (proteins) that directly play role in ALS and related processes. While checking on disease associations of TDP-43, only two types of diseases were found. One type that is related to muscles and the other that is related to the nervous system. From anxiety and depression to nerve degeneration and frontotemporal lobar degeneration, from muscle cramps and general muscle weakness to neurogenic muscular atrophy and ALS, in all these diseases TDP-43 has been found to play some role or the other. This also coincides with our expression analysis results as TDP-43 was found to be expressed in almost the entire brain, hence the relation to neurological and psychological disorders and higher/more prominent expression level in the cerebellum which governs voluntary movements hence TDP-43 malfunction can lead to muscular diseases. The involvement of multiple transcription factors and miRNA with TDP-43 proves that it is a gene of biological significance. The numerous drug and chemical interactions show that it is an important gene from a pharmacology point of view it has been targeted before and it still has unexplored potential as a key gene for neuromuscular diseases. At last, we used the KEGG database to establish TDP-43s exact role in biological pathways, there we found that it plays a role in four major pathways, biological pathways RNA transport, and mRNA surveillance. The two other pathways are disease-related and of our interest, one being ALS and the other a neurodegeneration pathway for multiple diseases. It is marked as a disease-causing gene and 7 times in the whole pathway. All these things solidify our initial assumption about TDP-43/TDP-43 being a gene of importance in ALS and other neurodegenerative diseases. Hence, further research with TDP-43 knockout and inhibited models should be done to explore the ways that this gene can be exploited in our cause.

Acknowledgment

S.K.S. is financially supported by University Grants Commission (UGC), New Delhi, India.

References

1. Arai T, et al. TDP-43 is a component of ubiquitin-positive tau-negative inclusions in frontotemporal lobar degeneration and amyotrophic lateral sclerosis. *Biochem Biophys Res Commun* 2006;**351**:602–11.
2. Hasegawa M, et al. Phosphorylated TDP-43 in frontotemporal lobar degeneration and amyotrophic lateral sclerosis. *Ann Neurol* 2008;**64**:60–70.
3. Neumann M, Kwong LK, Sampathu DM, Trojanowski JQ, Lee VM-Y. TDP-43 proteinopathy in frontotemporal lobar degeneration and amyotrophic lateral sclerosis: protein misfolding diseases without amyloidosis. *Arch Neurol* 2007;**64**:1388–94.

4. Kumar V, et al. Protein aggregation and neurodegenerative diseases: from theory to therapy. *Eur J Med Chem* 2016;**124**:1105–20.
5. Higashi S, et al. Concurrence of TDP-43, tau and alpha-synuclein pathology in brains of Alzheimer's disease and dementia with Lewy bodies. *Brain Res* 2007;**1184**:284–94.
6. Nakashima-Yasuda H, et al. Co-morbidity of TDP-43 proteinopathy in Lewy body related diseases. *Acta Neuropathol (Berl)* 2007;**114**:221–9.
7. Schwab C, Arai T, Hasegawa M, Yu S, McGeer PL. Colocalization of transactivation-responsive DNA-binding protein 43 and huntingtin in inclusions of Huntington disease. *J Neuropathol Exp Neurol* 2008;**67**:1159–65.
8. Lagier-Tourenne C, Polymenidou M, Cleveland DW. TDP-43 and FUS/TLS: emerging roles in RNA processing and neurodegeneration. *Hum Mol Genet* 2010;**19**:R46–64.
9. Casafont I, Bengoechea R, Tapia O, Berciano MT, Lafarga M. TDP-43 localizes in mRNA transcription and processing sites in mammalian neurons. *J Struct Biol* 2009;**167**:235–41.
10. Gregory RI, et al. The Microprocessor complex mediates the genesis of microRNAs. *Nature* 2004;**432**:235–40.
11. Sreedharan J, et al. TDP-43 mutations in familial and sporadic amyotrophic lateral sclerosis. *Science* 2008;**319**:1668–72.
12. Ayala YM, Misteli T, Baralle FE. TDP-43 regulates retinoblastoma protein phosphorylation through the repression of cyclin-dependent kinase 6 expression. *Proc Natl Acad Sci USA* 2008;**105**:3785–9.
13. Sephton CF, et al. TDP-43 is a developmentally regulated protein essential for early embryonic development. *J Biol Chem* 2010;**285**:6826–34.
14. Aulas A, Vande Velde C. Alterations in stress granule dynamics driven by TDP-43 and FUS: a link to pathological inclusions in ALS? *Front Cell Neurosci* 2015;**9**.
15. Prakash A, Kumar V, Meena NK, Hassan MI, Lynn AM. Comparative analysis of thermal unfolding simulations of RNA recognition motifs (RRMs) of TAR DNA-binding protein 43 (TDP-43). *J Biomol Struct Dyn* 2019;**37**:178–94.
16. Ling S-C, Polymenidou M, Cleveland DW. Converging mechanisms in ALS and FTD: disrupted RNA and protein homeostasis. *Neuron* 2013;**79**:416–38.
17. Cm D, et al. TDP-43 is directed to stress granules by sorbitol, a novel physiological osmotic and oxidative stressor. *Mol Cell Biol* 2010;**31**:1098–108.
18. Colombrita C, et al. TDP-43 is recruited to stress granules in conditions of oxidative insult. *J Neurochem* 2009;**111**:1051–61.
19. Molliex A, et al. Phase separation by low complexity domains promotes stress granule assembly and drives pathological fibrillization. *Cell* 2015;**163**:123–33.
20. Prakash A, Kumar V, Banerjee A, Lynn AM, Prasad R. Structural heterogeneity in RNA recognition motif 2 (RRM2) of TAR DNA-binding protein 43 (TDP-43): clue to amyotrophic lateral sclerosis. *J Biomol Struct Dyn* 2021;**39**:357–67.
21. Kumar V, et al. Exploring the aggregation-prone regions from structural domains of human TDP-43. *Biochim Biophys Acta Proteins Proteom* 2019;**1867**:286–96.
22. Buratti E. TDP-43 post-translational modifications in health and disease. *Expert Opin Ther Targets* 2018;**22**:279–93.
23. Buratti E, Baralle FE. Multiple roles of TDP-43 in gene expression, splicing regulation, and human disease. *Front Biosci J Virtual Libr* 2008;**13**:867–78.
24. Buratti E, et al. TDP-43 binds heterogeneous nuclear ribonucleoprotein A/B through its C-terminal tail: an important region for the inhibition of cystic fibrosis transmembrane conductance regulator exon 9 splicing. *J Biol Chem* 2005;**280**:37572–84.

25. Ayala YM, et al. Structural determinants of the cellular localization and shuttling of TDP-43. *J Cell Sci* 2008;**121**:3778–85.
26. Moisse K, et al. Divergent patterns of cytosolic TDP-43 and neuronal progranulin expression following axotomy: implications for TDP-43 in the physiological response to neuronal injury. *Brain Res* 2009;**1249**:202–11.
27. Moisse K, et al. Cytosolic TDP-43 expression following axotomy is associated with caspase 3 activation in NFL-/- mice: support for a role for TDP-43 in the physiological response to neuronal injury. *Brain Res* 2009;**1296**:176–86.
28. Farooqui A, et al. Assessment of the key regulatory genes and their Interologs for Turner Syndrome employing network approach. *Sci Rep* 2018;**8**:10091.
29. Alam A, et al. Identification and classification of differentially expressed genes and network meta-analysis reveals potential molecular signatures associated with tuberculosis. *Front Genet* 2019;**10**.
30. Consortium TG. The Genotype-Tissue Expression (GTEx) pilot analysis: multitissue gene regulation in humans. *Science* 2015;**348**:648–60.
31. Sunkin SM, et al. Allen Brain Atlas: an integrated spatio-temporal portal for exploring the central nervous system. *Nucleic Acids Res* 2013;**41**:D996–1008.
32. Guo L, Lin W, Zhang Y, Li W, Wang JBEST. A web server for brain expression Spatio-temporal pattern analysis. *BMC Bioinf* 2019;**20**:632.
33. Wroe R, Butler AW-L, Andersen PM, Powell JF, Al-Chalabi, Alsod A. The amyotrophic lateral sclerosis online database. *Amyotroph Lateral Scler* 2008;**9**:249–50.
34. Szklarczyk D, et al. STRING v11: protein-protein association networks with increased coverage, supporting functional discovery in genome-wide experimental datasets. *Nucleic Acids Res* 2019;**47**:D607–13.
35. Shannon P, et al. Cytoscape: a software Environment for integrated models of biomolecular interaction networks. *Genome Res* 2003. https://doi.org/10.1101/gr.1239303.
36. Cytoscape App Store - MCODE. https://apps.cytoscape.org/apps/mcode.
37. Bader GD, Hogue CWV. An automated method for finding molecular complexes in large protein interaction networks. *BMC Bioinf* 2003;**4**:2.
38. Zhou G, et al. NetworkAnalyst 3.0: a visual analytics platform for comprehensive gene expression profiling and meta-analysis. *Nucleic Acids Res* 2019;**47**:W234–41.
39. Cotto KC, et al. DGIdb 3.0: a redesign and expansion of the drug-gene interaction database. *Nucleic Acids Res* 2018. https://doi.org/10.1093/nar/gkx1143.
40. Kanehisa M, Goto S. KEGG: Kyoto encyclopedia of genes and genomes. *Nucleic Acids Res* 2000;**28**:27–30.
41. Doncheva NT, Morris JH, Gorodkin J, Jensen LJ. Cytoscape StringApp: network analysis and visualization of proteomics data. *J Proteome Res* 2019;**18**:623–32.
42. Wickham H. *ggplot2: elegant graphics for data analysis*. Incorporated: Springer Publishing Company; 2009.
43. KEGG PATHWAY: Amyotrophic lateral sclerosis - Reference pathway. https://www.genome.jp/kegg-bin/show_pathway?map05014.
44. Koziol LF, Budding DE, Chidekel D. From movement to thought: executive function, embodied cognition, and the cerebellum. *Cerebellum* 2012;**11**:505–25.
45. Cerebellum (Section 3, Chapter 5) Neuroscience Online: An Electronic Textbook for the Neurosciences | Department of Neurobiology and Anatomy - The University of Texas Medical School at Houston. https://nba.uth.tmc.edu/neuroscience/m/s3/chapter05.html.

46. Mompeán M, Romano V, Pantoja-Uceda D, Stuani C, Baralle FE, Buratti E, et al. The TDP-43 N-terminal domain structure at high resolution. *FEBS J* 2016;**283**:1242–60. https://doi.org/10.1111/febs.13651.
47. Prasad A, Bharathi V, et al. Molecular mechanisms of TDP-43 misfolding and pathology in amyotrophic lateral sclerosis. *Front Mol Neurosci* 2019. https://doi.org/10.3389/fnmol.2019.00025. https://www.frontiersin.org/articles/10.3389/fnmol.2019.00025/full.

Chapter 8

TDP-43 proteinopathy mechanisms from non-mammalian model systems

Vidhya Bharathi, Amandeep Girdhar, Basant K. Patel
Department of Biotechnology, Indian Institute of Technology Hyderabad, Kandi, Sangareddy, Telangana, India

Introduction

Amyotrophic lateral sclerosis (ALS) is a neurodegenerative disease with loss of function of motor neurons.[1] ALS causes paralysis of the patients and eventually leads to their death within a few years of the disease symptom onset. The TDP-43 protein occurs as inclusions in about 97% of the ALS patients, therefore, it is a key protein in understanding the mechanism of the ALS pathogenesis.[2,3] Several mammalian and nonmammalian model systems have been developed for the purpose of unraveling the TDP-43-induced pathology and to test potential therapeutic options.[4] The mammalian model systems offer close parallel to the human physiology for studying neuronal diseases due to their similar complex biochemical processes and immunological response. However, due to their simplistic, yet informative nature, several non-mammalian model systems have also been established for studying neurodegeneration as the key cytotoxicity mechanisms are conserved.[5,6] As the fundamental cellular mechanisms are conserved from the single cell eukaryote yeast to humans, the usage of the non-mammalian alternatives to model the human diseases rapidly facilitates identification of the implicated pathways. These models also help in drug screening and drug target validation in a simple and cost-effective manner. Also, these simple systems allow for easy genetic manipulations and the mechanistic similarity between the key cellular pathways facilitates the extrapolation of the findings to the mammalian system.

Conserved cellular mechanisms, easy handling, and rapid genetic manipulations have enabled the usage of the yeast *Saccharomyces cerevisiae* to study several proteinopathies relevant to neurodegeneration. The striking discovery

TDP-43 and Neurodegeneration. https://doi.org/10.1016/B978-0-12-820066-7.00002-3

of endogenous yeast "prions" that mimic the amyloid-like template-driven aggregation behavior and "protein-only" infectivity similar to as proposed for the human prion PrP^{Sc} and the observations of aggregation of several neurodegenerative disorder–associated human proteins such as Aβ-42 and polyglutamine, etc., in the yeast cell have catapulted yeast as a popular host for investigating the mechanistic details and cytotoxicities of numerous amyloid-like aggregation-prone human proteins.[7–9] In fact, the expressions in yeast cell of numerous aggregating human proteins with links to neurodegenerative diseases can manifest cytotoxicities and pathology mechanisms similar to as observed in neurons such as: mitochondrial dysfunction, defects in protein trafficking and proteasomal impairment etc.[10] Notably, a yeast model established to examine the molecular mechanism of the Huntington's disease by expressing the polyglutamine repeat protein led to uncovering of several mechanisms of toxicity and also identified several potential synthetic and natural therapeutic drug-like molecules.[11–13] Likewise, the Alzheimer's disease-linked human Aβ-42 protein-expressing yeast model has also been established. Strikingly, when Aβ-42 was directed to the secretory pathway, it could cause cellular toxicity in yeast that helped in the identification of several Aβ-42's toxicity modifying genes.[14] Furthermore, studies on the inhibition of the Aβ-42 oligomer formation have also been carried out using the yeast model towards developing new potential therapeutic molecules.[15,16] Several RNA-binding proteins linked to ALS pathology including TDP-43 were also found to successfully aggregate in yeast and cause cytotoxicity. Therefore, many facets of the molecular mechanisms of their cytotoxicities have been elucidated using the yeast model.

Another non-mammalian model system *Drosophila melanogaster* (fruit fly) has also become an invaluable tool to model and study neurodegenerative diseases. This is owing to the detailed understanding of its several complex processes such as embryogenesis and neuronal development in addition to its rapid life cycle and the ability to produce large number of progeny.[17] Remarkably, about two-third of the genes of *Drosophila* also have homologs in humans which further underpins the relevance and usefulness of this model.[18] Therefore, *Drosophila* models for the Huntington's, Parkinson's, and ALS diseases have been developed which have helped unearth and understand several pathogenic mechanisms associated with these human neurodegenerative diseases.[19,20] Notably, the expression of the human TDP-43 protein in *Drosophila* caused several pathologies that are also found in the ALS patients such as neuronal degeneration and reduced life span etc.[21,22]

The translucent nematode worm *Caenorhabditis elegans* manifests several complex physiological processes similar to that of the mammals and therefore it has also been used for modeling the learning and memory impairments usually associated with the neurodegenerative disorders.[23,24] As the in vivo modeling of neurons is easier in *C. elegans*, it often presents a preferable platform for preliminary screening of potential therapeutic medications before investigating them in the transgenic rodent models. Several devastating neurodegenerative diseases such as the Parkinson's disease, ALS, Alzheimer's disease and Huntington's disease have also been modeled in *C. elegans* and several toxic and pathogenic mechanisms have been elucidated.[25–27]

Recently, the zebrafish *Danio rerio* is also being employed as a non-mammalian vertebrate model system to study neurodegeneration. Zebrafish offers higher fecundity rate and an ease of visualization in addition to also providing a cost-effective host system for large-scale drug and genetic screenings.[28] Most of the human proteins have a functional homolog in zebrafish and they also exhibit extraordinary sequence similarities. Model systems of several neurodegenerative diseases, for example, the Huntington's and ALS diseases, have been established in zebrafish to investigate the mechanistic pathological details of these diseases.[4,29,30]

Taken together, these non-mammalian model systems offer an ease of genetic manipulations, simple cultivation, and easy imaging all of which have also facilitated successful establishment of the TDP-43 proteinopathy models to examine the ALS pathology. Thus, here we discuss the molecular mechanisms of the TDP-43 proteinopathy elucidated from these non-mammalian model systems. Also, we present various genetic insights of the TDP-43 proteinopathy mechanisms obtained from these model systems. Furthermore, small molecule and genetic modifiers of the TDP-43 proteinopathy screened out using these model systems have also been discussed.

The yeast *Saccharomyces cerevisiae* model of the TDP-43 proteinopathy

Even though there is a lack of a neuronal system in the yeast *S. cerevisiae*, several physiological insults inflicted in neurodegeneration such as the mitochondrial dysfunction, oxidative stress, formation of the intracellular proteinaceous aggregates and apoptosis etc., can also be investigated in the yeast cells.[31–35] The ease of effective genetic manipulations due to the well-characterized genome coupled with the fast growth rate has established yeast as a favoured organism for gaining mechanistic insights into protein aggregation and executing high-throughput drug screening.[36]

The yeast *S. cerevisiae* does not contain any apparent ortholog of the human TDP-43 protein. When TDP-43 was first expressed in the yeast cells using a low copy number plasmid encoding the human TDP-43 fused with GFP, it was found to localize to nucleus and many cells also showed a few intranuclear punctate foci suggesting of its nuclear aggregation.[34] In order to examine if the ALS-linked pathological features of TDP-43, such as its cytoplasmic mislocalization and toxicity can also be observed in yeast, TDP-43 was then overexpressed from a multicopy 2μ yeast plasmid using the tightly regulated and galactose-inducible, *GAL1* promoter.[34] Strikingly, the overexpressed TDP-43 exhibited a cytoplasmic movement from nucleus and also manifested the formation of "punctate foci" in cytoplasm, thereby mimicking an ALS-like TDP-43 proteinopathy. This high-expression of TDP-43 also caused a dramatic increase in the toxicity to the yeast cells. However, these TDP-43 inclusions in the yeast model could not be stained with the amyloid-specific dye, Thioflavin-T, and also did not form SDS-insoluble aggregates thereby suggesting a lack of typical amyloid-like architecture contrary to as previously observed for the yeast prions and other neurodegenerative disease–linked proteins expressed in yeast.[34]

The yeast model has also given insights into the importance of the different domains of TDP-43 toward its aggregation and toxicity. The TDP-43 protein contains two RNA recognition motifs termed RRM1 and RRM2 and an N-terminal domain which contains a nuclear localization signal (NLS) and a C-terminal glycine-rich prion-like domain.[2,3] As several TDP-43 C-terminal fragments (CTFs) have also been found to aggregate along with the full-length TDP-43 in the ALS patients, thus it is useful to understand the toxicity and aggregation capability of these CTFs.[37,38] In the yeast model, while the N-terminal fragments of TDP-43 containing NLS were exclusively targeted to nucleus and did not aggregate, the CTFs containing RRM2 domain formed cytoplasmic foci. The CTFs having partial or total deletion of RRM2, though cytoplasmic, did not form aggregates. Although the CTFs that resemble the caspase-dependent aberrant fragmentation products of TDP-43 showed toxicity in the yeast model, it is noteworthy that these CTFs exhibited far less toxicity compared to the full-length TDP-43.[34] Consistent with the reduced toxicity observed in the yeast model, in several higher in vivo models also CTFs did not induce any motor or behavioral dysfunctions characteristics of the ALS patients.[37] Therefore, CTFs may not be the primary inducers of the proteinopathy in ALS.[37] Furthermore, seven ALS-linked mutants of TDP-43, namely G294A, M337V, Q343R, N345K, N390D, Q331K, and R361S, were also expressed in the yeast model to examine their aggregation properties and cytotoxicities.[35] Two of these TDP-43 mutants, Q331K and M337V, were found to form more number of aggregates per yeast cell than the wild-type TDP-43. Furthermore, the respective recombinantly purified mutant proteins also aggregated more rapidly compared to the wild-type protein in vitro. Thus, the yeast model suggested that the point mutations in TDP-43 led to a higher extent of aggregation which could potentially enhance the toxicity.[35]

Additionally, in order to unearth the cellular components and pathways affecting the TDP-43 toxicity, overexpression and deletion genetic screens were also carried out in the yeast TDP-43 proteinopathy model.[22,39] So far, several suppressor and enhancer yeast genes have been identified to affect the TDP-43 toxicity when either deleted or overexpressed. An overexpression genetic screen identified that several genes of yeast can suppress the TDP-43 toxicity including the transcriptional corepressor protein Cyc8 (which is already known to form the [OCT^{+}] yeast prion), the protein kinases Nnk1 and Rim15, the stress granule and P-body component proteins Tis1 and Vts1, and

DNA double-strand break repair protein Xrs2.[22] On the contrary, certain other proteins such as the protein kinases Kin3, Sak1, Vhs1, and Yck2, and the stress granule and P-body component proteins Hrp1 and Kem1 and the ATP binding protein required for the DNA replication Cdc6, were found to enhance the TDP-43 toxicity when overexpressed.[22] Furthermore, a deletion genetic screen in yeast also identified several suppressors of the TDP-43 toxicity such as the histone deacetylase complex protein Set3, the actin assembly protein Siw14, and the ribosomal subunit protein Rpl16b.[39] Also, certain enhancers of the TDP-43 toxicity were found from the gene deletion screen including the Na^+/H^+ and K^+/H^+ exchanger Nhx1 and the transcriptional activator Msn2.[39]

Of high significance, the deletion of PBP1 (Pab1 binding protein 1), a human Ataxin-2 (ATXN2) ortholog, resulted in reduction of the TDP-43-mediated toxicity, whereas its overexpression enhanced the TDP-43 toxicity.[21] These findings from the yeast model could also be replicated in a *Drosophila* model, and moreover, the TDP-43 and ATXN2 proteins were found to interact physically depending on the presence of RNA in the mammalian cell cultures. Subsequently, when the neurons from the spinal cord of the ALS patients were examined, ATXN2 was observed to be mislocalized.[21] Furthermore, genetic analysis showed that an increase in the number of glutamine residues in ATXN2 was significantly associated with a small percentage of the ALS patients.[21] Additionally, another gene, *DBR1*, identified through the genetic screen in yeast, which codes for an RNA lariat debranching enzyme, also modulated the TDP-43 toxicity in the rodents and the neuronal cell lines.[39] In a recent study, a large-scale genetic screen was carried out to identify the modifiers of wild-type and M337V and Q331K mutant TDP-43 toxicities, where surprisingly little overlap was observed between the modifiers of the wild-type and mutant toxicities.[133] Succinate dehydrogenase (SDH) and lysine N-methyltransferase (EFM4) were found to be modifiers of the wild-type TDP-43 toxicity whereas, mitochondrial porin protein (POR2) and ATP-dependent mitochondrial chaperone HSC82 were found to be modifiers of the M337V and Q331K mutant TDP-43 toxicities. Recently, Nucleolin protein has been shown to reduce the toxicity of TDP-43 in the yeast and mammalian cell models.[134] Furthermore, the yeast model has shown that the TDP-43 aggregation affects several cellular processes causing toxicity *via* diverse mechanisms some of which have been discussed here in detail in the following sections.

TDP-43 induces oxidative stress and mitochondria-dependent cell death in yeast

When the human TDP-43 protein was expressed in yeast, higher levels of reactive oxygen species (ROS) and elevated oxidative stress were observed.[32,33,40] Furthermore, the elevated oxidative stress caused by several amyloid-like aggregation-prone proteins including TDP-43 could be analyzed in yeast by a simple red to white color conversion assay analogous to the widely used red to white color change assay of the [*PSI*$^+$] prion.[33,40] The expression of TDP-43 caused white color conversion of the *ade1* or *ade2* mutant yeast which are inherently red colored due to the reduced glutathione-dependent transport of an adenine biosynthesis intermediate metabolite to the yeast vacuoles. Accumulation of oxidative stress converts the reduced glutathione (GSH) to its oxidized form (GSSG) which thwarts the translocation of the intermediate metabolite to vacuole which therefore prevents the red color development. Thus, the oxidative stress due to the TDP-43 aggregation causes white coloration of the *ade1* or *ade2* mutant yeast which is reversible to red color upon turning-off the TDP-43 expression.[40] Notably, depletion of glutathione in the primary cortical neurons and the NSC cells *via* ethacrynic acid has been shown to induce serine phosphorylation, cytoplasmic mislocalization, C-terminal fragmentation and aggregation of TDP-43 thereby suggesting of an important role of ROS towards the TDP-43 proteinopathy-mediated ALS pathogenesis.[41]

Furthermore, an increased cell death associated with the TDP-43 expression in yeast has been linked to its effects on the mitochondria and their respiratory capability.[32] Initially, TDP-43 toxicity was found to be markedly decreased in the yeast cells lacking the mitochondrial DNA. Surprisingly however, deletions of none of the mitochondrial programmed cell death regulators such as AIF1, NUC1, CYC1, or CYC7 were found to mitigate the TDP-43 toxicity.[32] In fact, decreased respiration by inhibition of the respiratory chain complexes was found to relieve the TDP-43 cytotoxicity.[31,32] Notably, TDP-43-induced oxidative stress-mediated cell death in yeast was recently found to depend on the Cyclin C protein one function of which is to mediate the mitochondrial hyper-fission-mediated cell death.[135] Furthermore, the deletions of a mitochondrial fission gene, *DNM1*, and a pro-apoptotic gene *YBH3* also resulted in the rescue of the TDP-43 toxicity.[135] Recently, it has been reported from the other model systems that TDP-43 localizes to mitochondria, binds to the respiratory complex mRNAs, and impairs their function.[42] Strikingly, inhibition of the mitochondrial localization of TDP-43 was found to relieve the TDP-43 toxicity.[42]

Recently, respiring versus non-respiring yeast models of the TDP-43 aggregation have been examined for the relative TDP-43 toxicity.[43] Strikingly, the growth of yeast on carbon sources that promote respiration allowed for enhanced TDP-43 toxicity, whereas the carbon sources that inhibit respiration diminished the TDP-43 aggregation-induced cytotoxicity. Importantly, even in the absence of functional mitochondrial, TDP-43 was found to cause some residual toxicity thereby hinting that certain additional pathways other than the respiration also contribute in mediating the TDP-43 toxicity in yeast. Strikingly, an enhanced oxidative stress *via* hydrogen peroxide treatment was also found to enhance the TDP-43 toxicity in yeast, whereas a supplementation with an antioxidant molecule, N-acetyl cysteine (NAC), markedly decreased the TDP-43 toxicity thereby establishing oxidative stress as an important mediator of the TDP-43 toxicity.

TDP-43 causes disruption of the proteasomal degradation machinery in yeast

Pathological cellular inclusions in the ALS patients contain ubiquitinated TDP-43 and the abnormal ubiquitination has been reported as one of the major causes leading to the neuronal cell death.[44,45] Mutations in the ubiquitin degradation pathway genes such as Ubiquilin-2 and Sequestosome-1 are also known to affect the TDP-43 aggregation in the FTLD and ALS patients.[45] In the yeast model, expression of TDP-43 was found to severely affect the functioning of ubiquitin proteasome system (UPS). A UPS function reporter protein CG* which is a secretion-deficient mutant of carboxypeptidase Y fused to EGFP, degradation of which signifies a proper functioning of UPS, failed to degrade and rather accumulated in the yeast cell which were also co-overexpressing TDP-43.[46] Notably however, although TDP-43 causes the proteasomal dysfunction, the inhibition of proteasome did not affect the TDP-43's stability or turnover in the yeast model.[47]

TDP-43 interferes with the endolysosomal pathways of clearance in yeast

Recently, it was shown that TDP-43 is degraded in the yeast cell *via* non-selective autophagy and also in the yeast vacuoles *via* non-autophagic mechanisms.[47] In yeast, there are two non-autophagic mechanisms of protein degradation, endocytosis and the Golgi-vacuolar trafficking, both of which are regulated by the conserved enzyme classes of phosphoinositide 3-kinase (PI3K) and Vps34. Remarkably, the deletion of *VPS34* gene in yeast showed enhanced sensitivity to the TDP-43 expression thereby suggesting that the Vps34 kinase activity is essential for mitigating the TDP-43 toxicity. Also, TDP-43 localizes with the endocytic proteins and the absence of these proteins enhances toxicity, foci formation, as well as the TDP-43 protein stability.[47] The TDP-43's expression, in particular that of an ALS-linked TDP-43 familial mutant Q331K, resulted in the impairment of endocytosis and enhancing the endocytosis caused a reduction in the TDP-43 levels in yeast.[47]

Furthermore, mutations in Valosin-containing protein (VCP) have been associated with the TDP-43's cytoplasmic localization and toxicity in the ALS patients and also in the ALS disease models.[48,49] VCP is an essential protein involved in endosomal trafficking and ER-associated degradation (ERAD).[50] Recently, the yeast homolog of VCP, the Cdc48 protein, was shown to localize with TDP-43 and regulate its degradation. Inhibition of Cdc48 and its endocytosis promoting cofactor Ubx3 in yeast led to an increased accumulation of TDP-43 and its decreased degradation.[51]

Recently, it was shown that the yeast vacuoles, which are organelles analogous to the human lysosomes, partly aided in the clearance of TDP-43.[52] Expression of TDP-43 in yeast resulted in an increased vacuolar fragmentation, impaired vacuolar function, and decreased cell survival (Fig. 8.1).[52] Also, deletion of the gene that encodes for the vacuolar t-SNARE, Vam3, which facilitates the fusion of vesicles to the vacuoles, resulted in an increase in the TDP-43-mediated growth defects.[52] Similarly, the inactivation of the vacuolar machinery *via* gene deletions increased the non-degraded TDP-43 levels in yeast.[53] TDP-43 was also found to co-localize with various endocytic proteins. Also, vacuolar degradation of the endocytic substrates was affected by the TDP-43 expression and consequently the TDP-43's degradation was also decreased.

TDP-43 is predominantly cleared in yeast *via* the endolysosomal pathway.[52] Transport of the protease Pep4/CatD into the vacuoles is generally mediated by the endosomal−vacuolar pathway, and the improper functioning of the endosomal−vacuolar pathway due to the TDP-43 pathology can significantly reduce the vacuolar proteolytic activity.[52] This may be a contributory factor to the overall TDP-43-mediated cytotoxicity as the proper functioning of the vacuoles has house-keeping roles for the cell survival. In turn, activation of the endolysosomal pathway causes clearance of TDP-43 and rescues the TDP-43 toxicity in the yeast cell.[52] On the contrary, the induction of autophagy *via* chemicals resulted in an increase in the TDP-43-mediated growth defects.[52]

TDP-43 interferes with the RNA metabolism in yeast

Professor Gitler and coworkers have shown that the TDP-43 toxicity is rescued in a *dbr1Δ* yeast.[39] In absence of the activity of the RNA lariat debranching enzyme, Dbr1, in the *dbr1Δ* yeast, the intronic lariats generated during splicing accumulate in the cytoplasm that may act as sites for the sequestration of TDP-43. This would prevent the TDP-43's interference in the RNA metabolism which would require its binding to the essential RNAs and RNA-binding proteins.[39] As decreasing the Dbr1 activity relieved the TDP-43 toxicity, it is also proposed to have therapeutic implications for ALS.[39]

Yeast chaperones can modulate the TDP-43 toxicity

In a yeast gene overexpression screen, the chaperone Hsp104 was found to relieve the TDP-43 toxicity.[39] Hsp104 has been established to be a protein disaggregase that mediates the propagation of several yeast prions by fragmenting the large aggregates to generate new prion seeds that can thereafter recruit more monomers into the growing prion aggregates.[54−56] The potential of Hsp104 chaperone in modulating the TDP-43 aggregation and toxicity has been therefore extensively investigated.[57] The Hsp104 protein contains several

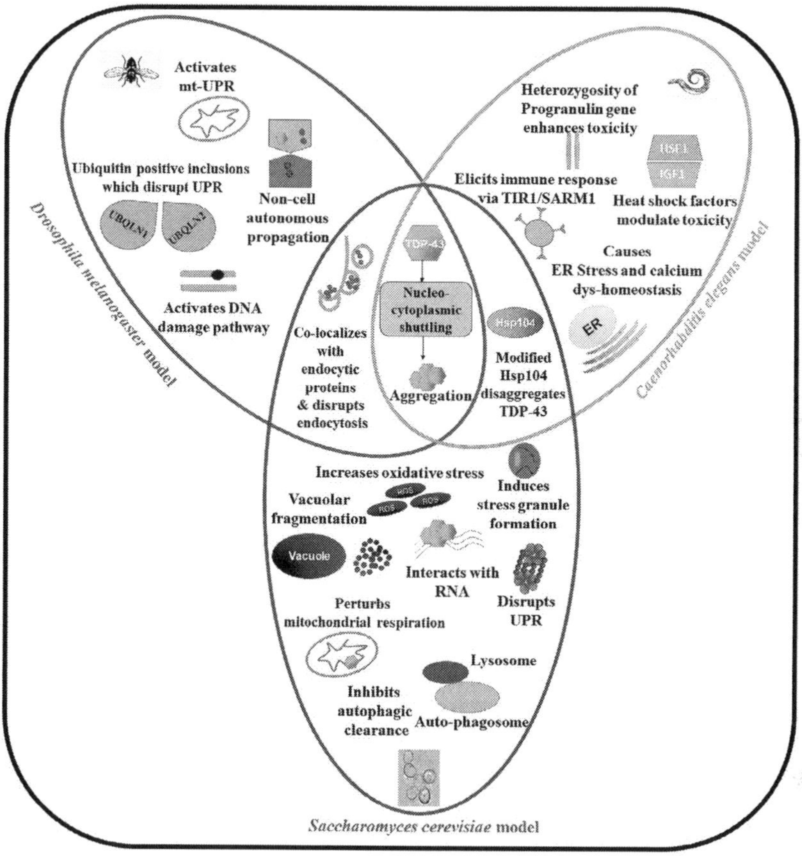

FIGURE 8.1 Cellular processes linked to TDP-43 proteinopathy elucidated from the non-mammalian models. The expression of the human TDP-43 protein in the non-mammalian models such as the yeast *Saccharomyces cerevisiae*, the fly *Drosophila*, and the worm *Caenorhabditis elegans* recapitulated several features of the ALS-associated TDP-43 pathology including its cytoplasmic mislocalization and aggregation. The expression of TDP-43 in *Drosophila* resulted in the deposition of ubiquitinated and hyper-phosphorylated TDP-43 aggregates along with the UPR-associated proteins such as Ubiquilin1 and Ubiquilin2 and also activated the mitochondria-dependent unfolded protein response (mt-UPR). The *Drosophila* model of the TDP-43 aggregation exhibited activation of the DNA damage and cell death signal response that was transmitted to the neighboring cells similar to as observed *via* gypsy retroviruses. Expression of TDP-43 in *C. elegans* led to the disruption of the ER-mediated calcium homeostasis and activated immune response *via* TIR1/SARM1 signaling. Progranulin and heat shock factor-1 (HSF-1) were also found to modulate the TDP-43 toxicity in the worm model. The expression of TDP-43 in the yeast model led to the increased oxidative stress, decreased autophagic and proteasomal clearance, and enhanced stress granule formation. It also affected the mitochondrial and vacuolar functions. TDP-43 colocalizes with the endocytic proteins and causes disruption of endocytosis in the fly and yeast models. Modified yeast Hsp104 chaperone reduced the aggregation in the worm and yeast models of the TDP-43 proteinopathy thereby promising of potential therapeutic applications.

domains including two nucleotide-binding domains (NBD1 and NBD2) and a middle domain (MD).[58–60] The Hsp104 protein is conserved in the eubacteria and the eukaryotes except for the metazoans that lack an Hsp104 homolog.[61,62] The wild-type Hsp104 chaperone can reverse the TDP-43 toxicity in yeast.[58] Professor Shorter and coworkers have identified potentiated Hsp104 middle domain (aa: 411–538) variants, by incorporating point mutations, that rescue the TDP-43 toxicity better than the wild-type Hsp104.[58] The enhanced rescue of the TDP-43 cytotoxicity by the variants of Hsp104 may be due to several reasons such as enhanced ATPase activity, enhanced unfoldase activity, enhanced amyloid disaggregase activity, independence from the Hsp40 and Hsp70 chaperones to perform the disaggregation or due to their faster substrate translocation ability than the wild-type Hsp104.[58] In fact, the $Hsp104^{A503V}$ and $Hsp104^{A503S}$ variants were observed to be better at disaggregating the preformed fibrils.[58] Notably, these functionally enhanced Hsp104 MD variants eradicate the TDP-43 aggregates and also restore the TDP-43 protein back to the nucleus without influencing the TDP-43 expression levels.[58] One shortcoming of the Hsp104 MD variants was found to be their limited substrate specificity that may cause off-target toxicity.[58] Thus, NBD1 and NBD2 potentiated variants of Hsp104 were generated that are found to be non-toxic and manifest specificity towards the TDP-43 disaggregation and its toxicity rescue.[63] Among the generated NBD1 Hsp104 variants, $Hsp104^{I230N}$, $Hsp104^{L355A}$, and $Hsp104^{E360K}$ were the most potent rescuer of the TDP-43 toxicity.[63] The NBD1 variants displayed better effect against the cytoplasmic mislocalization of TDP-43 thereby restoring it into the nucleus.[63] Additionally, another Hsp104 variant harboring mutations in the NBD2 domain was also found to rescue the TDP-43 toxicity in yeast.[63] In summary, Hsp104 variants are emerging as useful tools with potential toward TDP-43 disaggregation and toxicity rescue.

Another yeast chaperone, Hsp40 Sis1, which is a co-chaperone of the yeast Hsp70 chaperone SSA1, has been examined by Professor Liebman and coworkers for affecting the TDP-43 aggregation and toxicity in the yeast TDP-43 proteinopathy model.[46] Upon overexpression of Sis1, a reduction in the TDP-43-induced cytotoxicity was observed.[46] In fact, the Sis1 overexpression increased the survival of the yeast cells and also reversed their abnormal elongated cellular morphology that resulted from the TDP-43 toxicity in the yeast cells containing an endogenous prion [PIN^{+}]. Strikingly, Sis1 co-overexpression also restored the proteasomal function that was compromised upon the TDP-43 expression and aggregation.[46] In a recent study, Hsp90 chaperone and its co-chaperone sti1 have been shown to modify the TDP-43 aggregation and toxicity.[136]

In addition to giving insights into the molecular events that are important for mediating the TDP-43 toxicity, the yeast model has also been used to find small organic molecules that can potentially inhibit the TDP-43 aggregation or toxicity.[3,64,141]

The *Caenorhabditis elegans* model of the TDP-43 proteinopathy

The nematode worm, *C. elegans*, has a well-defined nervous system, short life span, and short generation time which make it a fruitful model system to investigate the mechanisms pertaining to neurodegeneration.[24] The RNA-binding protein TDP-43 is conserved in evolution across the humans, *Drosophila* and *C. elegans* and the homologs show extraordinary similarity in the nucleic acid–binding specificity. The *C. elegans* TDP-43 (known as TDP-1), much alike the human TDP-43, specifically binds to the UG-rich sequences of the RNA molecules.[65] TDP-1 also contains identical caspase cleavage sites and an NLS similar to that of the human TDP-43. Of note, the C-terminal glycine-rich segment found in the human TDP-43, which is important for the protein aggregation propensity and also for its interaction with other proteins,[66–68] is absent in the *C. elegans* TDP-43.[65] However, the absence of the C-terminal region does not affect the splicing activity of TDP-1 toward CFTR mRNA thus reinforcing the notion of TDP-1 being a functional homolog of the human TDP-43.[69] Human TDP-43, when expressed in neurons of *C. elegans*, resulted in synapse abnormalities in the motor neurons and the worms exhibited an "uncoordinated" phenotype analogous to the observations in the ALS patients. Similarly, overexpression of the worm TDP-1 also induced an abnormal neuronal activity, whereas a genetic deletion of TDP-1 did not result in any effects on the motor neurons and the worm movement.[69] In addition to the wild-type TDP-43 expression-induced abnormalities in *C. elegans*, several ALS-associated TDP-43 familial mutants such as A315T, G290A, and M337V also resulted in movement abnormalities, paralytic phenotype, and a reduced life span.[70] Furthermore, when the structural integrity of the GABAergic D-type motor neurons expressing the human TDP-43 bearing ALS mutations was examined, there was observable degeneration of the dopaminergic neurons coupled with the axonal loss.[70] Surprisingly this phenotype was not observed when the wild-type human TDP-43 was expressed. Importantly, TDP-43 formed detergent-resistant aggregates which were found to be phosphorylated in *C. elegans*. Also, inhibition of the Serine-409/410 phosphorylation site led to a reversal of the familial ALS mutant TDP-43-mediated motor neuron defects.[70] In contrast, the wild-type TDP-43 overexpression-associated motor neuron phenotype remained the same despite the ablation of the phosphorylation site.[70] In fact, phosphorylation of TDP-43 in *C. elegans* was found to be mediated *via* the cell division cycle kinase 7 (CDC7) and an inhibition of the CDC7 activity by a small molecule was found to rescue the effects of the TDP-43 toxicity.[71] Another study found that when a CTF of TDP-43 (25 kDa) was expressed in *C. elegans*, this CTF also manifested a propensity to aggregate and caused neurodegeneration.[69] As ALS is an adult-onset neurodegenerative disorder, several aging pathways have also been studied in association with the TDP-43 expression. When the insulin/insulin-like growth factor (IGF-1) pathway, which is conserved in the evolution from the worms to humans, was examined in relation to the TDP-43 aggregation, it was found that the deficiency of IGF-1 in *C. elegans* resulted in reduction in the soluble as well as aggregated TDP-43 protein and the neurotoxicity caused by the TDP-43 aggregation was significantly ameliorated.[72] This was further confirmed in the mammalian system where a deficiency of IGF-1 also affected the TDP-43 protein aggregation[72] (Table 8.1). Furthermore, the effect of another aging modulator protein, HSF-1, which regulates the expression of several heat shock proteins, was also examined in relation to the TDP-43 toxicity. It was found that the human TDP-43's aggregation as well as its neurotoxicity were enhanced with an increasing thermal stress on *C. elegans*[72]. Also, a loss of function of the *HSF-1* gene caused neurotoxicity thereby highlighting the neuroprotective effects of HSF-1 against the TDP-43-mediated toxicity.[72] Furthermore, it was found that the

TABLE 8.1 Genetic modifiers of the TDP-43 toxicity screened from the *Caenorhabditis elegans* model.

Gene	Overexpression or deletion	Effects on TDP-43 toxicity	Protein function	References
IGF1	Deficiency	Suppressor	Regulates aging	72
HSF1	Loss-of-function	Enhancer	Protective role in protein folding and distribution	72
CRT1	Null mutation	Suppressor	Regulates ER-calcium homeostasis	77
CNX1	Deletion	Suppressor	Regulates ER-calcium homeostasis	77
UNC-68	Deletion	Suppressor	ER Ca^{2+} release channel ryanodine receptor	77
ITR1	Deletion	Suppressor	ER Ca^{2+} release channel inositol triphosphate receptor	77
TRA-3	Mutations	Suppressor	Calcium-regulated calpain protease	77
ASP-4	Mutations	Suppressor	Calcium-regulated aspartyl protease	77
HSE-5	Loss-of-function	Suppressor	Heparan sulfate–modifying system	75

worm TDP-43 homolog, TDP-1, was in fact a gene modulating the stress response by the insulin/IGF-1 signaling pathway that can regulate oxidative stress and the life span of *C. elegans*.[72] Moreover, the mutant TDP-43 caused oxidative stress in *C. elegans* which in turn resulted in an upregulation of TDP-1 that negatively affected the life span.[73] Also, the mutant TDP-43's expression in TDP-1 deletion background failed to cause neurodegeneration.[74] Genome-wide loss-of-function screen for suppressors of TDP-43 in *C. elegans* revealed 46 candidate genes that significantly reduced the TDP-43 toxicity (Table 8.1).[75] Notably, a loss-of-function of the heparan sulfate–modifying enzyme, hse-5, that is involved in cell migration and growth, has been found to reduce the TDP-43 accumulation and the associated motor deficits.[75] As a validation of this genetic screen, the human hse-5 homolog, glucuronic acid epimerase's (GLCE) knockdown also exhibited protection against the oxidative stress–induced TDP-43 accumulation in the cultured human cells.[75] Noticeably, an analysis of the frontal cortex of the FTLD-TDP-43 patients revealed a decrease in the protein levels of GLCE which might potentially represent a protective response.[75]

To identify any paralytic effects of the TDP-43 expression, which would usually take 7–12 days on the solid culture plates, a liquid culture assay was developed in order to quickly identify the paralysis phenotype that could enable a faster genetic and pharmacological screening in *C. elegans*.[73] This assay helped in a screen that identified methylene blue as a molecule that reduces the paralytic effect induced by the mutant TDP-43 protein expression in the *C. elegans* model.[73] Subsequently, it was found that methylene blue also reduces the oxidative stress which originates as a result of the unfolded protein response (UPR) pathway activated by the TDP-43 aggregation in *C. elegans*.[76] As endoplasmic reticulum (ER) is the storage center for calcium, whose regulation can affect the functions of ER, the role of calcium signaling in mediating the TDP-43 toxicity was also studied using the *C. elegans* model. It was found that the introduction of a null mutation in the calreticulin (CRT-1) gene, which functions in the regulation of the ER Ca^{2+} homeostasis, was capable of suppressing the paralysis and degeneration of the neurons caused by the mutant TDP-43 expression in *C. elegans*.[77] Deletions of several genes involved in the calcium regulation such as calnexin (CNX1), ER calcium release channel (UNC-68), and ITR-1 were found to supress the TDP-43 toxicity (Table 8.1).[77] Furthermore, mutations in the Ca^{2+}-regulated calpain protease TRA-3 and the aspartyl protease ASP-4 also caused suppression of the toxicity of TDP-43.[78] Notably, the Ca^{2+} release triggered from ER by thapsigargin successfully restored the toxicity of TDP-43 in the CNX1 deleted nematode.[78] Additionally, phosphorylation of eIf2α, that indicates the activation of the protein kinase RNA-like ER kinase (PERK) pathway, was found to be protective against the TDP-43-induced neurodegeneration.[78] These findings highlight the importance of the ER calcium storage and its role in mediating the TDP-43 toxicity-induced neurodegeneration. Expression of the mutant TDP-43 in the *C. elegans* motor neurons was also found to induce the innate immune response *via* the microbial infection response TIR1/SARM1 pathway, which in turn can result in the activation of the tissue repair pathways (Fig. 8.1).[78] Furthermore, heterozygosity and not the complete loss of progranulin, that normally helps in neuronal survival, was found to exacerbate the TDP-43 toxicity in the *C. elegans* model.[79] Recently, granulins have been proposed to modulate the liquid−liquid phase separation and aggregation behavior of TDP-43.[80]

Deletion of TDP-1 in *C. elegans* resulted in the accumulation of double-stranded RNA (dsRNA)[81] and this was later replicated in the M17 human neuroblastoma cell lines when TDP-43 was knocked down using small interfering RNA (siRNA).[81] The worm TDP-1 and heterochromatin protein HPL-2, were found to interact directly and the loss of TDP-1 could significantly alter the association of HPL-2 with chromatin[82]. This hints that a depletion of the nuclear TDP-43 could alter the RNA metabolism. Indeed, TDP-43 pathology in *C. elegans* was found to correlate with the DNA damage and the DNA damage response activation similar to that of the TDP-43 pathology was found in the ALS patients.[83] A recent study has shown that the deletion of TDP-1 significantly reduces human α-synuclein-associated toxicity in the transgenic *C. elegans* model.[84] Also, the co-expression of TDP-1 along with α-synuclein led to an increased locomotive defect and shorter life span as compared to when these proteins were expressed singly.[84] Notably, the co-expression of Ubiquilin-2 and TDP-43 exacerbated the extent of neurodegeneration in the *C. elegans* model of ALS.[139]

Furthermore, the *C. elegans* model facilitates the screening of molecules that can modulate the TDP-43 aggregation and toxicity. Several compounds which extend the life span of *C. elegans* were tested for their potential to rescue the TDP-43-induced toxicity. Some chemicals such as resveratrol, reserpine, rolipram, propyl gallate, trolox, and ethosuximide, exhibited protective effects against the mutant TDP-43-induced neuronal toxicity.[85] Also, maple syrup, which contains several active molecules such as sugars and phenols, caused a reduction in the age-dependent motor neuron defects exhibited upon the expression of an ALS-associated mutant, TDP-43-A315T.[86] Another compound, α-methyl-α-phenylsuccinimide (MPS), relieved the locomotion defects caused by the mutant TDP-43 expression.[87] An anti-psychotic compound, pimozide, was found to reduce the mutant TDP-43-induced defects in the *C. elegans* as well as in the zebrafish model of ALS.[88] Recently, inhibitors against cyclin-dependent kinase, CDC7, have been shown to reduce the phosphorylation of TDP-43 in the *C. elegans* model suggesting potential for therapeutic applications.[140] Various genetic and chemical screens carried out thus far using the *C. elegans* model of the TDP-43 toxicity have provided important mechanistic insights as well as potential therapeutic targets and leads for ALS.

The *Drosophila* model of the TDP-43 proteinopathy

The fruit fly, *Drosophila melanogaster*, has been used to examine the toxicity mechanisms of several neurodegenerative diseases.[89] The rapid life-cycle of *Drosophila* as well as the availability of tools for genetic manipulations have greatly facilitated its usage as a model organism for investigating the mechanisms relevant to the ALS pathogenesis. *Drosophila* has an ortholog of TDP-43 known as TBPH which has structural and functional similarity to the human protein[65]. The human TDP-43 and the *Drosophila* TDP-43 exhibit about 77% sequence similarity and have conserved RNA recognition motifs that preferentially bind to the UG-rich sequences. Although the C-terminal regions of the human and *Drosophila* TDP-43 are divergent, the *Drosophila* TDP-43, alike to the human TDP-43, also regulates exon splicing for which the C-terminal glycine-rich domain is a pre-requisite.[65] Also, similar to that of the human TDP-43, the C-terminal region of the *Drosophila* TDP-43 interacts with various hnRNPs.[90,91] Notably, the expression of the human wild-type TDP-43 in the *Drosophila* eyes leads to a marked reduction in their life span and the flies also exhibit motor neuron deficits in a dose-dependent manner.[92] As the human TDP-43 predominantly localized to the nuclei and did not form any detectable cytosolic aggregates, the observed adverse effects suggested of the presence of certain alternate toxicity mechanisms. The modulation of the TBPH expression in *Drosophila* has shown that TDP-43 promotes the formation and growth of the neuromuscular synapses.[93]

Furthermore, when the human TDP-43 was overexpressed in the mushroom bodies of *Drosophila*, which have varied roles in memory and learning, TDP-43 was found to aggregate in the cell bodies and axons.[94] Also, enhanced axon swelling and axonal and neuronal loss were observed.[95] Moreover, the expression of an ALS-causing M337V mutation in TDP-43 caused redistribution of the human TDP-43 from nucleus to cytoplasm and could recapitulate the neuronal degeneration.[48] Furthermore, expression of TDP-43 bearing another ALS mutation, A315T, markedly increased the neuronal loss in comparison with the wild-type TDP-43 and also caused more frequent fly deaths.[96] Notably, the coexpression of two ALS implicated proteins, FUS and TDP-43, containing ALS-associated mutations, synergistically enhanced the neurodegeneration in *Drosophila*.[97] A defective axonal transport has also been reported in several *Drosophila* models of the ALS pathology.[98]

Of note, the *Drosophila* model has also been used to study the RNA-binding and splicing activities of the TDP-43 protein. The deletion of RRM1 and a substitution mutation W113A have been shown to abrogate the RNA-binding ability of TDP-43 and also reduce its toxicity.[65]

Notably, a distinctive characteristics of the TDP-43-mediated ALS pathogenesis is the presence of several CTFs of TDP-43 in the pathological TDP-43 inclusions.[95] Several CTFs of TDP-43 with deleted NLS were expressed in *Drosophila*, that led to cytoplasmic insoluble protein aggregation with high molecular weight species similar to the pathological inclusions found in the ALS patients.[99] It was also reported that the hyper-phosphorylation and ubiquitination occur temporally after the aggregation of CTFs and that the hyper-phosphorylation confers the pathogenesis by TDP-43.[99] Additionally, alike to the ALS disease-specific phosphorylation of the Ser409/Ser410, the human TDP-43 also undergoes phosphorylation at S409/410 when expressed in *Drosophila*.[100]

Furthermore, the *Drosophila* TDP-43 ortholog, TBPH, was found to change the expression of the microtubule associated protein, Futsch (human ortholog, MAP1B).[99] Futsch has a vital role in microtubule integrity during expansion of the neuromuscular junction. Modulation of the Futsch expression may mediate the observed synaptic defects and cytoskeletal abnormalities associated with the TDP-43 pathogenesis.[99] Similarly, the *Drosophila* TBPH depletion caused silencing of the histone deacetylase 6 (HDAC6) mRNA and reduced the protein levels which caused accumulation of an HDAC6 substrate, acetyl-tubulin[101]. Recently, a downregulation of a poly-(ADP-ribose) polymerase (PARP), tankyrase, was observed to reduce the cytoplasmic accumulation of TDP-43 that also reduced neurodegeneration (Table 8.2).[102] Also, an overexpression of the *Drosophila* fragile X mental retardation protein, dFMRP, was able to relieve the TDP-43-mediated toxicity in the *Drosophila* model of ALS (Table 8.2).[103] In a recent study, an overexpression of omega class of glutathione S-transferase, which protects against oxidative damage, has been shown to relieve the motor neuron degeneration in the *Drosophila* model of ALS.[104]

The failure in the clearance of the misfolded TDP-43 protein *via* the ubiquitin proteasome response as well as the autophagy pathway has been observed in the ALS models.[105,106] Also, the TDP-43 aggregates are found to be hyper-phosphorylated and ubiquitinated and are frequently found to be deposited with the UPR-associated proteins such as Ubiquilin1 and Ubiquilin2.[44,45] Several mutations in the VCP, that segregates ubiquitinated substrates from the multimeric protein complexes, are known to enhance the mutant TDP-43 toxicity in *Drosophila* and are sufficient to cause the cytoplasmic mislocalization of TDP-43.[48] In fact, the *Drosophila* model has shown that VCP interacts physically and genetically with TDP-43.[48] Countering the effects of the dysfunctional proteostasis caused by the depletion of Ubiquilin2 by increasing the *Drosophila* VCP protein restored the TBPH levels to the normal and also resulted in a reduction in the levels of toxic soluble TBPH.[107] Another UPR-associated protein, Ubiquilin1's coexpression with TDP-43, was found to increase the TDP-43 toxicity even though there was a marked increase in the clearance of TDP-43.[92] Recently, depletion of the *Drosophila* Ubiquilin2 protein has been shown to cause cytoplasmic mislocalization of the *Drosophila* TDP-43 (TBPH), which then partitions into insoluble ubiquitin-positive inclusions and causes moderate toxicity. Furthermore, upregulation of the autophagy gene, ATG7, has been found to relieve the motor neuron

TABLE 8.2 Genetic modifiers of the TDP-43 toxicity from the *Drosophila* model.

Gene	Overexpression or deletion	Effect on TDP-43 toxicity	Protein function	References
TNKS	Downregulation	Suppressor	Poly-ADP-ribose (PAR) polymerase	102
FMR1	Overexpression	Suppressor	Cognitive development and reproductive function	103
VCP	Mutations	Enhancer	Transitional endoplasmic reticulum ATPase	48
UBQLN-1	Overexpression	Enhancer	Regulation of UPS, autophagy, and ERAD pathways	92
ATG7	Upregulation	Suppressor	Regulates autophagy and CVT pathway	108
LON	Downregulation	Enhancer	Mitochondrial protease	110
PLD1	Downregulation	Suppressor	Phospholipase D	124
GSTO2	Overexpression	Suppressor	Glutathione S-transferase Omega-2, antioxidant	104

deficits in the transgenic fly depleted for TBPH thereby signifying the importance of the autophagy-mediated clearance in ALS.[108]

Notably, TDP-43 has been found to abnormally localize to mitochondria and an inhibition of this localization was found to decrease the neuronal toxicity.[42,109] In the *Drosophila* model of ALS, an increased TDP-43 expression was also found to be associated with the mitochondrial dysfunction, increased ROS generation, as well as decreased mitochondrial membrane potential.[110] Importantly, the TDP-43 expression in flies also activated the mitochondrial UPR, downregulated the mitochondrial protease Lon (human homolog LonP1), and exacerbated the TDP-43's mitochondrial localization and toxicity (Fig. 8.1).[110] Also, as an indication of the elevated ROS levels, the expression of TDP-43 in *Drosophila* resulted in an increased protein carbonylation.[111] The TDP-43's turnover and toxicity was found to be dependent on endocytosis in the yeast model of ALS, likewise, in *Drosophila* the TDP-43-induced locomotor dysfunction was found to be enhanced or suppressed by the respective impairment or enhancement of the endocytosis machinery.[47]

Furthermore, TDP-43 was found to also interfere with the nucleosome dynamics.[112,113] The abnormal chromatin dynamics exerted upon the TDP-43 expression was modulated by the expression of the chromodomain helicase DNA-binding protein 1 (CHD1).[72] Recently, in the *Drosophila* model, expression of the human TDP-43 in a focally small group of glial cells caused death of the adjacent neurons possibly through replication and/or *via* the expression of the endogenous retroviruses RTE/ERV, which can transmit the DNA damage and cell death signal to the neighboring cells.[114]

Several studies aimed towards therapeutic modulation of the TDP-43's toxicity have also been carried out using the transgenic fly models of TDP-43. Inhibition of the aggregation of TDP-43 and its disease-associated CTF of 25 kDa have been attempted *via* pharmacological upregulation of the heat shock response pathway.[112] Of high interest, genetic upregulation of a small Hsp, CG14207, was found to cause partial reduction in the neurotoxicity and enhancement of the TDP-43's clearance from brain[115]. PARP inhibitors have also been proposed to inhibit the TDP-43's liquid–liquid phase separation and hence reduce the TDP-43 toxicity.[102] Metabolomic profiling upon the TDP-43 overexpression in the *Drosophila* model found an evidence of perturbed mitochondrial lipid metabolism *via* an increase in the fatty acid conjugates and decrease in beta-hydroxybutyrate.[116] Increasing the dietary fatty acids corrected the associated locomotor dysfunction.[117] Small molecule inhibitors of stress granule formation have been proposed to prevent the RNA-dependent recruitment of TDP-43 into stress granules thereby preventing the toxicity.[118] Furthermore, a small molecule inhibitor, rTRD01, which could bind to the RRM1/RRM2 domains of TDP-43 in the in silico studies, was also found to improve the larval turning and neuromuscular coordination in a mutant TDP-43 *Drosophila* model.[119] Heterocyclic inhibitors of casein kinase, CK1, displayed potency to inhibit the TDP-43 phosphorylation thereby increasing the life-span of the fly model of ALS.[120] A recent study has shown that reducing the histone deacetylase 1 (HDAC1) levels by HDAC1 inhibitors or by siRNA significantly reduces the TDP-43 toxicity thereby highlighting the importance of the TDP-43 acetylation as a potential therapeutic target against ALS.[121]

The *D. melanogaster* family of sHSPs contains 11 proteins having role in regulation of the protein homeostasis.[122] An upregulation of the chaperone Dmel\CG14207 (also known as HspB8) reduces the neurotoxicity induced by the full-length TDP-43 and its N-terminally truncated disease-associated fragment, TDP-25.[115] In another study, it was found that an overexpression/upregulation of HSP67Bc, a Dm functional ortholog of HspB8, protects against the toxicity of various mutant forms of TDP-43 such as an NLS mutant TDP-43, TDP-35, and an ALS-causing point mutant, TDP-43-M337V.[123] Recently, a genome-wide screen has revealed that the down-regulation of the phospholipase D1/2 pathway genes cause protection against the TDP-43-M337V-induced pathology (Table 8.2).[124] Notably, the overexpression of the long non-coding RNA, NEAT1 isoform, NEAT1_1, was found to rescue the TDP-43 toxicity in the *Drosophila* and yeast models of ALS.[137]

The zebrafish model of the TDP-43 proteinopathy

The zebrafish (*Danio rerio*) is a vertebrate model also used for investigating the neurodegenerative diseases as it has significantly conserved physiology and maintenance of the nervous system akin to humans.[28] The zebrafish offers the salient features of a good model organism such as rapid life-cycle, high reproduction rate, and a striking transparency that allows for non-invasive imaging at the cellular level. The zebrafish TDP-43 protein, TARDBP, has about 72% amino acid similarity with that of the human TDP-43 and a knock-down of the zebrafish TDP-43 gene results in the motor neuron deficits alike to that of the ALS patients.[125] Two orthologs of human TDP-43 namely tardbp and tardbpl are present in zebrafish. A transient knock-down of tardbp resulted in shortened and disorganized axons in the motor neurons, whereas a knock-down of the other TDP-43 ortholog, tardbpl, did not result in any observable phenotypic defects.[125] However, the knockdown of both tardbp and tardbpl resulted in the locomotion defects, muscular degeneration, vascular dysfunction, reduced outgrowth of the motor neuron axons, and early death of zebrafish.[126] This finding hinted that a loss of function of TDP-43 might lead to the ALS-associated pathogenesis.[127]

Strikingly, when the human wild-type TDP-43 was expressed in the zebrafish knocked-down for *tardbp*, the motor neuron deficits associated with *tardbp* knockdown were rescued, however, the expression of ALS mutant TDP-43 could not rescue these motor neuron deficits.[125] Similarly, an over-expression of the wild-type human TDP-43 in zebrafish did not cause any significant motor neuron deficits, whereas the overexpression of the mutant human TDP-43 resulted in reduced length of axons of the motor neurons with premature and increased branching and the zebrafish exhibited swimming deficits.[125] Recently, a transgenic zebrafish model for both the wild-type as well as the ALS-associated mutant TDP-43 was generated and a transcriptome analysis was performed to identify the candidate genes that may be linked to the TDP-43 proteinopathy in zebrafish.[128]

Furthermore, the mutant TDP-43 expression in zebrafish resulted in the neuronal degeneration and also induced oxidative stress and a treatment with methylene blue significantly lowered the oxidative stress and the neuronal deficits.[73] Also, the expression of a mutant TDP-43 in the zebrafish larvae led to functional alterations in the glutamatergic synapses in the spinal cord.[129] In a recent study, reduction in the TDP-43 ortholog expression in zebrafish resulted in the reduction of the expression of acetylcholinesterase which is commonly found in the neuromuscular junctions.[138]

Several compounds which modulate the UPR and oxidative stress rescued the mutant TDP-43 toxicity in the zebrafish model.[76] Also, an overexpression of progranulin was shown to rescue the TDP-43 axonopathy in the zebrafish model.[130] Furthermore, an inhibitor of the T-type calcium channels was found to rescue the TDP-43 toxicity by stabilizing the neuromuscular transmission.[88] Recently, the zebrafish models with partial or complete loss of the *tardbp* allelic function were created using the CRISPR-Cas9 system which can be used for drug screening.[131] Using an optogenetic TDP-43 variant, study has shown that the TDP-43 oligomerization has differential short-term and long-term effects, therefore, this model can facilitate the study of time progression of the ALS-like pathology in zebrafish.[132]

Conclusions and future directions

Several mammalian and non-mammalian models are available for studying the TDP-43-associated pathogenic mechanisms relevant to the neurodegeneration in ALS. Valuable insights from the non-mammalian models are being obtained with regard to the various genes and pathways that affect or mediate the toxicity associated with the TDP-43 dyshomeostasis (Fig. 8.1). The ease of genetic manipulations and the large-scale drug screening has made the unicellular yeast model an important tool for investigating the TDP-43-associated pathogenic mechanisms. Furthermore, some of the non-mammalian models such as *C. elegans*, zebrafish, and *Drosophila* themselves contain endogenous homologs of TDP-43 with sequence as well as functional similarities to the human TDP-43. Therefore, these models have become useful for examining the neuronal health upon the TDP-43 dyshomeostasis. These models are also being used to validate potential therapeutic hits obtained from the yeast, in silico or in vitro experiments. As a clear role of TDP-43 in the pathogenesis of ALS/FTLD is only beginning to be elucidated, the non-mammalian model systems continue to be helpful in unearthing and understanding the pertinent molecular mechanisms. Recently, genome-scale genetic screens using a library of cloned human genes have been developed for studying the toxicity of neurodegenerative disorder proteins in the yeast model. This can potentially help in identifying more *bona fide* suppressors of the TDP-43 toxicity.

Acknowledgments

Vidhya Bharathi thanks DBT, Govt. of India, for research fellowship. Amandeep Girdhar thanks Ministry of education, Govt. of India, for research fellowship. Basant K Patel thanks SERB-DST, Govt. of India for research support (Grant no: SERB/EMR/2016/006327).

References

1. Rowland LP, Shneider NA. Amyotrophic lateral sclerosis. *N Engl J Med* 2001;**344**(22): 1688–700.

2. Ling SC, Albuquerque CP, Han JS, Lagier-Tourenne C, Tokunaga S, Zhou H, et al. ALS-associated mutations in TDP-43 increase its stability and promote TDP-43 complexes with FUS/TLS. *Proc Natl Acad Sci USA* 2010;**107**(30):13318–23.
3. Prasad A, Bharathi V, Sivalingam V, Girdhar A, Patel BK. Molecular mechanisms of TDP-43 misfolding and pathology in amyotrophic lateral sclerosis. *Front Mol Neurosci* 2019;**12**:25.
4. Liu Y-C, Chiang P-M, Tsai K-J. Disease animal models of TDP-43 proteinopathy and their pre-clinical applications. *Int J Mol Sci* 2013;**14**(10):20079–111.
5. Verduyckt M, Vignaud H, Bynens T, Van den Brande J, Franssens V, Cullin C, et al. Yeast as a model for Alzheimer's disease: latest studies and advanced strategies. *Methods Mol Biol* 2016;**1303**:197–215.
6. Prüßing K, Voigt A, Schulz JB. *Drosophila melanogaster* as a model organism for Alzheimer's disease. *Mol Neurodegener* 2013;**8**:35.
7. Patel BK, Gavin-Smyth J, Liebman SW. The yeast global transcriptional co-repressor protein Cyc8 can propagate as a prion. *Nat Cell Biol* 2009;**11**(3):344.
8. Wickner RB. [URE3] as an altered URE2 protein: evidence for a prion analog in *Saccharomyces cerevisiae*. *Science* 1994;**264**(5158):566–9.
9. Patel BK, Liebman SW. "Prion-proof" for [PIN+]: infection with *in vitro*-made amyloid aggregates of Rnq1p-(132-405) induces [PIN+]. *J Mol Biol* 2007;**365**(3):773–82.
10. Rencus-Lazar S, DeRowe Y, Adsi H, Gazit E, Laor D. Yeast models for the study of amyloid-associated disorders and development of future therapy. *Front Mol Biosci* 2019;**6**(15).
11. Krobitsch S, Lindquist S. Aggregation of huntingtin in yeast varies with the length of the polyglutamine expansion and the expression of chaperone proteins. *Proc Natl Acad Sci USA* 2000;**97**(4):1589–94.
12. Walter GM, Raveh A, Mok SA, McQuade TJ, Arevang CJ, Schultz PJ, et al. High-throughput screen of natural product extracts in A yeast model of polyglutamine proteotoxicity. *Chem Biol Drug Des* 2014;**83**(4):440–9.
13. Ruetenik AL, Ocampo A, Ruan K, Zhu Y, Li C, Zhai RG, et al. Attenuation of polyglutamine-induced toxicity by enhancement of mitochondrial OXPHOS in yeast and fly models of aging. *Microbial Cell* 2016;**3**(8):338.
14. Treusch S, Hamamichi S, Goodman JL, Matlack KE, Chung CY, Baru V, et al. Functional links between Aβ toxicity, endocytic trafficking, and Alzheimer's disease risk factors in yeast. *Science* 2011;**334**(6060):1241–5.
15. Park S-K, Ratia K, Ba M, Valencik M, Liebman SW. Inhibition of Aβ42 oligomerization in yeast by a PICALM ortholog and certain FDA approved drugs. *Microbial Cell* 2016;**3**(2):53.
16. Park S-K, Pegan SD, Mesecar AD, Jungbauer LM, LaDu MJ, Liebman SW. Development and validation of a yeast high-throughput screen for inhibitors of $A\beta_{42}$ oligomerization. *Dis Model Mech* 2011;**4**(6):822–31.
17. McGurk L, Berson A, Bonini NM. *Drosophila* as an *in vivo* model for human neurodegenerative disease. *Genetics* 2015;**201**(2):377–402.
18. Rubin GM, Yandell MD, Wortman JR, Gabor GL, Nelson CR, Hariharan IK, et al. Comparative genomics of the eukaryotes. *Science* 2000;**287**(5461):2204–15.
19. Warrick JM, Paulson HL, Gray-Board GL, Bui QT, Fischbeck KH, Pittman RN, et al. Expanded polyglutamine protein forms nuclear inclusions and causes neural degeneration in *Drosophila*. *Cell* 1998;**93**(6):939–49.
20. Wittmann CW, Wszolek MF, Shulman JM, Salvaterra PM, Lewis J, Hutton M, et al. Tauopathy in *Drosophila*: neurodegeneration without neurofibrillary tangles. *Science* 2001;**293**(5530):711–4.

21. Elden AC, Kim H-J, Hart MP, Chen-Plotkin AS, Johnson BS, Fang X, et al. Ataxin-2 intermediate-length polyglutamine expansions are associated with increased risk for ALS. *Nature* 2010;**466**(7310):1069.
22. Kim H-J, Raphael AR, LaDow ES, McGurk L, Weber RA, Trojanowski JQ, et al. Therapeutic modulation of eIF2α phosphorylation rescues TDP-43 toxicity in amyotrophic lateral sclerosis disease models. *Nat Genet* 2014;**46**(2):152.
23. Alexander AG, Marfil V, Li C. Use of *Caenorhabditis elegans* as a model to study Alzheimer's disease and other neurodegenerative diseases. *Front Genet* 2014;**5**:279.
24. Wolozin B, Gabel C, Ferree A, Guillily M, Ebata A. Watching worms whither: modeling neurodegeneration in *C. elegans*. *Prog Mol Biol Transl Sci* 2011;**100**:499−514.
25. Kraemer BC, Zhang B, Leverenz JB, Thomas JH, Trojanowski JQ, Schellenberg GD. Neurodegeneration and defective neurotransmission in a *Caenorhabditis elegans* model of tauopathy. *Proc Natl Acad Sci USA* 2003;**100**(17):9980−5.
26. Lakso M, Vartiainen S, Moilanen AM, Sirvio J, Thomas JH, Nass R, et al. Dopaminergic neuronal loss and motor deficits in *Caenorhabditis elegans* overexpressing human alpha-synuclein. *J Neurochem* 2003;**86**(1):165−72.
27. Vartiainen S, Pehkonen P, Lakso M, Nass R, Wong G. Identification of gene expression changes in transgenic *C. elegans* overexpressing human alpha-synuclein. *Neurobiol Dis* 2006;**22**(3):477−86.
28. Xi Y, Noble S, Ekker M. Modeling neurodegeneration in zebrafish. *Curr Neurol Neurosci Rep* 2011;**11**(3):274−82.
29. Henshall TL, Tucker B, Lumsden AL, Nornes S, Lardelli MT, Richards RI. Selective neuronal requirement for huntingtin in the developing zebrafish. *Hum Mol Genet* 2009;**18**(24):4830−42.
30. Newman M, Ebrahimie E, Lardelli M. Using the zebrafish model for Alzheimer's disease research. *Front Genet* 2014;**5**(189).
31. Braun RJ. Mitochondrion-mediated cell death: dissecting yeast apoptosis for a better understanding of neurodegeneration. *Front Oncol* 2012;**2**:182.
32. Braun RJ, Sommer C, Carmona-Gutierrez D, Khoury CM, Ring J, Buttner S, et al. Neurotoxic 43-kDa TAR DNA-binding protein (TDP-43) triggers mitochondrion-dependent programmed cell death in yeast. *J Biol Chem* 2011;**286**(22):19958−72.
33. Bharathi V, Girdhar A, Prasad A, Verma M, Taneja V, Patel BK. Use of ade1 and ade2 mutations for development of a versatile red/white colour assay of amyloid-induced oxidative stress in *Saccharomyces cerevisiae*. *Yeast* 2016;**33**(12):607−20.
34. Johnson BS, McCaffery JM, Lindquist S, Gitler AD. A yeast TDP-43 proteinopathy model: exploring the molecular determinants of TDP-43 aggregation and cellular toxicity. *Proc Natl Acad Sci Unit States Am* 2008;**105**(17):6439−44.
35. Johnson BS, Snead D, Lee JJ, McCaffery JM, Shorter J, Gitler AD. TDP-43 is intrinsically aggregation-prone, and amyotrophic lateral sclerosis-linked mutations accelerate aggregation and increase toxicity. *J Biol Chem* 2009;**284**(30):20329−39.
36. Miller-Fleming L, Giorgini F, Outeiro TF. Yeast as a model for studying human neurodegenerative disorders. *Biotechnol J* 2008;**3**(3):325−38.
37. Berning BA, Walker AK. The pathobiology of TDP-43 C-terminal fragments in ALS and FTLD. *Front Neurosci* 2019;**13**:335.
38. Tsuji H, Arai T, Kametani F, Nonaka T, Yamashita M, Suzukake M, et al. Molecular analysis and biochemical classification of TDP-43 proteinopathy. *Brain J Neurol* 2012;**135**(Pt 11):3380−91.

39. Armakola M, Higgins MJ, Figley MD, Barmada SJ, Scarborough EA, Diaz Z, et al. Inhibition of RNA lariat debranching enzyme suppresses TDP-43 toxicity in ALS disease models. *Nat Genet* 2012;**44**(12):1302–9.
40. Bharathi V, Girdhar A, Patel BK. A protocol of using white/red color assay to measure amyloid-induced oxidative stress in *Saccharomyces cerevisiae. Bio-Protocol* 2017;**7**(15).
41. Iguchi Y, Katsuno M, Takagi S, Ishigaki S, Niwa J, Hasegawa M, et al. Oxidative stress induced by glutathione depletion reproduces pathological modifications of TDP-43 linked to TDP-43 proteinopathies. *Neurobiol Dis* 2012;**45**(3):862–70.
42. Wang W, Wang L, Lu J, Siedlak SL, Fujioka H, Liang J, et al. The inhibition of TDP-43 mitochondrial localization blocks its neuronal toxicity. *Nat Med* 2016;**22**(8):869–78.
43. Park SK, Park S, Liebman SW. Respiration enhances TDP-43 toxicity, but TDP-43 retains some toxicity in the absence of respiration. *J Mol Biol* 2019;**431**(10):2050–9.
44. Neumann M, Sampathu DM, Kwong LK, Truax AC, Micsenyi MC, Chou TT, et al. Ubiquitinated TDP-43 in frontotemporal lobar degeneration and amyotrophic lateral sclerosis. *Science* 2006;**314**(5796):130–3.
45. Deng HX, Chen W, Hong ST, Boycott KM, Gorrie GH, Siddique N, et al. Mutations in UBQLN2 cause dominant X-linked juvenile and adult-onset ALS and ALS/dementia. *Nature* 2011;**477**(7363):211–5.
46. Park SK, Hong JY, Arslan F, Kanneganti V, Patel B, Tietsort A, et al. Overexpression of the essential Sis1 chaperone reduces TDP-43 effects on toxicity and proteolysis. *PLoS Genet* 2017;**13**(5):e1006805.
47. Liu G, Coyne AN, Pei F, Vaughan S, Chaung M, Zarnescu DC, et al. Endocytosis regulates TDP-43 toxicity and turnover. *Nat Commun* 2017;**8**(1):2092.
48. Ritson GP, Custer SK, Freibaum BD, Guinto JB, Geffel D, Moore J, et al. TDP-43 mediates degeneration in a novel *Drosophila* model of disease caused by mutations in VCP/p97. *J Neurosci* 2010;**30**(22):7729–39.
49. Koppers M, van Blitterswijk MM, Vlam L, Rowicka PA, van Vught PW, Groen EJ, et al. VCP mutations in familial and sporadic amyotrophic lateral sclerosis. *Neurobiol Aging* 2012;**33**(4). 837.e7-13.
50. Bug M, Meyer H. Expanding into new markets–VCP/p97 in endocytosis and autophagy. *J Struct Biol* 2012;**179**(2):78–82.
51. Liu G, Byrd A, Warner AN, Pei F, Basha E, Buchanan A, et al. Cdc48/VCP and endocytosis regulate TDP-43 and FUS toxicity and turnover. *Mol Cell Biol* 2020;**40**(4). e00256-19.
52. Leibiger C, Deisel J, Aufschnaiter A, Ambros S, Tereshchenko M, Verheijen BM, et al. Endolysosomal pathway activity protects cells from neurotoxic TDP-43. *Microbial Cell* 2018;**5**(4):212–4.
53. Leibiger C, Deisel J, Aufschnaiter A, Ambros S, Tereshchenko M, Verheijen BM, et al. TDP-43 controls lysosomal pathways thereby determining its own clearance and cytotoxicity. *Hum Mol Genet* 2018;**27**(9):1593–607.
54. Chernoff YO, Lindquist SL, Ono B, Inge-Vechtomov SG, Liebman SW. Role of the chaperone protein Hsp104 in propagation of the yeast prion-like factor [psi+]. *Science* 1995;**268**(5212):880–4.
55. Shorter J, Lindquist S. Hsp104 catalyzes formation and elimination of self-replicating Sup35 prion conformers. *Science* 2004;**304**(5678):1793–7.
56. Higurashi T, Hines JK, Sahi C, Aron R, Craig EA. Specificity of the J-protein Sis1 in the propagation of 3 yeast prions. *Proc Natl Acad Sci USA* 2008;**105**(43):16596–601.

57. Tariq A, Lin J, Noll MM, Torrente MP, Mack KL, Murillo OH, et al. Potentiating Hsp104 activity via phosphomimetic mutations in the middle domain. *FEMS Yeast Res* 2018;**18**(5):foy042.
58. Jackrel ME, DeSantis ME, Martinez BA, Castellano LM, Stewart RM, Caldwell KA, et al. Potentiated Hsp104 variants antagonize diverse proteotoxic misfolding events. *Cell* 2014;**156**(1–2):170–82.
59. Parsell DA, Sanchez Y, Stitzel JD, Lindquist S. Hsp104 is a highly conserved protein with two essential nucleotide-binding sites. *Nature* 1991;**353**(6341):270–3.
60. Schirmer EC, Glover JR, Singer MA, Lindquist S. HSP100/Clp proteins: a common mechanism explains diverse functions. *Trends Biochem Sci* 1996;**21**(8):289–96.
61. Duennwald ML, Echeverria A, Shorter J. Small heat shock proteins potentiate amyloid dissolution by protein disaggregases from yeast and humans. *PLoS Biol* 2012;**10**(6): e1001346.
62. Shorter J. Hsp104: a weapon to combat diverse neurodegenerative disorders. *Neurosignals* 2008;**16**(1):63–74.
63. Tariq A, Lin J, Jackrel ME, Hesketh CD, Carman PJ, Mack KL, et al. Mining disaggregase sequence space to safely counter TDP-43, FUS, and alpha-synuclein proteotoxicity. *Cell Rep* 2019;**28**(8):2080–20895 e6.
64. Girdhar A, Bharathi V, Tiwari VR, Abhishek S, Deeksha W, Mahawar US, et al. Computational insights into mechanism of AIM4-mediated inhibition of aggregation of TDP-43 protein implicated in ALS and evidence for *in vitro* inhibition of liquid-liquid phase separation (LLPS) of TDP-432C-A315T by AIM4. *Int J Biol Macromol* 2020;**147**:117–30. https://doi.org/10.1016/j.ijbiomac.2020.01.032.
65. Ayala YM, Pantano S, D'Ambrogio A, Buratti E, Brindisi A, Marchetti C, et al. Human, *Drosophila*, and C.elegans TDP43: nucleic acid binding properties and splicing regulatory function. *J Mol Biol* 2005;**348**(3):575–88.
66. Fuentealba RA, Udan M, Bell S, Wegorzewska I, Shao J, Diamond MI, et al. Interaction with polyglutamine aggregates reveals a Q/N-rich domain in TDP-43. *J Biol Chem* 2010;**285**(34):26304–14.
67. Budini M, Buratti E, Stuani C, Guarnaccia C, Romano V, De Conti L, et al. Cellular model of TAR DNA-binding protein 43 (TDP-43) aggregation based on its C-terminal Gln/Asn-rich region. *J Biol Chem* 2012;**287**(10):7512–25.
68. Prasad A, Sivalingam V, Bharathi V, Girdhar A, Patel BK. The amyloidogenicity of a C-terminal region of TDP-43 implicated in Amyotrophic Lateral Sclerosis can be affected by anions, acetylation and homodimerization. *Biochimie* 2018;**150**:76–87.
69. Ash PEA, Zhang Y-J, Roberts CM, Saldi T, Hutter H, Buratti E, et al. Neurotoxic effects of TDP-43 overexpression in *C. elegans*. *Hum Mol Genet* 2010;**19**(16):3206–18.
70. Liachko NF, Guthrie CR, Kraemer BC. Phosphorylation promotes neurotoxicity in a *Caenorhabditis elegans* model of TDP-43 proteinopathy. *J Neurosci* 2010;**30**(48): 16208–19.
71. Liachko NF, McMillan PJ, Guthrie CR, Bird TD, Leverenz JB, Kraemer BC. CDC7 inhibition blocks pathological TDP-43 phosphorylation and neurodegeneration. *Ann Neurol* 2013;**74**(1):39–52.
72. Zhang T, Mullane PC, Periz G, Wang J. TDP-43 neurotoxicity and protein aggregation modulated by heat shock factor and insulin/IGF-1 signaling. *Hum Mol Genet* 2011;**20**(10):1952–65.

73. Vaccaro A, Patten SA, Ciura S, Maios C, Therrien M, Drapeau P, et al. Methylene blue protects against TDP-43 and FUS neuronal toxicity in *C. elegans* and *D. rerio*. *PLoS One* 2012;**7**(7):e42117.
74. Vaccaro A, Tauffenberger A, Ash PEA, Carlomagno Y, Petrucelli L, Parker JA. TDP-1/TDP-43 regulates stress signaling and age-dependent proteotoxicity in *Caenorhabditis elegans*. *PLoS Genet* 2012;**8**(7):e1002806.
75. Liachko NF, Saxton AD, McMillan PJ, Strovas TJ, Keene CD, Bird TD, et al. Genome wide analysis reveals heparan sulfate epimerase modulates TDP-43 proteinopathy. *PLoS Genet* 2019;**15**(12):e1008526.
76. Vaccaro A, Patten SA, Aggad D, Julien C, Maios C, Kabashi E, et al. Pharmacological reduction of ER stress protects against TDP-43 neuronal toxicity *in vivo*. *Neurobiol Dis* 2013;**55**:64–75.
77. Aggad D, Vérièpe J, Tauffenberger A, Parker JA. TDP-43 toxicity proceeds via calcium dysregulation and necrosis in aging *Caenorhabditis elegans* motor neurons. *J Neurosci* 2014;**34**(36):12093–103.
78. Vérièpe J, Fossouo L, Parker JA. Neurodegeneration in *C. elegans* models of ALS requires TIR-1/Sarm1 immune pathway activation in neurons. *Nat Commun* 2015;**6**(1):7319.
79. Salazar DA, Butler VJ, Argouarch AR, Hsu T-Y, Mason A, Nakamura A, et al. The progranulin cleavage products, granulins, exacerbate TDP-43 toxicity and increase TDP-43 levels. *J Neurosci* 2015;**35**(25):9315–28.
80. Bhopatkar AA, Uversky VN, Rangachari V. Granulins modulate liquid-liquid phase separation and aggregation of TDP-43 C-terminal domain. *bioRxiv* 2019:812115.
81. Saldi TK, Ash PE, Wilson G, Gonzales P, Garrido-Lecca A, Roberts CM, et al. TDP-1, the *Caenorhabditis elegans* ortholog of TDP-43, limits the accumulation of double-stranded RNA. *EMBO J* 2014;**33**(24):2947–66.
82. Saldi TK, Gonzales P, Garrido-Lecca A, Dostal V, Roberts CM, Petrucelli L, et al. The *Caenorhabditis elegans* ortholog of TDP-43 regulates the chromatin localization of the heterochromatin protein 1 homolog HPL-2. *Mol Cell Biol* 2018;**38**(15). e00668-17.
83. Mitra J, Guerrero EN, Hegde PM, Liachko NF, Wang H, Vasquez V, et al. Motor neuron disease-associated loss of nuclear TDP-43 is linked to DNA double-strand break repair defects. *Proc Natl Acad Sci USA* 2019;**116**(10):4696–705.
84. Shen L, Wang C, Chen L, Leung KL, Lo E, Lakso M, et al. TDP-1/TDP-43 potentiates human α-Synuclein (HASN) neurodegeneration in *Caenorhabditis elegans*. *Biochim Biophys Acta - Mol Basis Dis* 2020;**1866**(10):165876.
85. Tauffenberger A, Julien C, Parker JA. Evaluation of longevity enhancing compounds against transactive response DNA-binding protein-43 neuronal toxicity. *Neurobiol Aging* 2013;**34**(9):2175–82.
86. Aaron C, Beaudry G, Parker JA, Therrien M. Maple syrup decreases TDP-43 proteotoxicity in a *Caenorhabditis elegans* model of amyotrophic lateral sclerosis (ALS). *J Agric Food Chem* 2016;**64**(17):3338–44.
87. Wong SQ, Pontifex MG, Phelan MM, Pidathala C, Kraemer BC, Barclay JW, et al. α-Methyl-α-phenylsuccinimide ameliorates neurodegeneration in a *C. elegans* model of TDP-43 proteinopathy. *Neurobiol Dis* 2018;**118**:40–54.
88. Patten SA, Aggad D, Martinez J, Tremblay E, Petrillo J, Armstrong GA, et al. Neuroleptics as therapeutic compounds stabilizing neuromuscular transmission in amyotrophic lateral sclerosis. *JCI Insight* 2017;**2**(22).

89. Marsh JL, Thompson LM. *Drosophila* in the study of neurodegenerative disease. *Neuron* 2006;**52**(1):169–78.
90. D'Ambrogio A, Buratti E, Stuani C, Guarnaccia C, Romano M, Ayala YM, et al. Functional mapping of the interaction between TDP-43 and hnRNP A2 *in vivo*. *Nucleic Acids Res* 2009;**37**(12):4116–26.
91. Romano M, Feiguin F, Buratti E. *Drosophila* answers to TDP-43 proteinopathies. *J Amino Acids* 2012;**2012**:13.
92. Hanson KA, Kim SH, Wassarman DA, Tibbetts RS. Ubiquilin modifies TDP-43 toxicity in a *Drosophila* model of amyotrophic lateral sclerosis (ALS). *J Biol Chem* 2010;**285**(15):11068–72.
93. Strah N, Romano G, Introna C, Klima R, Marzullo M, Ciapponi L, et al. TDP-43 promotes the formation of neuromuscular synapses through the regulation of Disc-large expression in *Drosophila* skeletal muscles. *BMC Biol* 2020;**18**(1):34.
94. Li Y, Ray P, Rao EJ, Shi C, Guo W, Chen X, et al. A *Drosophila* model for TDP-43 proteinopathy. *Proc Natl Acad Sci USA* 2010;**107**(7):3169–74.
95. Lagier-Tourenne C, Polymenidou M, Cleveland DW. TDP-43 and FUS/TLS: emerging roles in RNA processing and neurodegeneration. *Hum Mol Genet* 2010;**19**(R1):R46–64.
96. Guo W, Chen Y, Zhou X, Kar A, Ray P, Chen X, et al. An ALS-associated mutation affecting TDP-43 enhances protein aggregation, fibril formation and neurotoxicity. *Nat Struct Mol Biol* 2011;**18**(7):822.
97. Lanson Jr NA, Maltare A, King H, Smith R, Kim JH, Taylor JP, et al. A *Drosophila* model of FUS-related neurodegeneration reveals genetic interaction between FUS and TDP-43. *Hum Mol Genet* 2011;**20**(13):2510–23.
98. Baldwin KR, Godena VK, Hewitt VL, Whitworth AJ. Axonal transport defects are a common phenotype in *Drosophila* models of ALS. *Hum Mol Genet* 2016;**25**(12):2378–92.
99. Godena VK, Romano G, Romano M, Appocher C, Klima R, Buratti E, et al. TDP-43 regulates *Drosophila* neuromuscular junctions growth by modulating Futsch/MAP1B levels and synaptic microtubules organization. *PLoS One* 2011;**6**(3):e17808.
100. Miguel L, Frébourg T, Campion D, Lecourtois M. Both cytoplasmic and nuclear accumulations of the protein are neurotoxic in *Drosophila* models of TDP-43 proteinopathies. *Neurobiol Dis* 2011;**41**(2):398–406.
101. Fiesel FC, Voigt A, Weber SS, Van den Haute C, Waldenmaier A, Gorner K, et al. Knockdown of transactive response DNA-binding protein (TDP-43) downregulates histone deacetylase 6. *EMBO J* 2010;**29**(1):209–21.
102. McGurk L, Gomes E, Guo L, Mojsilovic-Petrovic J, Tran V, Kalb RG, et al. Poly(ADP-Ribose) prevents pathological phase separation of TDP-43 by promoting liquid demixing and stress granule localization. *Mol Cell* 2018;**71**(5):703–717 e9.
103. Coyne AN, Yamada SB, Siddegowda BB, Estes PS, Zaepfel BL, Johannesmeyer JS, et al. Fragile X protein mitigates TDP-43 toxicity by remodeling RNA granules and restoring translation. *Hum Mol Genet* 2015;**24**(24):6886–98.
104. Cha SJ, Han YJ, Choi H-J, Kim H-J, Kim K. Glutathione S-transferase rescues motor neuronal toxicity in fly model of amyotrophic lateral sclerosis. *Antioxidants* 2020;**9**(7).
105. Scotter EL, Vance C, Nishimura AL, Lee Y-B, Chen H-J, Urwin H, et al. Differential roles of the ubiquitin proteasome system and autophagy in the clearance of soluble and aggregated TDP-43 species. *J Cell Sci* 2014;**127**(6):1263.

106. Lamark T, Johansen T. Autophagy: links with the proteasome. *Curr Opin Cell Biol* 2010;**22**(2):192–8.
107. Jantrapirom S, Lo Piccolo L, Yoshida H, Yamaguchi M. Depletion of Ubiquilin induces an augmentation in soluble ubiquitinated *Drosophila* TDP-43 to drive neurotoxicity in the fly. *Biochim Biophys Acta - Mol Basis Dis* 2018;**1864**(9, Part B):3038–49.
108. Donde A, Sun M. *Upregulation of ATG7 attenuates motor neuron dysfunction associated with depletion of TARDBP/TDP-43*. 2019. p. 1–11.
109. Gao J, Wang L, Yan T, Perry G, Wang X. TDP-43 proteinopathy and mitochondrial abnormalities in neurodegeneration. *Mol Cell Neurosci* 2019;**100**:103396.
110. Wang P, Deng J, Dong J, Liu J, Bigio EH, Mesulam M, et al. TDP-43 induces mitochondrial damage and activates the mitochondrial unfolded protein response. *PLoS Genet* 2019;**15**(5):e1007947.
111. Zhan L, Xie Q, Tibbetts RS. Opposing roles of p38 and JNK in a *Drosophila* model of TDP-43 proteinopathy reveal oxidative stress and innate immunity as pathogenic components of neurodegeneration. *Hum Mol Genet* 2015;**24**(3):757–72.
112. Berson A, Sartoris A, Nativio R, Van Deerlin V, Toledo JB, Porta S, et al. TDP-43 promotes neurodegeneration by impairing chromatin remodeling. *Curr Biol* 2017;**27**(23):3579–90. e6.
113. Berson A, Nativio R, Berger SL, Bonini NM. Epigenetic regulation in neurodegenerative diseases. *Trends Neurosci* 2018;**41**(9):587–98.
114. Chang Y-H, Dubnau J. The gypsy endogenous retrovirus drives non-cell-autonomous propagation in a *Drosophila* TDP-43 model of neurodegeneration. *Curr Biol* 2019;**29**:3135–52. https://doi.org/10.1016/j.cub.2019.07.071.
115. Gregory JM, Barros TP, Meehan S, Dobson CM, Luheshi LM. The aggregation and neurotoxicity of TDP-43 and its ALS-associated 25 kDa fragment are differentially affected by molecular chaperones in *Drosophila*. *PLoS One* 2012;**7**(2):e31899.
116. Manzo E, O'Conner AG, Barrows JM, Shreiner DD, Birchak GJ, Zarnescu DC. Medium-chain fatty acids, beta-hydroxybutyric acid and genetic modulation of the carnitine shuttle are protective in a *Drosophila* model of ALS based on TDP-43. *Front Mol Neurosci* 2018;**11**:182.
117. Su TT. Drug screening in *Drosophila*; why, when, and when not? *Wiley Interdiscip Rev Dev Biol* 2019;**8**(6):e346.
118. Fang MY, Markmiller S, Vu AQ, Javaherian A, Dowdle WE, Jolivet P, et al. Small-molecule modulation of TDP-43 recruitment to stress granules prevents persistent TDP-43 accumulation in ALS/FTD. *Neuron* 2019;**103**(5):802–19. e11.
119. François-Moutal L, Felemban R, Scott DD, Sayegh MR, Miranda VG, Perez-Miller S, et al. Small molecule targeting TDP-43's RNA recognition motifs reduces locomotor defects in a *Drosophila* model of amyotrophic lateral sclerosis (ALS). *ACS Chem Biol* 2019;**14**(9):2006–13.
120. Salado IG, Redondo M, Bello ML, Perez C, Liachko NF, Kraemer BC, et al. Protein kinase CK-1 inhibitors as new potential drugs for amyotrophic lateral sclerosis. *J Med Chem* 2014;**57**(6):2755–72.
121. Sanna S, Esposito S, Masala A, Sini P, Nieddu G, Galioto M, et al. HDAC1 inhibition ameliorates TDP-43-induced cell death *in vitro* and *in vivo*. *Cell Death Dis* 2020;**11**(5):369.

122. Vos MJ, Carra S, Kanon B, Bosveld F, Klauke K, Sibon OC, et al. Specific protein homeostatic functions of small heat-shock proteins increase lifespan. *Aging Cell* 2016;**15**(2):217–26.
123. Crippa V, Cicardi ME, Ramesh N, Seguin SJ, Ganassi M, Bigi I, et al. The chaperone HSPB8 reduces the accumulation of truncated TDP-43 species in cells and protects against TDP-43-mediated toxicity. *Hum Mol Genet* 2016;**25**(18):3908–24.
124. Kankel MW, Sen A, Lu L, Theodorou M, Dimlich DN, McCampbell A, et al. Amyotrophic lateral sclerosis modifiers in *Drosophila* reveal the phospholipase D pathway as a potential therapeutic target. *Genetics* 2020;**215**(3):747–66.
125. Kabashi E, Lin L, Tradewell ML, Dion PA, Bercier V, Bourgouin P, et al. Gain and loss of function of ALS-related mutations of TARDBP (TDP-43) cause motor deficits *in vivo*. *Hum Mol Genet* 2010;**19**(4):671–83.
126. Schmid B, Hruscha A, Hogl S, Banzhaf-Strathmann J, Strecker K, van der Zee J, et al. Loss of ALS-associated TDP-43 in zebrafish causes muscle degeneration, vascular dysfunction, and reduced motor neuron axon outgrowth. *Proc Natl Acad Sci USA* 2013;**110**(13):4986–91.
127. Vanden Broeck L, Callaerts P, Dermaut B. TDP-43-mediated neurodegeneration: towards a loss-of-function hypothesis? *Trends Mol Med* 2014;**20**(2):66–71.
128. Lissouba A, Liao M, Kabashi E, Drapeau P. Transcriptomic analysis of zebrafish TDP-43 transgenic lines. *Front Mol Neurosci* 2018;**11**(463).
129. Petel Légaré V, Harji ZA, Rampal CJ, Allard-Chamard X, Rodríguez EC, Armstrong GAB. Augmentation of spinal cord glutamatergic synaptic currents in zebrafish primary motoneurons expressing mutant human TARDBP (TDP-43). *Sci Rep* 2019;**9**(1):9122.
130. Laird AS, Van Hoecke A, De Muynck L, Timmers M, Van den Bosch L, Van Damme P, et al. Progranulin is neurotrophic *in vivo* and protects against a mutant TDP-43 induced axonopathy. *PLoS One* 2010;**5**(10):e13368.
131. Bose P, Armstrong GAB, Drapeau P. Neuromuscular junction abnormalities in a zebrafish loss-of-function model of TDP-43. *J Neurophysiol* 2019;**121**(1):285–97.
132. Asakawa K, Handa H, Kawakami K. Optogenetic modulation of TDP-43 oligomerization accelerates ALS-related pathologies in the spinal motor neurons. *Nat Commun* 2020;**11**(1):1004.
133. Kim J-H, Rahman MH, Park D, Jo M, Kim H-J, Suk K. Identification of genetic modifiers of TDP-43: inflammatory activation of astrocytes for neuroinflammation. *Cells* 2021;**10**:676.
134. Peggion C, Massimino ML, Stella R, Bortolotto R, Agostini J, Maldi A, et al. Nucleolin rescues TDP-43 toxicity in yeast and human cell models. *Front. Cell. Neurosci.* 2021;**15**:1. https://doi.org/10.3389/fncel.2021.625665. 625665.
135. Bharathi V, Girdhar A, Patel BK. Role of CNC1 gene in TDP-43 aggregation-induced oxidative stress-mediated cell death in *S. cerevisiae* model of ALS. *Biochim. Biophys. Acta, Mol. Cell Res.* 2021;**1868**:1–16. 118993.
136. Lin LT, Razzaq A, Di Gregorio SE, Hong S, Charles B, Lopes MH, et al. Hsp90 and its co-chaperone Sti1 control TDP-43 misfolding and toxicity. *FASEB J* 2021;**35**:e21594.
137. Matsukawa K, Kukharsky MS, Park S-K, Park S, Watanabe N, Iwatsubo T, et al. Long non-coding RNA NEAT1_1 ameliorates TDP-43 toxicity in in vivo models of TDP-43 proteinopathy. *RNA BIOLOGY* 2021:1–9. https://doi.org/10.1080/15476286.2020.1860580.

138. Campanari M-L, Marian A, Ciura S, Kabashi E. TDP-43 regulation of AChE expression can mediate ALS-like phenotype in zebrafish. *Cells* 2021;**10**:221. https://doi.org/10.3390/cells10020221.

139. Saxton AD, Kraemer BC. Human Ubiquilin 2 and TDP-43 copathology drives neurodegeneration in transgenic *Caenorhabditis elegans. G3* 2021;**11**(8):jkab158. https://doi.org/10.1093/g3journal/jkab158.

140. Rojas-Prat E, Martinez-Gonzalez L, Gonzalo-Consuegra C, Liachko NF, Perez C, Ramírez D, et al. Targeting nuclear protein TDP-43 by cell division cycle kinase 7 inhibitors: a new therapeutic approach for amyotrophic lateral sclerosis. *Eur. J. Med. Chem.* 2021;**210**:112968. https://doi.org/10.1016/j.ejmech.2020.112968.

141. Prasad A, Raju G, Sivalingam S, Girdhar A, Verma M, Vats A, et al. An acridine derivative, [4,5-bis{(N-carboxy methyl imidazolium)methyl}acridine] dibromide, shows anti-TDP-43 aggregation effect in ALS disease models. *Scientific Reports* 2016;**6**:39490. https://doi.org/10.1038/srep39490.

Chapter 9

New opportunities for treatment of neurodegenerative disease through the modulation of TDP-43

Pasha Apontes
Department of Psychiatry, Icahn School of Medicine at Mount Sinai, New York, NY, United States

Ah, tell me not that memory
Sheds gladness o'er the past;
What is recalled by faded flowers,
Save that they did not last

Letitia Elizabeth Landon.

Introduction

Overview of neurodegenerative disorders and rationale for targeting of TDP-43

Essential to all aspects of life is the ability to produce and retain memory. The deterioration of memory, awareness, learning, and cognitive function epitomizes most neurodegenerative disorders. Through the years, many neuronal disease—associated proteins have been discovered and found to undergo conformational changes and to generate insoluble aggregates within the nervous system. There is some disagreement as to whether misfolded proteins and aggregates associated with neurological disorders contribute to, help to counteract, or are neutral bystanders with respect to the diseases with which they are associated, but these may still providing potentially useful markers for the diagnosis and treatment monitoring of patients with neurodegenerative afflictions.[1,2] TDP-43 has risen to the forefront worldwide as an important protein implicated in many neurological disorders. It is a DNA/RNA-binding protein expressed during early stages of embryonic development and is vital to

TDP-43 and Neurodegeneration. https://doi.org/10.1016/B978-0-12-820066-7.00008-4

the central nervous system, as well as to other tissues in the body. Neumann et al. identified TDP-43 as a misfolded, ubiquitinated protein found in inclusions of patients with frontotemporal dementia (FTD) and amyotrophic lateral sclerosis (ALS).[3] Early studies revealed it to be a factor that inhibits transcription of human immunodeficiency syndrome type 1 at the long terminal repeat sequence called TAR, with the help of the cofactor RNA Helicase DDX5 (p68), which is part of the Drosha complex involved in microRNA synthesis.[4] TDP-43 was later cloned by Ou et al.[5,6] Eventually it was implicated in many other neuropathologies, e.g., in more than 90% of patients with sporadic ALS (sALS), ~75% of patients with severe Alzheimer's disease (AD), and a large number (~45%) of patients with FTD have TDP-43-positive inclusions and other abnormalities, such as having TDP-43 aggregates which may spread from one cell to another in a prion-like manner.[7,8] In neuropathology, TDP-43's normal cellular functions can be abrogated and new functions gained in association with changes to its tertiary structure, to its posttranslational modifications, and to its resultant oligomers and aggregate formation.[3,9] But there is still much to be learned, and the causal role of aggregates has been called into question by Hergesheimer et al.[1]

Along with TDP-43, there are several well-known proteins that likewise undergo conformational changes, form self-aggregates or inclusions in various cell compartment, which has become implicated in neurodegenerative disorders, although, as mentioned, there is debate as to the causal roles played.[2] That is, it is somewhat uncertain as to whether self-aggregates and inclusion bodies are compensatory mechanisms resulting from cellular stresses to actually avoid death of neurons, astrocytes, or microglial cells, as opposed to reflecting causal mechanisms that necessarily contribute to neurodegeneration.[2] Other well-known proteinopathies include Rho Guanine Nucleotide Exchange Factor (RGNEF) in Parkinson's disease (PD), α-synuclein in Lewy bodies dementia (LBD), and multiple system atrophy (MSA) and β-amyloid and hyper-phosphorylated tau, which form plaques and neurofibrillary tangles, respectively, in AD.

A mutational study using yeast to investigate the relationship between TDP-43 aggregates, condensates and neurotoxicity, analyzed more than 50,000 mutations in the prion-like domain of TDP-43 to unravel the functional consequences of mutations in this domain.[10] Bolognesi et al., found that there are mutant variants that cause hydrophobic aggregates to form which exert a protective role arising during molecular stresses, while there are other mutant variants that cause liquid, hydrophilic condensates to form, which appear to exert a cytotoxic role.[10]

In another study in an ALS transgenic murine model, in which mice expressed endogenous levels of either human wildtype or one of two ALS-linked TDP-43 mutations, those expressing mutations manifested aberrant RNA splicing, along with degeneration of motor neuron axons and motor neuron death. Surprisingly, however, was that, despite the motor neuron pathology, TDP-43 was enriched in the nucleus and there were no insoluble cytoplasmic aggregates observed.[11] These findings point to the importance of functional splicing ability of nuclear TDP-43, rather than merely correct nuclear localization, and suggests that the commonly observed aggregate TDP-43 is not either required for or not an early causal event in ALS.

Although the causative mechanisms underlying TDP-43's precise role in contributing to neurodegeneration require further study, the focus of this chapter will be on updating the reader on preclinical and clinical trials geared at combating motoneuron and cognitive disorders and discussing what is known about the way in which these various interventions modulate TDP-43.

The physiological functions of TDP-43

The *TARDBP* gene encodes TDP-43, a ubiquitous and highly conserved nuclear protein that binds to both DNA and RNA.[12] TDP-43 is a critical protein with pleiotropic DNA- and RNA-processing activities.[13] It is transcribed early during embryonic development, and is ubiquitously expressed in a variety of tissues. It is important in mRNA splicing, transcription, regulation of noncoding RNA, miRNA biogenesis, RNA stability, nucleocytoplasmic RNA trafficking, and autoregulation of its own protein synthesis through feedback mechanisms.[14,15] In addition, TDP-43 has been shown to be involved in the maintenance of genomic integrity and is a component of the nonhomologous end-joining DNA damage—response pathway, which is activated in response to DNA double-stranded breaks.[12,16] In addition to its role in a wide range of DNA and RNA-regulatory processes described above, TDP-43 also has important functions in neurite outgrowth. Neurite projections, axons and dendrites, emanate from neuron cell bodies, and TDP-43 is involved in the transport of ribosomal RNA from nuclei to these distant projections.[15]

TDP-43 structure and motifs associated with TDP proteopathies

TDP-43 is comprised of 414 amino acids, encoded on chromosome 1, and consists of the following main structural motifs and domains: two RNA recognition motifs (RRM1 and RRM2); five mitochondrial localization motifs comprised of hydrophobic residues; an N-terminal domain (NTD); a nuclear localization signal (NLS); a nuclear export signal (NES); and a C-terminal domain (CTD, also called a low complexity domain), which contains an intrinsically disordered region (IDR), a prion-like domain and a glycine-rich region, which is important for TDP-43 interactions with other heterogeneous nuclear ribonucleoprotein[9,13] and is also involved in phase separation and stress granule formation.[17]

The dynamic structure of TDP-43 NTD has been determined to be solenoid in shape and there is a physical separation between the C- and N-terminal domains. The glycine-rich and prion-like regions are also implicated in

TDP-43 aggregate proteinopathy in neurons and glia,[18] its 3D conformation, and, interestingly, nearly half of all ALS-associated TDP-43 mutations occur in the CTD.[19,20] The CTD is important for the normal physiological function of the protein and has roles in aggregate accumulation.[12,21]

Both the NTD and NLS are important regions for the nuclear localization of the protein as deletion of these regions results in its mislocalization, whereby the protein translocates from the nucleus, where its ordinarily enriched, to various cytoplasmic regions where inclusions may form, although nucleus include may occur as well (for example, in subtypes of AD of FTD).[22] Importantly, using algorithms to predict the ability of peptide sequences to self-aggregate, domains of TDP-43 were analyzed and the highest proclivity to self-aggregate and misfold was found to be the RRM2 and NTD peptide sequences, which should prove very important in efforts to screen for peptide inhibitors based upon the interaction with these domains.[23]

Residues within the CTD (residues 320–343) produce a partial α-helix which is important for TDP-43's formation of liquid–liquid phase-separated compartments, as well as its capacity to splice RNA.[24] Phase separation occurs when a homogenous solution of molecules, such as TDP-43, forms a membraneless phase-separated compartment, providing a local environment in which to facilitate the processing of nascent RNA.[24] Some patients with ALS have a G335D mutation which prevents TDP-43 from self-dimerizing. Additional residues in the NTD (Leu71 and Val72) are involved in the reversible TDP-43 self-assembly. Self-dimerization of TDP-43 appears to be functionally important as it has been reported to prevent aggregates formation as well as being necessary for carrying out its normal physiological functions in alternative splicing.[25] It has been shown that engineering an alanine at the G335 position (G335A) increases NTD self-dimerization and RNA splicing.[24] The scientific elucidation of these key structural components of TDP-43 may be important for opening new therapeutic windows for neurodegenerative diseases.

TDP-43 proteinopathy

The role of TDP-43 in neurodegenerative disease

TDP-43 is considered as a crucial player in neurodegenerative disease. The most well-known TDP-43-linked disorder is ALS. ALS is a rapidly progressive and fatal group of motor neuron diseases that leads to upper and lower motor neuron (UMN and LMN, respectively) degeneration and cell death, resulting in loss of the brain's communication with muscles. Ultimately, patients are unable to move voluntarily and, eventually, this leads to the loss of the ability to breath, which accounts for ALS's high fatality rate. ALS occurs most often in late middle life. Classic ALS leads to the progressive death of motor neurons in the cerebral motor cortex and spinal cord, and the gradual loss of respiratory ability, muscle paralysis, and atrophy. Treatment consists of supportive care for respiratory difficulties, symptom relief, and nutritional needs. There are currently only two FDA-approved medications for ALS, riluzole, an antiglutamatergic compound and edaravone, an antioxidant, both of which only provide modest benefit in only some patients.[26] ALS often culminates in death within 2–5 years of clinical presentation, although there do exist variants that have a slower rate of progression and better prognosis.

Approximately 5%–10% of patients with ALS have a familial history (fALS), while the majority of cases (90%–95%) emerge sporadically in those without a family history, and these sALS forms arise from indeterminate genetic and environmental causes. Uncovering the genetic variants associated with sALS may be hampered by the methodological approaches employed. For instance, most genetic studies have relied upon short-read and whole exome sequencing platforms, approached more conducive to detecting diseases attributed to one or more single nucleotide polymorphisms. This misses larger structural variants such as deletions, inversions, or tandem repeats, which appear likely to play a fundamental role in ALS pathology.

In ALS, TDP-43 forms aberrant aggregates in both neurons and glia.[27] Within the CTD is a glycine-rich region that is involved in protein–protein interactions and the binding of TDP-43 to different heterogeneous nuclear ribonucleoproteins. It is in this region that the majority of ALS mutations occur. However, TDP-43 NTD was found to form a homodimer in solution and as a homodimer the formation of cytoplasmic aggregates is prevented, and, instead, the mRNA splicing functions of TDP-43 in the cell nucleus is enhanced, showing the importance of various domains of the proteins.[13]

During cell stress, TDP-43 relocates to cytoplasmic stress granules. Cell stress granules form in the cytoplasm and are composed of mRNA- and RNA-binding proteins following a stress to the cell, such as in the presence of toxins or high temperature. In normal physiology, cell stress granules are dynamic structures that can be assembled and disassembled when the cell stress trigger subsides. In ALS, stress granules are persistent in motor neurons and this culminates in the degeneration of the motor neurons.

Although TDP-43 is normally present in the nucleus, it can become sequestered during neurodegeneration in the cytoplasm. It is also susceptible to cleavage into C-terminal fragments (CTFs) composed of a truncated RRM2 and the glycine-rich region, such as of ~25 and ~35 kilo Daltons (kDa) fragments (TDP-43-CTF25 and TDP-43-CTF35), which enters the cell cytoplasm and is prone to aggregate formation.[28,29] However, it has been argued that, while TDP-43 CTFs are hallmarks of ALS and FTD, these fragments do not appear to be primary causes of these diseases.[30] It is reported that TDP-43 CTFs do not reproduce TDP pathology consistently in cell lines, rarely in animal models and the observation that, while spinal cord of ALS patients largely lack CTFs, these regions are nevertheless main regions of neurodegeneration.[30,31] Instead it is reported that CTFs (mainly CTF25) are present in the frontal and temporal lobes and, also, that TDP-43 inclusions in LMN seem to be predominantly full length.[31] Therefore, TDP-43 CTFs may be present without cell death or absent with neurodegeneration and seemingly not a prerequisite for disease.[30]

In addition to ALS, TDP-43 is also implicated in other neuropathologies including progressive muscular atrophy (PMA), which affects primarily LMN and, in some patients, also UMN; primary lateral sclerosis (PLS), a slowly progressive motor neuron disease that affects upper neurons, and is more readily managed than ALS and also may not decrease overall lifespan; FTD, the second most common form of early-onset dementia, following AD, alters personality, behavior, language, and mental functions, and, with the first onset of symptoms, life expectancy is 6–8 years. TDP-43 aberrations are also known to occur in Perry syndrome, PD, LD, AD, Huntington's disease (HD), chronic traumatic encephalopathy (CTE), and limbic predominant age-related TDP-43 encephalopathy (LATE). Based on it being a cornerstone of neurodegenerative disorders, TDP-43 has become identified as an attractive target in the clinical setting.

Mutations and epigenetic variations of TDP-43 and other ALS-linked genes in familial and sporadic neurodegenerative diseases

There are at least 70 known mutations in the TDP-43 gene which have been identified in ALS and FTD, with most occurring in the largest exon (exon 6) of the TARDBP gene, 29 of which are missense mutation within the glycine-rich domain at the C-terminal.[19] Certain mutations in this gene are responsible for the development of several other neurodegenerative disorders. These include primary lateral sclerosis, progressive muscular atrophy, and FTD. Yet, TDP-43 mutations only account for an estimated 5%–10% of fALS,[32] while the remainders of fALS mutations occur in other ALS-linked genes.[1] The most well-characterized of ALS-linked genes are the superoxide dismutase (SOD1), chromosome 9 open reading frame 72 (C9ORF72), and fused in sarcoma (FUS) genes. For fALS-associated genes, all are inherited in an autosomal dominant fashion.

TDP-43 gain-of-function: mitochondrial dysfunction, persistent stress granules, and aggregate formation

In ALS there is a perturbation in TDP-43 trafficking between the nucleus and cytoplasm.[7] TDP-43 is predominantly localized to the nucleus. In disease states, it becomes sequestered in the cytoplasmic compartment, shifting the main locality to the cytoplasm. This leads to a commensurate loss of endogenous nuclear TDP-43. With an excessive amount of the protein in cytoplasm, insoluble, cytoplasmic aggregates develop.[3,19]

There are several reasons that account for why TDP-43 aggregates form. One explanation has to do with the NTD or NES of TDP-43. The NTD of two TDP-43 molecule homodimerize with each other, resulting in a conformational change that exposes the RRM2 which is necessary for TDP-43 to function in regulating RNA splicing and prevention of aggregates.[28,32] When experimental mutations are introduced into the NLS or the NES of TDP-43, cytoplasmic or nuclear aggregates form, respectively.[33] Besides the NLS, the RNA-binding region of TDP-43, specifically RRM1, is important as TDP-43 binding to GU-rich RNA and, when associated with RNA, oligomerization and aggregation of the protein is prevented.[34] It has also been shown that exogenous cytoplasmic TDP-43 accumulation is toxic to primary rat cortical neurons, which suggests that an overabundance of the protein in the cytoplasm may be a contributory factor in neuropathology.[1]

TDP-43 has also been reported to be sequestered in the cytoplasm by coaggregation with other proteins, such as poly(GR) and poly(PR).[15] Normally, TDP-43 binds to the 3′ UTR of its own pre-mRNA, which leads to its undergoing nonsense-mediated mRNA decay,[35] thereby decreasing the nuclear to cytoplasmic shuttling of the transcript as well as its corresponding translation into a functional protein.[36] Any perturbation of the endogenous autoregulation of TDP-43 compromises its normal function and disrupts its homeostasis. Loss of homeostatic nucleo-cytoplasmic shuttling triggers either nuclear or cytoplasmic TDP-43 aggregates, a salient feature in the pathology of all variants of ALS.[37] Interestingly, in different subtypes of FTD-TDP with differing severities of neurodegeneration, and in a study between subtypes A and C (A, high neurotoxicity and C, low neurotoxicity), it was shown that each subtype has distinct conformations of TDP-43 aggregates and distinct patterns of ubiquitination of hyperphosphorylated TDP-43, a “shapeshifting” attribute of TDP-43 which is reminiscent of prion diseases.[38,39] Altogether, these observations imply that clinical heterogeneity of TDP-43-related neuronal diseases is related to the specific morphological features and modifications of the TDP-43 protein.

In TDP-43-containing neuronal inclusions in the brain and spinal cord of patients with ALS and FTD, there are often aberrant posttranslational modifications. Specific TDP-43 modifications lead to certain alterations in the protein, and can cause it to be targeted for removal by autophagosomes.[32] Some examples of posttranscriptional modifications to TDP-43 include ubiquitination, which alters its solubility and aggregation; hyperphosphorylated which may lead to enrichment within the cytoplasm and mitochondria and aggregation; acetylation is associated with altered mitochondria function and RNA-binding impairment; PARylation may induce an increase in phase separation and persistent stress granules; cysteine oxidation within the NTD, RRM1, and RRM2 can induce oligomerization, seeding, and RRM1 aberrant self-dimers.[32] It also appears that some modifications of TDP-43 lead to others, such as acetylation leading to hyperphosphorylation and ubiquitination, and the occurrence of mitochondrial and inflammatory pathological changes.[40]

TDP-43 loss-of-function: loss of normal DNA and RNA processing functions

As described, TDP-43 is an important protein that is normally enriched in the nucleus, where it binds to both DNA and RNA. Some of the normal functions of the protein include the regulation of transcription, mRNA splicing, RNA stability, transport, and translation. The attenuation of these critical cell processes is implicated in a variety of neurological disorders, and it has been argued[41] that the loss of normally functioning TDP-43 is the primary cause of neuropathology, while the toxic formation of aggregates is a secondary pathological feature, arising from protein conformational changes. In addition, abnormal tertiary structural changes create oligomers of repeating units that result in higher-order structures and these accumulate and cluster in specific brain regions and in motor neurons.

A study by Polymenidou et al. revealed important TDP-43 direct target genes which are downregulated as a result of the loss of TDP-43. The authors reported that depletion of TDP-43, employing antisense oligonucleotides injected into mice striatum, triggered the downregulation most dramatically of genes with long introns, many of which were found to be enriched in synaptic activity.[42] For example, these are some of the transcripts which are downregulated in response to TDP-43 depletion: NMDA receptor, Grin2a, glutamate receptor 6, calcium-activated potassium channel alpha, Cacna1c, neuroligin, E3 ubiquitin ligase parkin, SET and MYND domain-containing protein 3, neurexin 1/3, Na^+/K^+, ataxin 1, and the imprinted ncRNA associated with the disorder Prader-Willi.[43,44]

These findings suggests that the cytoplasmic clustering of TDP-43 can lead to decreases in long pre-mRNA transcripts, particularly of synapse-related genes, which sheds light on how the diminishment of normal nuclear levels of TDP-43 can contribute to the development of ALS, or neuropathology in general.[43] The importance of TDP-43 loss-of-function is quite plausible especially given that it preferentially regulates many long genes and many of the longest genes are actually expressed in the nervous system across diverse species.[45] In addition to neurons, TDP-43 aggregates also form in glia in ALS and Wang and colleagues demonstrated, through oligodendrocyte knockdown of TDP-43 in mice, that functional TDP-43 is required for survival of mature oligodendrocytes, the production of myelination, and that the loss results in progressive motor neuron deterioration and a short lifespan.[46] This is further evidence highlighting the importance of the loss of TDP-43 in neurodegeneration.

TDP-43 as a therapeutic target

Modulation of TDP-43 appears to hold promise in making appreciable inroads in the treatment of neurodegenerative disorders, even in the overwhelming instances wherein a patient does not possess a mutation in the TARDBP gene. For example, the vast majority (~95%) of ALS patients (sporadic and familial) and a significant percentage of FTD patients (~45%) exhibit TDP-43 neuronal inclusions in their cortical and spinal cord neurons although only a mere 5%–10% possess mutations in the gene.[7] Since TDP-43 was first determined to be a major constituent of ubiquitin-positive inclusions in ALS patients,[47] many additional studies have led to its recognition as a hallmark of ALS and other neurodegenerative disorders.[3,17] Advances in our understanding of TDP-43's loss-of-function and gain-of-function changes in neuropathology have inspired new targeting strategies to treat these conditions.

For many neurocognitive disorders treatment options are sparse and, as monotherapies, patients mainly achieve a slowing of disease progression or alleviation of symptoms. These are some of the FDA-approved drugs for the various neuropathologies: 1 for HD (Xenazine/tetrabenazine); 2 for ALS (Rilutek/riluzole and Radicava/edaravone); 5 for AD (donepezil, galantamine, rivastigmine, memantine, a combination of donepezil and memantine and one that may soon be approved, aducanumab); and 13 for PD (Carbidopa-levodopa/Sinemet and long acting version, Rytary, Dopamine agonists: pramipexole, ropinirole, rotigotine patch, Apomorphine MAO-B inhibitors: selegiline, rasagiline, COMT inhibitors: entacapone, tolcapone, and Amantadine Trihexyphenidyl/Artane).

Standard-of-care medications in the treatment of ALS

As mentioned previously, there exist presently only two medications to treat patients with ALS, and the benefit they afford is at best modest.[26] The two medications are the antioxidant edaravone and the antiglutamatergic riluzole.

Edaravone (MCI-186, Mitsubishi Tanabe Pharma Corporation), antioxidant

Edaravone is a free radical scavenger that has been approved for the treatment of acute cerebral infarction in 2001.[48] Yoshino et al. conducted a 6 month open-label study of edaravone at two doses (30 mg, n = 5 and 60 mg n = 15) by IV drip once daily.[49] This study concluded that edaravone has potential in slowing ALS progression based upon the primary endpoint of change in ALSFRS-R score and the secondary endpoint of 3-nitrotyrosine, a marker of oxidative stress, in the cerebrospinal fluid.

In the first edaravone placebo-controlled, double-blind study, designed to test efficacy and safety over a 36-week period (with a 12 week pretreatment observational period to include only patients with an ALSFRS-R score that changed by −1 to −4), edaravone was not demonstrated to be slow ALS progression.[50] However, in a subsequent 24-week intervention trial, Abe et al. reported that there was a decrease in the progressive decline in ALS in the participants receiving edaravone, based on the ALSFRS-R (edaravone −5.70 ± 0.85, n = 102 and placebo, −6.35 ± 0.84, n = 104), which proved edaravone's superiority, without a difference in adverse effects.[51] The second phase-3 trial (*Efficacy and Safety Study of MCI-186 for Treatment of the Patients With Amyotrophic Lateral Sclerosis (ALS);* ClinicalTrials.gov *Identifier: 2NCT01492686*) was conducted based on the inclusion criteria that identified, through post-hoc analysis of the Abe study, a potential subpopulation of patients that may benefit from edaravone treatment.[48] The delay in the decline in ALSFRS-R score found in the post-hoc analysis was confirmed but the effects on long-term survival are still uncertain.[52,53]

Riluzole (Sanofi), glutamine antagonist, benzothiazole

Riluzole, the first FDA-approved medication for ALS, is an antiglutamatergic compound belonging to a class of medicines called benzothiazoles.[26,54] Lacomblez et al. (in 1996) reported that riluzole has neuroprotective activity and that it prolonged survival in patients in a dose-dependent manner, but with the best cost/benefit ratio afforded to those receiving 100 mg/day versus placebo (HR 0.55, 95% CI 0.36–0.83; log-rank $P = .037$). In a retrospective analysis, it was determined that the prolongation of survival in ALS with riluzole was derived with treatment in ALS stage 4, a late disease stage, just before death (stage 5) in which there is dysphagia-induced weight loss, and nutritional and respiratory failure.[54]

While riluzole's reduction in mortality in late-stage ALS is generally reported to be 2–3 months, in a real-world evidence (RWE) analyses, which includes patient registries or electronic health records, it was shown that survival benefits may be greater than 19 months overall in the different ALS populations, also accounting for patients who lived longer than the 18–21 months trial study follow-up durations (Table 9.1).[55]

Therapeutic modulation of TDP-43

Prognosis remains dismal for the many neurodegenerative diseases, especially the motor neuron disease ALS, a fatal neurodegenerative disorder that occurs most often in late middle life. Classic ALS leads to progressive death of motor neurons in the cerebral motor cortex and spinal cord, and the gradual loss of respiratory ability, muscle paralysis, and atrophy. Treatment consists of supportive care for respiratory difficulties, symptom relief, and nutritional support. ALS often culminates in death within 2–5 years of clinical presentation, although other variants exist which have a slower progression and better prognosis. As mentioned there are at this time, only two FDA-approved drugs indicated for the treatment of ALS: the antiglutamatergic drug riluzole, and an antioxidant, edaravone, which provides only modest benefit in some patients.[26]

TABLE 9.1 Summary of therapeutic Interventions for TDP-43-associated Neurodegenerative Diseases.

Trial name	Intervention	Study characteristics	Condition	Phase	Outcome measures	Results	References
Study of MCI-186 for Treatment of Amyotrophic Lateral Sclerosis: Edaravone	Edaravone (MCI-186)	20- to 75-year-old patients with ALS (n = 137) MCI-186/ edaravone or placebo in open label study	ALS	3	At 24 weeks, the study outcome measures were change from baseline in ALSFRS-R, % forced vital capacity, modified norris scale score, ALS assessment questionnaire, adverse events, and adverse drug reactions	Edaravone was demonstrated to slow functional decline	48
Riluzole: real-world evidence supports significant extension of median survival times in patients with amyotrophic lateral sclerosis	Riluzole	Real-world evidence (RWE) from analysis of 10 clinical ALS databases of patients treated with riluzole (n = ~6000)	ALS	RWE	The study outcome measure was survival	Survival benefits in ALS was 2–3 months, but in approximately 9% of patients, an additional up to 19 months median survival was achievable	55

Continued

TABLE 9.1 Summary of therapeutic Interventions for TDP-43-associated Neurodegenerative Diseases.—cont'd

Trial name	Intervention	Study characteristics	Condition	Phase	Outcome measures	Results	References
The Study of Tamoxifen Treatment in Patients With Motor Neuron Disease	Tamoxifen	ALS patients ($n = 18$; 10 in the tamoxifen group and 8 in the placebo group); 40 mg per day tamoxifen or placebo, with all participants receiving riluzole twice daily (100 mg/day total) for a period of 1, 3, or 12 months	ALS	2	Primary outcome measured was time to death or dependence on mechanical ventilation, while secondary outcomes were decline in ALSFRS-R score and change in forced vital capacity	At 1, 3, and 6 months, ALSFRS-R score, but not forced vital capacity, decreased less in the tamoxifen group than in the placebo	61
Induction of autophagy mitigates TDP-43 pathology and translational repression of neurofilament mRNAs in mouse models of ALS/FTD	IMS-088 (analogue of withaferin-A)	Transgenic mice expressing hTDP-43A315T or hTDP-43G348C were treated with IMS-088 at ~1 year (beginning of pathological stage) administered by gavage (30 mg/kg) twice a day for 8 weeks, while vehicle group received equal volumes of buffered saline for the same durations	ALS/FTD	Preclinical	In TDP-43 transgenic mice, study endpoints were correction of cognitive deficits (passive avoidance test and novel object recognition) and the quantification of cytoplasmic TDP-43 inclusions, autophagy, and neuronal translational profile	IMS-088 improved cognitive deficiency, decreased cytoplasmic TDP-43, and increased autophagy and neurofilament protein synthesis	63

A Randomized, Double Blind, Placebo-controlled Study to Evaluate the Impact of Bosutinib on Safety, Tolerability, Biomarkers and Clinical Outcomes in Dementia With Lewy Bodies (DLB)	Bosutinib	A 90-day trial of dementia with Lewy bodies patients treated with 100 mg daily bosutinib or placebo (n = 15)	DLB	2	The safety, tolerability and clinical efficacy will be monitored as the primary outcomes, while CSF and plasma levels of bosutinib and disease-related biomarkers will be determined as secondary measures	No interim results	ClinicalTrials.gov Identifier: NCT03888222
Phase 1/2a Study of Ropinirole Hydrochloride Extended-Release Tablets in Patients with Amyotrophic Lateral Sclerosis (ALS); ROPALS trial	Ropinirole hydrochloride (extended release)	Cohort is Japanese ALS patients (n = 20; n = 5 to receive the placebo and n = 15 assigned to receive the active drug). All patients receive standard ALS treatment of riluzole, along with either placebo or ropinirole hydrochloride, beginning with 2 mg and increasing to up to 16 mg, in an extended-release formulation (Requip CR) for 24 weeks	ALS	1/2a	The primary outcome measure is to assess drug safety and tolerability. The secondary outcome measures are ALSFRS-R, change in the combined assessment of function and survival (CAFS) and exploratory molecular effects on TDP-43 and neurofilament light chain (NfL) in blood and spinal fluid	No interim results	Japanese Clinical Trial Identifier: UMIN000034954 and JMA-IIA00397

Continued

TABLE 9.1 Summary of therapeutic Interventions for TDP-43-associated Neurodegenerative Diseases.—cont'd

Trial name	Intervention	Study characteristics	Condition	Phase	Outcome measures	Results	References
Efficacy, Safety and Tolerability Study of 1 mg Rasagiline in Patients With Amyotrophic Lateral Sclerosis (ALS) Receiving Standard Therapy (Riluzole)—An AMG Trial With a Market Authorized Substance	Rasagiline	ALS patients were followed for 2, 6, 12, and 18 months after randomization to active drug or placebo; n = 252 patients (rasagiline n = 127 or placebo n = 125); patients were treated with rasagiline at a dose of 1 mg/day for up to 18 months, as an add on to riluzole at 100 mg/day	ALS	2	The primary outcome measure was survival prolongation and the secondary outcome measure was ALSFRS	Rasagiline had good tolerability. The primary outcome of survival prolongation was not achieved, but for the secondary measure of ALSFRS, post-hoc analysis demonstrated a trend toward the slowing of disease progression	72
A Phase II Randomized, Placebo-Controlled, Double Blinded, Multi-Centre Clinical Trial of Pimozide in Patients With Amyotrophic Lateral Sclerosis; clinicaltrials.gov Identifier	Pimozide	ALS patients randomized to receive 2 mg/day (4 mg/day at study initiation) pimozide or placebo for 22 weeks (n = 100)	ALS	2	The primary outcome measure is ALSFRS-R, while the secondary outcome measures are Change in Slow Vital Capacity, decremental responses on repetitive nerve stimulation, changes in motor power, quality of life, frequency of cramps and adverse events	No interim results. In preclinical studies, synaptic transmission, paralysis, and neurodegeneration were corrected in the mutant TDP-43 roundworm, while aberrant swimming and motor neuron disease phenotypes were corrected in the mutant TDP-43 zebrafish	73 and Clinicaltrials.gov Identifier: NCT03272503

A Phase I Dose Finding Study of Intranasal Oxytocin in Frontotemporal Dementia, Protocol # FTDOXY10EF	Oxytocin (Syntocinon)	In patients diagnosed with probable behavioral variant frontotemporal dementia (bvFTD, n = 23) intranasal oxytocin or saline was administered two times a week (0, 24, 48, or 72 IU) for 1 week; telephone assessments were obtained after 1 week washout period	FTD (bvFTD)	1	At baseline and after 1 week of oxytocin treatment compared to placebo, the primary outcome measure was adverse events/safety and tolerability and the secondary outcome measures were behavioral ratings, frontal behavioral inventory, clinicians global impression of change, interpersonal reactivity index, and multi-faceted empathy test	All three doses of oxytocin were well tolerated and a significant improvement was observed, dose dependently, on the FBI Apathy subscale, with the greatest improvement obtained at 72 IU. Although the effect of oxytocin on TDP-43 is unknown, it is known that oxytocinergic nerves release oxytocin centrally to brain regions important in dementia with TDP-43 pathology	77–79 and Clinicaltrials.gov Identifier: NCT01386333
A Phase 2 Clinical Trial of Intranasal Oxytocin for Frontotemporal Dementia; FOXY	Oxytocin (Syntocinon)	Intranasal oxytocin (n = 112; n = 20 per one of three oxytocin dosage schedules and placebo controls). In stage one, participants will receive 72 IU intranasal oxytocin twice daily, on alternate days, or every third day. A washout period will follow (6 weeks) and crossover to different treatment schedule for 6 weeks. An additional	FTD (bvFTD, semantic subtype or Progressive Nonfluent Aphasia)	2	In stage one, the most encouraging dose of oxytocin will be determined using the change in neuropsychiatric inventory (NPI) apathy/indifference domain score (primary outcome). In stage two, new participants will receive the most efficacious dose. At up to 20 weeks, the	No interim results	ClinicalTrials.gov identifier: NCT03260920

Continued

TABLE 9.1 Summary of therapeutic Interventions for TDP-43-associated Neurodegenerative Diseases.—cont'd

Trial name	Intervention	Study characteristics	Condition	Phase	Outcome measures	Results	References
		40 participants will be treated with the most beneficial dose			primary and secondary outcome measures will be determined: changes in emotional facial expression recognition performance, revised self-monitoring scale score and modified clinicians global impression of change (apathy) scores		
Motor neuron preservation and decrease of in vivo TDP-43 phosphorylation by protein CK-1δ kinase inhibitor treatment	IGS-2.7	TDP-43 (A315T) transgenic mice ($n = 10$ per group), at 65 days old, were treated intraperitoneally with IGS-2.7 (1 mg/kg of body weight) or vehicle until sacrifice at the age of 90 days old. Human sALS lymphocytes were treated with 5 μM of IGS-2.7 for 24 h	ALS	Preclinical	The measured outcomes were CK-1-dependent phosphorylation of TDP-43, inflammation (by astrocyte and microglia immunoreactivity using GFAP and Iba-1 staining, respectively), modulation of TDP-43 cytosolic localization, and motor neuron survival	In mice, IGS-2.7 delayed weight loss, increased motor neurons survival in lumbar anterior horn, decreased inflammation and the level of TDP-43 phosphorylation in TDP-43 transgenic mice to that of wild-type animals. In the human sALS lymphocytes, treatment for 24 h with 5 μM of IGS-2.7 decreased phospho-TDP-43 as well as its cytosolic mislocalization	82

A Phase 2 Randomized, Placebo Controlled Trial of Tocilizumab in ALS Subjects	Tocilizumab	Patients (n = 14, treatment group, n = 8 placebo) received 8 mg/kg of IV tocilizumab every 4 weeks for 3 months via IV infusion	ALS	2	The primary outcome measures were all-cause mortality and safety and tolerability, while the secondary outcome measures were the mean change in ALSFRS-R total score, rate of decline in Slow Vital Capacity, and the change in peripheral blood mononuclear cells and cytokine levels	Currently the treatment and placebo group lost no patients to death, and decline in ALSFRS-R and the Rate of Decline in Slow Vital Capacity did not reach significance, while the fold change in two cytokine markers, CRP (decreased with treatment) and IL-6 (increased with treatment), did reach significance. The final results are not yet published	ClinicalTrials.gov identifier: NCT02469896
Rapamycin (Sirolimus) Treatment for Amyotrophic Lateral Sclerosis; The rapamycin for ALS trial (RAP-ALS)	Rapamycin (Sirolimus)	An 18-week study, with a total follow-up duration of 54 weeks in patients randomized to one of three groups of (n = 21 per group): (1) 1 mg/m2/d rapamycin + riluzole, (2) 2 mg/m2/d rapamycin + riluzole, or (3) placebo + riluzole, in addition to mainstay supportive care in accordance with established ALS guidelines	ALS	2	The primary endpoint is the number of T regulatory cells (Tregs) at baseline and 18 weeks. Secondary endpoints that will be quantified are safety and tolerability, minimum dose required for detection of rapamycin in the cerebrospinal fluid, modulation of inflammatory and mTOR markers, ALSFRS-R, and quality of life	No interim results. Through TDP-43 is not an indicated measured outcome in this study, yet rapamycin was shown (in N2A and SH-SY5Y cells) to induce autophagy along with decreasing TDP-43 aggregates, mislocalization. TDP-43's clearance by autophagy has also been demonstrated in murine and in human stem cell–derived neurons and astrocytes with mutant TDP-43	87–90 and ClinicalTrials.gov Identifier NCT03359538

Continued

TABLE 9.1 Summary of therapeutic Interventions for TDP-43-associated Neurodegenerative Diseases.—cont'd

Trial name	Intervention	Study characteristics	Condition	Phase	Outcome measures	Results	References
Colchicine for Amyotrophic Lateral Sclerosis: a Phase II, Randomized, Double Blind, Placebo Controlled, Multicenter Clinical Trial	Colchicine	54-week study included 54 patients were treated with Colchicine or placebo, along with Riluzole (n = 18 in the placebo + Riluzole 100 mg, and n = 18 in one of two different dosages of the colchicine: 0.01 mg/kg/day + Riluzole 100 mg or 0.05 mg/kg/day + Riluzole 100 mg)	ALS	2	The primary outcome measures is change in ALSFRS-R (at 8, 18, 30, and 54 weeks). The secondary outcome measures are: overall survival, adverse effects, Forced Vital Capacity, and in ALS patient fibroblasts and lymphoblasts: autophagy, stress granules properties, full length and truncated TDP-43, SQSTM1/p62, UBQLN, OPTN), in extracellular vesicles from plasma and CSF: TDP-43, hyperphosphorylated TDP-43, SQSTM1/p62, UBQLN, and OPTN and inflammatory biomarkers and in plasma various neurodegenerative biomarkers	No interim results	ClinicalTrials.gov Identifier: NCT03693781

Effect of Mexiletine on Cortical Hyperexcitability in Sporadic Amyotrophic Lateral Sclerosis (SALS)	Mexiletine	Patients (n = 20) treated with mexiletine (300 mg, 600 mg) or placebo by mouth each day for 4 weeks	sALS	2	The primary measured outcome is change in resting motor threshold. The secondary measured outcomes are effect on short-interval intracortical inhibition, motor evoked ootential amplitude, cortical silent period, strength duration time constant, depolarizing and hyperpolarizing threshold electrotonus, super/subexcitability, frequency of muscle cramps and fasciculations, and changes in ALSFRS-R and Slow Vital Capacity	No interim results. Mexiletine has been reported to prevent both motor neuron cell death and intracellular nitroxidative stress caused by exposure of wild-type primary spinal cord cells to conditioned medium obtained from astrocytes with different ALS motoneuron-linked mutations (including SOD1G93A, SOD1G86R, and or TDP43A315T)	91 and ClinicalTrials.gov Identifier: NCT02781454

Continued

TABLE 9.1 Summary of therapeutic Interventions for TDP-43-associated Neurodegenerative Diseases.—cont'd

Trial name	Intervention	Study characteristics	Condition	Phase	Outcome measures	Results	References
Trial of Oxaloacetate in ALS	Oxaloacetate (OAA)	The first 3 patients will receive 500 mg twice daily, and each subsequent cohort will receive an additional 500 mg per dose to a maximum of 2500 mg twice daily ($n = 24$). The study duration is 28 days	ALS	1	The primary outcome measure will be dose-limiting toxicities, and the secondary outcome measures will be OAA levels, serum platelet TDP-43 and indicators of mitochondrial dysfunction (Annexin V/Mitotracker) in lymphocytes and plasma, as well as MR spectroscopy of brain glutathione	No interim results	ClinicalTrials.gov Identifier: NCT04204889
Perampanel for Sporadic Amyotrophic Lateral Sclerosis (ALS): A Multicenter, Randomized, Double-blind, Placebo-controlled, Parallel-group Phase 2 Trials	Perampanel (FYCOMPA)	Patients ($n = 60$) were assigned to receive a placebo, 4 and 8 mg per day for 48 weeks perampanel	ALS	2	The primary outcome measure will be the change in ALS functional rating scale, and the secondary outcome measures will be manual muscle tests and percent-predicted Forced Vital Capacity. The outcome measures will be obtained at 12, 24, 36, and 48 weeks	No interim results. Loss of glutamate AMPA receptor increases Ca^{2+} permeability which is linked with TDP-43 pathology	98,99 and ClinicalTrials.gov Identifier: NCT03019419

A Randomized, Double-blind, Placebo-controlled Pilot Study to Evaluate the Biological Activity, Safety, and Pharmacokinetics of PU-AD in Subjects With Amyotrophic Lateral Sclerosis (ALS)	PU-AD	Patients (n = 30) will be randomized to receive either PU-AD, 30 mg q.d., or placebo. All patients will also receive the stable treatment they are already receiving at the beginning of this Open Label study. PU-AD or placebo will be given on an empty stomach (1 h prior to food or 2 h after)	ALS	2	The primary outcome measures (obtained at 12 and 24 weeks) are: ALSFRS-R, Slow Vital Capacity, Handgrip strength using hand-held dynamometry, 6 Minute Walk Test, Rasch-Built Overall Amyotrophic Lateral Sclerosis Disability Scale. The secondary outcome measures (obtained at 12 and 24 weeks) will be: in blood (NfL, pNfH and SOD1), in CSF (NfL, pNfH, tau, p-tau231 GFAP, TDP-43 and C9orf72) and in urine (p75ECD), all compared to baseline values	No interim results	ClinicalTrials.gov Identifier: NCT04505358
A Phase 2A Study to Evaluate Safety, Tolerability and Pharmacological Effects of PU AD in Subjects With Mild Alzheimer's Disease (AD) Dementia	PU-AD	Patients will be treated with active PU-AD (n = 75) or placebo SyrSpend (n = 75) on a 1:1 ratio q.d.	AD	2A	The outcome measures will be obtained at baseline and over a 6 month period. The primary outcome measures are FDG-PET and adverse events. The	No interim results	ClinicalTrials.gov Identifier: NCT04311515

Continued

TABLE 9.1 Summary of therapeutic Interventions for TDP-43-associated Neurodegenerative Diseases.—cont'd

Trial name	Intervention	Study characteristics	Condition	Phase	Outcome measures	Results	References
					secondary outcome measures are tau PET, ratio of p-Tau to t-tau in cerebrospinal fluid, Alzheimer's Disease composite score, Global Statistical Test and Pharmacokinetics of PU-AD		
GM604 Phase 2A Randomized Double-blind Placebo Controlled Pilot Trial in Amyotrophic Lateral Disease (ALS)	GM604	12 ALS patients diagnosed within 2 years of the trial, were treated with GM604 ($n = 8$) or placebo ($n = 4$). The study duration was 3 months and patients received bolus treatment 3 times per week for 2-week or a comparator placebo	ALS	2	The primary outcome measures (baseline, week 2, and week 12) were percent change in biomarkers in the CSF (TDP-43, tau and SOD1), safety and tolerability. The secondary outcome measures (baseline, week 2, week 6, week 12) were clinical progression as determined by ALSFRS-R, Forced Vital Capacity, time up and go, muscle strength, and mortality	No serious adverse reactions occurred, and when compared to historic controls to increase power, the GM604-treated patients showed significant improvements in ALSFRS-R ($P = .005$), respiratory function, and these ALS biomarkers: TDP-43, $P = .008$; tau, $P = .037$; SOD1, $P = .009$. A compassionately treated patient also experienced clinical improvement in speech, fluid intake, mouth suctioning and the ALS biomarkers	111,112 and ClinicalTrials.gov Identifier: NCT01854294

A Phase 2, Multi-Center, Double-Blind, Randomized, Dose-Ranging, Placebo-Controlled Study to Evaluate the Efficacy, Safety and Tolerability of CK-2127107 in Patients With Amyotrophic Lateral Sclerosis (ALS)	Reldesemtiv (CK-2127107)	Reldesemtiv oral tablets (150, 300, 450 mg) or placebo twice daily, every 12 h for 12 weeks (n = 458)	ALS	2	The primary outcome measure was percent predicted Slow Vital Capacity. The secondary outcome measures were ALSFRS-R and muscle strength mega-score. The outcome measures were reported as change from baseline and at week 12	The study endpoints were slow vital capacity, change in baseline of ALSFRS-R and slope of muscle strength, with *P*-values of 0.10, 0.01, and 0.20, respectively. Although statistical significance was not met, Cytokinetics reported that the drug was fast-tracked by the FDA for ALS because all three endpoints showed a trend in the direction of improvement and near statistical significance for ALSFRS-R, in addition to demonstrating good tolerability	116 and ClinicalTrials.gov Identifier: NCT03160898
A Single-Center, Open Label Study to Assess the Safety and Tolerability of Metformin in Subjects With C9orf72 Amyotrophic Lateral	Metformin hydrochloride sustained-release (SR)	Metformin hydrochloride sustained-release (SR) will be administered with a starting dose of 500 mg and increased incrementally by	C9orf72 ALS and FTD	2	The primary outcomes to be measured are adverse events and change in RAN protein levels in cerebrospinal fluid (CSF), at baseline and	No interim results	ClinicalTrials.gov Identifier: NCT04220021

Continued

TABLE 9.1 Summary of therapeutic Interventions for TDP-43-associated Neurodegenerative Diseases.—cont'd

Trial name	Intervention	Study characteristics	Condition	Phase	Outcome measures	Results	References
Sclerosis Over 24 Weeks of Treatment		500 mg every week to a maximal dosage of 2000 mg in divided doses twice daily for a duration of 24 weeks			through 24 weeks. The secondary outcome is the ALSFRS-R which will be determined at baseline and through 52 weeks,		
Evaluation of the Safety, Tolerability, Efficacy and Activity of AMX0035, a Fixed Combination of Phenylbutyrate (PB) and Tauroursodeoxycholic Acid (TUDCA), for the Treatment of ALS; CENTAUR trial	AMX0035	ALS patients diagnosed within the past 18 months ($n = 136$) received either placebo ($n = 48$) or AMX0035 ($n = 89$), in a 2:1 ratio of sodium phenylbutyrate (an aromatic fatty acid) to taurursodiol (a bile acid derivative) (3g of sodium phenylbutyrate to 1 g of taurursodiol), once daily for 3 weeks and then twice daily for the duration of the 6 month study	ALS	2/3	Primary outcomes were the ALSFRS-R slope, adverse events and discontinuation rate (at 24 weeks), while secondary outcomes were accurate Testing of Limb Isometric Strength, Slow Vital Capacity, Survival, tracheostomy and hospitalizations, and blood and imaging biomarkers (at 24 weeks)	At the completion of the 24 week trial, it was found that ALSFRS-R decline was significantly slowed in treated patients compared to placebo controls ($P = .03$), drug related side effects were primarily gastrointestinal, while secondary endpoints were not different between these groups	125 and ClinicalTrials.gov Identifier: NCT03127514

Open Label Extension Study of AMX0035 in Patients With ALS; CENTAUR-OLE	AMX0035	Participants in the CENTAUR trial will continue to receive AMX0035 twice daily (3g of sodium phenylbutyrate to 1 g of taurursodiol) and the study duration will be 30 months	ALS	2	The primary outcome is quantity of adverse events and secondary outcomes are number of hospitalizations, ALSFRS-R, ATLIS Strength Measurement, Slow Vital Capacity, number of gastric tubes placed and number of patients of permanent invasive ventilation	No interim results	ClinicalTrials.gov Identifier: NCT03488524
Phase II Study to Assess the Safety, Tolerability, and Target Engagement of AMX0035, a Fixed Combination of Sodium Phenylbutyrate and Tauroursodeoxycholic Acid for the Treatment of Alzheimer's Disease; PEGASUS	AMX0035	AMX0035 (n = 100), in a specific fixed ratio of sodium phenylbutyrate and tauroursodeoxycholic acid for a duration of 6 months	AD	2	The effect of AMX0035 on Adverse Events, and secondary outcomes are hippocampal and whole brain atrophy (as assessed by volumetric MRI), cognition (as measured by ADAS-Cog), Neuropsychiatric Inventory, Functional MRI Imaging and CSF and plasma biomarkers	No interim results	ClinicalTrials.gov Identifier: NCT03533257

Continued

TABLE 9.1 Summary of therapeutic Interventions for TDP-43-associated Neurodegenerative Diseases.—cont'd

Trial name	Intervention	Study characteristics	Condition	Phase	Outcome measures	Results	References
Safety and Tolerability Ultra-high-caloric Food Supplements in Amyotrophic Lateral Sclerosis (ALS) (TOLCAL); also identified as LIPCAL-ALS study	Ultra-high-caloric food supplement (Calogen)	Participants were randomized (n = 201) in an equal proportion to either a high caloric fatty beverage supplement (Calogen) (100% lipids, 405 kcal/90 mL/day, 45 g fat per day in addition to daily food intake and 100 mg riluzole) or placebo (drinking solution 8 kcal/ 90 mL/day in addition to daily food intake and 100 mg riluzole). The participants were followed for 3, 6, 9, 12, 15, and 18 months, and up to 28 months	ALS	Safety and tolerability	At 18 months the primary outcome measure was survival and the secondary outcomes were changes in ALSFRS-R, individual quality of life (SEIQoL), Slow Vital Capacity, time to tracheostomy or death, BMI, Appetite-Score, Council of Nutrition appetite questionnaire	In a subset of fast progressing ALS and motor neuron disease patients, post-hoc analysis indicated that the high calorie fatty diet group experienced a significant increase in survival, the primary outcome measure	139 and ClinicalTrials.gov Identifier: NCT02306590
Effects of a medium-chain triglyceride-based ketogenic formula on cognitive function in patients with mild-to-moderate Alzheimer's disease; Ketonformula®	Medium-chain triglyceride-based ketogenic formula (Ketonformula®)	Japanese patients with mild-to-moderate AD (n = 20) provided with 50 g of ketogenic formula daily, dissolved in hot water, for 12 weeks; the formula contained 20 g of MCTs and cognitive functions were evaluated at weeks 4, 8, and 12.	AD (mild-to-moderate)	Longitudinal, open-label, single arm	The tested measures, at baseline and at 4, 8, or12 weeks, were blood ketone levels (acetoacetate and β-hydroxybutyrate) and cognitive functions. The cognitive function tests consisted of logical memory (immediate and delayed), digital and spatial span (parts of the Wechsler	Acetoacetate and β-hydroxybutyrate did not meet significance compared to baseline (possibly due to time interval between blood collection and ketogenic formula intake). Statistically significant improvements in cognitive processing speed and verbal memory (immediate	133

					Memory Scale-Revised, WMS-R), block-design and digit-symbol coding tests (a part of the Wechsler Adult Intelligence Scale-3rd, WAIS-III), the Stroop test, and the Trail Making Test (TMT).	and delayed) with chronic intake of ketogenic formula, and 3 patients discontinued the trial due to gastrointestinal side effects	
	Standardized phytotherapic extracts rescue anomalous locomotion and electrophysiological responses of TDP-43 *Drosophila melanogaster* model of ALS	*Mucuna pruriens* (Mpe) and *Withania somnifera* (Wse) (commonly used in Ayurvedic medicine) were supplemented in food (0.1% w/w) in a partial TDP-43 protein depletion *Drosophila* model of ALS	TDP-43-loss-of-function in motoneurons in a *Drosophila* model of ALS	Preclinical	The study tested if Mpe or Wse could correct the TDP-43 mutant ALS *Drosophila* model aberrant phenotypes, including: Ambulation, climbing and sleeping behavior, and evoked postsynaptic potential (ePSP) of the dorsal longitudinal muscle (DLM) neuromuscular junction after electrical stimulation, or the PSP after exposure of wild-type and mutant flies to volatile anesthetics under basal conditions and of treated with Mpe or Wse	Both Mpe and Wse rescued, in part, the ALS disease phenotypes, improving the hyperactive anomalies in ambulation and climbing behavior, dysregulated sleep (in two age cohorts, 3- to 6- or 10- to 15-day-old male flies), and Mpe was able to correct the ePSP response of treated mutant flies, showing a significant decrease in amplitude and an increase in latency. In TDP-43 mutant flies treated with Mpe or Wse, the DLM contractions with anaesthetics was similar to that observed in GAL4 wild-type flies	140

Continued

TABLE 9.1 Summary of therapeutic Interventions for TDP-43-associated Neurodegenerative Diseases.—cont'd

Trial name	Intervention	Study characteristics	Condition	Phase	Outcome measures	Results	References
Metabolic enhancement for neurodegeneration (MEND)	MEND protocol (an individualized, combinatorial program); (Therapeutic System 1.0)	AD patient pilot study (n = 10; n = 9 ApoE4+) and expanded study (n = 100). Participants treated with individualized MEND protocol: low glycemic, low grain and antiinflammatory diet, 12 h nightly fast (3 h prior to bedtime), stress reduction, exercise (30–60 min/day), 8 h nightly sleep (treatment for sleep apnea, if present, 0.5 mg po qhs melatonin and, if still awakening, 500 mg po 3 times per week Tryptophan), evaluation and/or treating heavy metal toxicity and pathogens, brain stimulation, and the following: Ashwagandha, Bacopa monnieri, magnesium-L-threonate, Vitamins D3, K2, H. erinaceus or	Subjective cognitive impairment (SCI), mild cognitive impairment (MCI) and AD	Pilot and expanded case studies, single arm	The pilot MEND study assessed cognitive performance using the Montreal Cognitive Assessment or MemTrax tests and functional improvement (such as ability to resume work). The expanded MEND study, evaluated cognitive performance at baseline and after treatment using the Montreal Cognitive Assessment (MoCA), and/or Mental Status Questionnaire (MSQ), Saint Louis University Mental Status (SLUMS), Mini-Mental Status Examination (MMSE), Neurocognitive Index (NCI) and functional improvements	In the pilot study, there was reversal of some aspects of neurocognitive disease in AD and importantly, 6 of the 10 patients resumed work. In the expanded MEND study, 72 of the 100 had baseline and post-treatment MoCA, MMSE, or SLUMS evaluation. There was a mean improvement of 4.9 points (SD of 2.6) and a range of 1–12	152 and 153

		Acetyl-L-carnitine, Citicoline, DHA, prebiotics and probiotics, curcumin, and oxidative stress (Me–B12, MTHF, P5P); follow-up 2.5 years					
A Retrospective Study to Investigate the Additive Effectiveness of Chinese Herbal Medicine in Alzheimer's Disease	Chinese Herbal Medicine	Chinese ALS patients (n = 344) will receive standard of care only or standard of care plus Chinese herbal GRAPE granules (Rehmannia glutinosa, 30 g/day, and 10 g/day each of: Ginseng, Acorus tatarinowii, Polygala tenuifolia, Epimedium brevicornu, Cornus officinalis, Cistanche deserticola, Curcuma aromatica, Salvia miltiorrhiza, Angelica sinensis, Gastrodia elata, and berberine, dissolved in 150 mL warm water and taken orally twice daily	AD	Observational, retrospective study	The primary outcome was the MMSE and secondary outcomes, the Activities of Daily Living, Clinical Dementia Rating and Neuropsychiatric Inventory and mild (MMSE 21–26) and moderate (MMSE 10–20) AD patients were followed up to 24 months and assessed at approximately 3 month intervals	The herbal granules added to standard AD medications was shown to outperform standard medication alone, and the improvement was seen both in mild AD (2.60 point higher MMSE, $P = .012$) and moderate AD (2.12 points higher MMSE, $P = .025$). One of the components of the GRAPE formula, berberine, has been reported to inhibit mTOR/p70S6K pathway and activate autophagy, which reduced insoluble TDP-43 and inhibited TDP-43 aggregate formation in a cell culture model of FTD and ALS	154,155 and ClinicalTrials.gov Identifier: NCT03221894

New therapeutics which enable a better quality of life and long-term survival are very urgently sought after for the ALS spectrum of disorders and also for other devastating neurological conditions, many of which exhibit TDP-43 pathological changes.

Preclinical and clinical studies: interventions which modulate TDP-43

Between June 20, 2020 and January 14, 2021, we conducted searches on Clinical Trials.gov (https://clinicaltrials.gov) for clinical trials for patients with ALS, neurodegeneration, and based upon TDP-43 as a measured outcome. A search of ALS and TDP-43 yielded 692 Studies and after applying the following filters to limit studies to those that were further along in clinical development stages, to phase 2, 3, and 4 trials. Limiting studies to the terms TDP-43, Active, not recruiting, Completed, Unknown status Studies, Studies With Results and Phase 2, 3, or 4 yielded 39 studies. We also conducted a search for clinical trials with the TDP-43 as a measured marker using the following search terms: "Recruiting, Not yet recruiting, Active, not recruiting, Completed, Enrolling by invitation Studies, Interventional Studies, TDP-43, Phase Early Phase 1, 1, 2, 3, 4" and five results were obtained, with one being a new diagnostic instrument and report the four pharmacological interventions. In addition, an NCBI Pubmed Central search was carried out on TDP-43 with the filters "Clinical Trial, Meta-Analysis and Randomized Controlled Trial" which yielded 19 results. The preclinical and clinical trials selected for this chapter will provide information for clinicians and patients in search of new therapies in development for motor neuron disease and dementias, many of which provide evidence for a mechanistic role involving the modulation of TDP-43.

Therapeutics that target DNA double-strand break signaling and DNA repair mechanisms

Mitra et al. show that depletion of TDP-43 is associated with an increase in double-strand break (DSB) signaling.[12] TDP-43 serves as a scaffold for DNA DSB factors, XRCC4/DNA Ligase 4, and the aberrant mislocalization of this protein feasibly contributes to genomic instability and the pronounced motor-neuron cell death that is a prominent hallmark evidenced in ALS. The authors report that TDP-43 is a part of the nonhomologous end joining DNA double-stranded damage response pathway.[16] TDP-43 is attracted to sites of DNA DSBs and serves as a scaffold for other factors that repair the damage. The authors prove this because when they employed shRNA or CRISPR/Cas9 to conditionally deplete TDP-43, they observed an increase in DSB, genomic instability and stress, which are signature genomic changes of sALS patients.

Strategies aimed at treating neuropathologies based on improving DNA damage responses seem to be meritorious given the ubiquitous nature of genomic instability in neurological disorders, and the finding of the importance of TDP-43 in DSB repair. In addition to TDP-43, it has been shown that another ALS-linked protein, FUS/TLS, recruits XRCC1/DNA Ligase IIIα to locations with DNA single-strand breaks to induce DNA repair.[56] Also, survivin, a protein that promotes apoptosis resistance and is a key member of the Chromosome Passenger Complex required for faithful cell division, has been shown, like TDP-43, to play a role in DSB and like TDP-43, forms a complex with Ku70 when the DNA damage response is triggered.[16] One may imaginatively conjecture that TDP-43 and survivin nervous system variants work together to ensure genomic fidelity and survival of neuronal and glial cells by mechanisms involving DNA repair, antiapoptosis or precision replication, as well as that neuronal-cell-specific survivin-agonists may be formulated for neurodegenerative diseases. In fact, survivin and modulation of apoptosis, such as caspase inhibition, has been suggested as a target for neurodegenerative disorders[57,58]. Therefore, restoring TDP-43's normal cellular functions may be efficacious in the treatment of neurological disorders by improving DNA repair response and genomic integrity as well as splicing and miRNA biogenesis abnormalities in neurons, glia, and astrocytes. Additional preclinical studies may help to advance our knowledge of the merits of this approach.[59]

Autophagy induction

Tamoxifen (Nolvadex, AstraZeneca), competitive estrogen inhibitor; autophagy upregulation as a treatment in patients with motor neuron disease

Tamoxifen has been reported to decrease TDP-43 aggregates in animal models of ALS via tamoxifen-induced autophagy.[60] In a small clinical trial, 18 ALS patients without mutations in SOD-1 or FUS gene were randomized to receive either 40 mg/day tamoxifen or placebo, with all participants receiving riluzole twice daily (100 mg/day total) for a period of 1, 3, or 12 months (*The Study of Tamoxifen Treatment in Patients With Motor Neuron Disease;* ClinicalTrials.gov *Identifier*: NCT02166944).[61]

Disappointedly, primary endpoints of time to death or need for mechanical ventilation was not statistically different between the tamoxifen and placebo group at any of the time points. However, at a 1, 3, and 6 month evaluation, one of the two secondary endpoints, ALSFRS-R score, but not forced vital capacity, decreased less in the tamoxifen group, while there were no statistically significant differences of any endpoint observed at 12 months.

The authors concluded that larger studies would be required to confirm the modest, although transient, improvement in ALSFRS-R scores. It would be important to know if, in the tamoxifen-treated group, enhanced autophagy brought about a decrease in TDP-43 aggregates, and whether it played a role in ALSFRS-R improvements, and the value of administering at the earliest stages of the disease.

IMS-088, novel withaferin-A (Wse) analogue (active compound from Withania somnifera, *Ashwagandha, Indian winter cherry/ginseng, ImStar), induces autophagy and an antagonist of NF-κB*

It has been reported that the RRM1 region of TDP-43 interacts with a subunit of NF-kB (p65), and is associated with the excessive activation of NF-kB and its toxic inflammatory effects on neurons and glia.[62] In a TDP-43^{G348C} mutant mouse model of ALS, a model of neuroinflammation, a monoclonal single chain antibody generated against the RRM1-p65 interaction (VH7Vk9 scFv), downregulated NF-kB, reduced TDP-43 mislocalization and cytoplasmic aggregates by proteasome and autophagy mechanisms, and decreased cognitive, motor, and inflammatory phenotypes.[62] IMS-088 is a semisynthetic analogue of the ayurvedic compound derived from Ashwagandha, withaferin-A, which has also been demonstrated to inhibit NF-kB and promote autophagy.[63] It is presently in preclinical testing as a drug with applications for the treatment of ALS and FTD. In hTDP-43^{A315T} transgenic mice, a model for ALS/FTD, oral administration of IMS-088 was demonstrated to induce autophagy, decrease TDP-43 pathological changes, and restore neurofilament protein synthesis as well as cognitive performance.[63]

Drug discovery: pluripotent stem cells, transgenic model organisms, and ALS-derived iPSC motor neurons

Pluripotent stem cell screens for new ALS drugs: ropinirole, retigabine, and bosutinib identified

In drug discovery and repurposing screens using ALS-induced pluripotent stem cells (ALS-iPSC), several new candidate ALS treatments have been identified.[64] Three potential anti-ALS drugs were identified using ALS-iPSCs: ropinirole (a dopamine receptor agonist approved for PD) was identified in a neuropathology phenotype screen of 1232 FDA-approved drugs using ALS-iPSC of human spinal cord motor neurons from patients with fALS, with mutations in TDP-43 or FUS, or with sALS;[65,66] retigabine (a Kv7 or KCNQ voltage-gated potassium channel activator used in epilepsy) was identified due to its ability to inhibit hyperexcitability in an ALS-iPSC motor neurons with mutations in SOD1, FUS, or C9orf72 repeats[67] bosutinib (a Src/c-Abl inhibitor) indicated for chronic myelocytic leukemia, reduced phosphorylated TDP-43 (pTDP-43) in human ALS-derived iPSC motor neurons.[68,69] Bosutinib was also able to increase survival and autophagy of these cells, while decreasing the misfolded mutant SOD1 protein and improved the ALS-associated altered expression of mitochondrial genes.[68,69]

These preclinical trials indicated that these drugs may have therapeutic efficacy in treating sALS and warranted the commencement of clinical studies, as described for each drug individually below.

Bosutinib (Bosulif, SKI-606, Pfizer), Src/c-Abl inhibitor

There is currently a phase II double-blind, placebo-controlled RCT of bosutinib for dementia with lewy bodies (*A Randomized, Double Blind, Placebo-controlled Study to Evaluate the Impact of Bosutinib on Safety, Tolerability, Biomarkers and Clinical Outcomes in Dementia With Lewy Bodies (DLB);* ClinicalTrials.gov *Identifier:* NCT03888222). It is a 90-day study during which 15 patients will receive 100 mg daily of bosutinib or placebo. The safety, tolerability, and clinical efficacy will be monitored as the primary outcomes, while CSF and plasma levels of bosutinib and disease-related biomarkers (S100B, neuron-specific enolase, phosphorylated neurofilaments, HVA, DOPAC, Abeta40/42, total tau, phosphorylate tau231/181, and total and oligomeric levels of alpha-synuclein) will be determined as secondary endpoints. The study completion date is reported to be in April 2021, and no interim study results have yet been posted.

Ropinirole hydrochloride (Requip XL and Requip, GlaxoSmithKline Co., Ltd), dopamine receptor agonist

Currently underway in Japan is a phase I/IIa double-blind, randomized, placebo-controlled, clinical trial to study ropinirole in patients with ALS as an add on to riluzole (*Phase 1/2a Study of Ropinirole Hydrochloride Extended-Release Tablets in Patients with Amyotrophic Lateral Sclerosis (ALS); https://www.umin.ac.jp/english/; Identifier: UMIN000034954 and JMA-IIA00397*). Referred to as the ROPALS trial, this study was initiated following the in vitro findings of superiority of ropinirole to the current ALS medications riluzole and edaravone.[66,67] The study consists of 20 Japanese ALS patients, without a family history of SOD-1 mutation (among other exclusion criteria), 15 assigned to receive the active drug and 5 to receive the placebo. The primary study aim will be to assess drug safety and tolerability. Patients receive the standard ALS treatment of riluzole, along with placebo or ropinirole hydrochloride, beginning with 2 mg and increasing to a maximum dosage of 16 mg, in an extended-release formulation (Requip CR) for 24 weeks. The secondary endpoints of ALSFRS-R, change in the Combined Assessment of Function and Survival (CAFS) and exploratory molecular effects on TDP-43 and neurofilament light chain (NfL) on in blood and spinal fluid will also be accessed using either single molecule arrays or an immunomagnetic reduction assay. The secondary endpoint will provide an indication of the efficacy of ropinirole as a therapy for ALS. Of great interest will be the efficacy of this intervention, as well as the ability to modulate TDP-43 and NfL. The study is reported to have concluded in November of 2020 and the results as yet are unavailable but anticipated in early 2021.

Rasagiline (TEVA Pharmaceuticals Co., Ltd), monoamine oxidase B inhibitor

Rasagiline is a drug FDA-approved for the treatment of PD that is known to have a good safety profile, and it has antioxidant, mitochondrial-modulatory, and antiapoptotic properties that may be important for its therapeutic activity.[70]

A clinical trial was conducted to ascertain rasagiline's efficacy in patients with ALS in a double placebo, RCT of 80 ALS patients assigned in a 3:1 ratio to treatment with 2 mg/day rasagiline or placebo (rasagiline n = 60, placebo n = 20), stratified by riluzole use and bulbar onset (speech or swallowing problems) at 10 testing centers in the United States (*Phase 2 Study of Rasagiline for Treatment of Amyotrophic Lateral Sclerosis;* ClincalTrials.gov *Identifier:* NCT01786603).

The primary outcome measured was the decline in ALSFRS, and secondary measures were the vital capacity and decline in vital capacity since these predictors of survival, ALS quality of life (ALSQOL; 0 = very bad to 10 = excellent). An exploratory biomarker tested was TDP-43, measured indirectly (instead of in CNS) as a peripheral biomarker in platelet cytoplasm. It was found that TDP-43, collected at baseline and 12 months, did not differ between treated and placebo, although there was a decrease in both. Since rasagiline was reported to increase mitochondrial membrane potential, decrease oxidative stress, and increase the antiapoptotic: proapoptotic protein ratio, the authors measured mitochondrial and oxidative stress pathology as biomarkers that are important in ALS neuropathology. At baseline, month 6 and month 12, they measured the following peripheral biomarkers: mitochondrial membrane potential (JC-1 and MitoTracker Red); oxidative stress (15-F_{2t}-isoprostane); antioxidant potential Oxygen Radical Antioxidant Capacity assay (ORAC) and apoptosis (Bcl2/Bax mRNA Ratio and Annexin V). The analysis revealed inconsistent trends between rasagiline and placebo groups.

However, regrettably, no functional differences were found between the treated and untreated groups after a study period of 12 months duration. In addition, while in the placebo group one death occurred (5%), there were eight deaths in the rasagiline group (13.3%, reported to have not reached statistically significance) although 30 of the original 80 patients discontinued the study (60% in the placebo and 30% in the rasagiline group). The authors note that there was no significant difference between rasagiline-treated and placebo-treated platelet TDP-43, or survival at 12 months between the rasagiline, placebo, or (to increase study power) historical controls (n = 177), defined as a comparison between a new treatment in a current patient cohort with an old treatment in a previous cohort.[71] The inability of the drug, at the concentration given (2 mg/day) and during the 12 months study to modulate the levels of the

TDP-43 may have a role in the drugs failure to impact survivability, although there are clearly other possibilities, such as potentially the power or duration of the study.

There is some indication of this as there was another double-blind, phase II multicenter RCT which reported a slowing of disease progression in patients with ALS (*Efficacy, Safety and Tolerability Study of 1 mg Rasagiline in Patients With Amyotrophic Lateral Sclerosis (ALS) Receiving Standard Therapy (Riluzole)—An AMG Trial With a Market Authorized Substance;* ClinicalTrials.gov *identifier:* NCT01879241). ALS patients were treated with rasagiline at a dose of 1 mg/day for up to 18 months, as an add on to riluzole at 100 mg/day, with the primary endpoint being survival prolongation, which was not achieved, and a secondary endpoint of ALSFRS, which post-hoc analysis demonstrated a trend toward the slowing of disease progression.[72]

Pimozide (D2-like antagonist, Tocris Bioscience)

The clinical presentation of ALS patients include muscle fatigue, weakness, and neuromuscular conduction defects. A high-throughput drug discovery screen was developed to test for drug activity in two ALS TDP-43 transgenic model organisms (the roundworm/*C. elegans*, $TARDBP^{A315T}$ and the zebrafish/*Danio rerio*, $TARDBPG^{348C}$), both of which exhibit neuromuscular defects.[73] The authors screened 3850 small molecules for efficacy using the amelioration of neuromuscular junction deficits as a read-out of drug activity. Pimozide was identified as the lead small molecule based on its ability to rescue the motor deficits observed in these ALS model organisms. Synaptic transmission, paralysis, and neurodegeneration were corrected in the mutant TDP-43 roundworm, while aberrant swimming and motor neuron disease phenotypes were corrected in the mutant TDP-43 zebrafish.[73] Pimozide is a neuroleptic drug used in the treatment of patients with Tourette syndrome to suppress the occurrence of motor and vocal tics, and its mechanism of action includes antagonizing dopamine D2 signaling and blocking Ca^{2+} channel activity.[73]

There is presently underway a double-blind, phase II RCT to test the effects of pimozide in 100 ALS patients randomized to receive 2 mg/day (4 mg/day at study initiation) pimozide or placebo for 22 weeks (*A Phase II Randomized, Placebo-Controlled, Double Blinded, Multi-Centre Clinical Trial of Pimozide in Patients With Amyotrophic Lateral Sclerosis;* Clinicaltrials.gov *Identifier:* NCT03272503). The primary endpoint will be the ALSFRS-R, while secondary endpoints will be Change in Slow Vital Capacity, decremental responses on repetitive nerve stimulation, and changes in motor power, quality of life, frequency of cramps, and adverse events. Reports are not posted at this time, but are anticipated at the end of 2020 when the study is completed.

Oxytocin (Syntocinon nasal spray, Novartis), neuropeptide hormone

Oxytocin is a neuropeptide that is important for prosocial behavior. For example, it improves empathy, emotions, trust, and cooperation. Simultaneously, it decreases aggression and apathy. Aversive behaviors are notable in dementias such as AD and a subtype of FTD called behavioral variant (bvFTD), and can pose an immense hardship on caregivers, as well as impacting self-care, and personal relationships.

The hypothalamic–pituitary–adrenal (HPA) axis, part of the neuroendocrine system involved in basal homeostasis, diurnal circadian rhythms, and stress responses, may contribute to the development of AD.[74] Oxytocin is produced mainly in the hypothalamic bilateral supraoptic (SON), paraventricular (PVN), and accessory nuclei, primarily in magnocellular neurons and is released either systemically (via the pituitary gland) or into the CNS.[75] In addition, oxytocin-producing neurons extend projections into the pituitary, amygdala, nucleus accumbens, and medial prefrontal cortex, regions affected in neurodegenerative diseases.[76] Diodati et al. did not identify TDP-43 pathological inclusions in the nuclei from the SON or PVN from LBD, AD, or TDP-43-FTD patients. However, the authors mention that the brain regions that oxytocinergic neurons directly project into (such as the pituitary, amygdala, nucleus accumbens, and medial prefrontal cortex) are known to exhibit TDP-43 and other neuropathological changes, such as tau and alpha-synuclein. TDP-43 pathology in neurodegenerative disease has been reported to occur in the amygdala,[77] hippocampus, entorhinal cortex, and temporal and frontal cortices.[78] Elderly nursing home residents, with a varying degrees of dementia, as determined by clinical and cognitive performance tests, but without a diagnosis of AD or FTD, had TDP-43 inclusions in the amygdala (43.7%, stage 1), hippocampus, or entorhinal cortex (40%, stage 2) and temporal and frontal cortices (16.3%, stage 3).[78]

It is conceivable that TDP-43 pathological changes could be ameliorated with oxytocin in view of its role in FTD. In a short-term (2 week) RCT of intranasal oxytocin (0, 24, 48, or 72 IU) conducted in patients diagnosed with probable bvFTD (n = 23) (*A Phase I Dose Finding Study of Intranasal Oxytocin in Frontotemporal Dementia, Protocol # FTDOXY10EF;* ClinicalTrials.gov *Identifier:* NCT01386333), intranasal oxytocin was administered two times a week for 1 week and telephone assessments after 1 week washout period.[79] The authors found that all three doses of oxytocin were well tolerated and a significant improvement was observed, in a dose-related manner, on the FBI Apathy subscale, with the greatest improvement obtained at 72 IU.

There is an additional oxytocin crossover RCT in FTD/Pick's Disease (bvFTD, FTD-semantic subtype or FTD-Progressive Nonfluent Aphasia), which is recruiting patients with a starting Neuropsychiatric inventory (NPI) apathy/indifference score (primary outcome measure) ≥ 2 (*A Phase 2 Clinical Trial of Intranasal Oxytocin for Frontotemporal Dementia;* ClinicalTrials.gov *identifier:* NCT03260920). The study has two stages. In the first stage, the most efficacious of three doses of oxytocin (n = 20 per treatment arm compared to placebo controls) will be determined based on the NPI. Participants will receive 72 IU intranasal oxytocin (or placebo) on either twice daily, on alternate days, or every third day basis for 6 weeks. A washout period will follow (6 weeks) and crossover to different treatment schedule for 6 weeks. During the second stage, additional participants (n = 40) will be recruited and started at the most beneficial dosage schedule for 6 weeks, followed by a washout period for 6 weeks and a crossover period for 6 weeks. At different intervals, the primary outcome will be measured (NPI), as well as the following secondary outcomes: changes in emotional facial expression recognition performance, Revised Self-Monitoring Scale score, and Modified Clinicians Global Impression of Change (apathy) scores.

The study is expected to be completed in September 2023 and the results are currently not available. This study and additional studies to evaluate the clinical usefulness of oxytocin in FTD, PD, AD, and other neurological disorders as well as the potential modulatory effects on TDP-43 and other proteinopathies appear well-justified.

Modulation of TDP-43 phosphorylation

Preclinical IGS-2.7 (synthesized in the laboratory of Dr. Eva de Lago), benzothiazole-based CK-1δ inhibitor

The TDP-43 protein is susceptible to a variety of posttranslational modifications which include hyperphosphorylation at the serine 409/serine 410 site, and ubiquitination and cleavage into CTFs by caspases or calpain at specific, susceptible amino acids residues.[80] The CTF are very prone to aggregate formation and inclusion bodies can result as these fragments associate with hyperphosphorylated, polyubiquitinated TDP-43 at the CTD, and also, since they lack an NTD, these fragments become partitioned in the cytoplasm.[59,80] These modifications deplete nuclear TDP-43 and promote its translocation into the cytoplasm. That is where TDP-43 aggregates often tend to form and the residues prone to aggregate are within the glycine-rich prion-like domain (PrLD) of the protein.[80]

The phosphorylation of TDP-43, which occurs widely in brain inclusions of ALS and FTD/FTD-TDP patients, may represent a targetable modification. Several protein kinases that induce phosphorylation of TDP-43 have been identified. One is protein casein kinase-1 (CK-1) which can phosphorylate TDP-43 at 29 different sites in vivo, 18 of which are within the C-terminal glycine-rich region involved in aberrant cytosolic compartmentalization and aggregation.[81,82] CK-1-dependent phosphorylation of TDP-43 is induced during cellular stresses and the inhibition of CK-1 may have therapeutic value in neurodegenerative disorders.[83] In virtually all TDP-43 proteinopathies, serine 409, and serine 410 sites are phosphorylated, triggering TDP-43 aggregation, a process that can become irreversible and has been implicated in cellular degeneration.[84]

In ALS the upper and lower motor neurons in the CNS and spinal cord degenerate and die and TDP-43 plays a major role in both sALS and fALS. In TDP-43 (A315T) transgenic mice and human lymphoblasts derived from sALS patients, the hyperphosphorylation of TDP-43 was targeted with IGS-2.7. In mice, IGS-2.7 delayed weight loss, increased motor neurons survival in lumbar anterior horn, decreased inflammation as shown by a decrease in astrocyte and microglia immunoreactivity shown using GFAP and Iba-1 staining, respectively, and decreased the level of TDP-43 phosphorylation in TDP-43 transgenic mice to that of wild-type animals.[82] In the human sALS lymphocytes, treatment for 24 h with 5 μM of IGS-2.7 decreased phospho-TDP-43 as well as its cytosolic mislocalization.

These preclinical study results suggest that IGS-2.7 has neuroprotective activity and that it is also expected to be blood—brain barrier permeable, and future clinical trials appear to be warranted to better evaluate the benefits in ALS, or other motor neurodegenerative diseases.

Immune modulation and TDP-43

Tocilizumab (Actemra/RoActemra, Roche), anti—interleukin-6 receptor monoclonal antibody

Tocilizumab is an FDA-approved drug indicated for the treatment of rheumatoid arthritis.[85] It is an antiinflammatory monoclonal antibody that works by competitively inhibiting the proinflammatory cytokine interleukin-6 (IL-6), which prevents the binding of IL-6 to its receptor (IL-6R). This decreases IL-6 proinflammatory signaling to mediators, such as the activation of B and CD4 + T cells.[85] Inflammation is known to play a role in motor neuron diseases, including ALS.

In a multicenter RCT of Tocilizumab in ALS subjects (*A Phase 2 Randomized, Placebo Controlled Trial of Tocilizumab in ALS Subjects;* ClinicalTrials.gov *identifier:* NCT02469896), tocilizumab was tested for efficacy in ALS patients from the ages of 18—75 in a study that included both genders. The trial included patients who were receiving a stable dose of riluzole, those not taking riluzole, or women of child-bearing age, to enable broader implications of the study. Patients (n = 14, treatment group, n = 8 placebo) received 8 mg/kg of i.v. tocilizumab every 4 weeks for 3 months via i.v. infusion. The primary endpoints included all-cause mortality, safety, and tolerability, while secondary endpoints included efficacy as determined by the mean change in ALSFRS-R total score, rate of decline in Slow Vital Capacity, and the change in peripheral blood mononuclear cells and cytokine levels.

At this time, both the treatment and placebo group lost no patients to death, and decline in ALSFRS-R and the Rate of Decline in Slow Vital Capacity did not reach significance, while the fold change in two cytokine markers, CRP (decreased with treatment) and IL-6 (increased with treatment), did reach significance. The final results are not yet published and TDP-43 was not measured as a biomarker in this study.

Rapamycin (Sirolimus, Pfizer), mTOR inhibitor

Rapamycin inhibitor of the Mechanistic/Mammalian Target of Rapamycin (mTOR) is a clinically approved drug prescribed in cancer therapeutics,[86] and for the prevention of organ rejection through immunomodulatory activity (that is, both stimulating the immune system and suppressing age-related excessive hyper-active immunity), antiinflammation, antisenescence, and activation of autophagy.[87,88] Autophagy, an important cell process for protein turnover and degradation of aberrant proteins, is upregulated by rapamycin.[88] Rapamycin also downregulates protein synthesis, inhibits inflammatory responses, and has demonstrated efficacy in animal models of neurodegenerative disorders. Rapamycin was shown to induce autophagy in vitro in N2A and SH-SY5Y cells, decrease TDP-43 aggregates, and correct the mislocalization of TDP-43 from the cytoplasmic aggregation to restore it to the nucleus.[89] The clearance of TDP-43 by autophagy has been demonstrated in murine and in human stem cell—derived neurons and astrocytes with mutant TDP-43.[90] In addition, it has been reported that rapamycin increases T lymphocytes and that these increases in these cells is associated with slower disease progression in ALS.[88]

In eight ALS treatment centers in Italy, the rapamycin for ALS trial (RAP-ALS) is being conducted for an 18-week period, with a total follow-up duration of 54 weeks (*Rapamycin (Sirolimus) Treatment for Amyotrophic Lateral Sclerosis*; ClinicalTrials.gov *Identifier* NCT03359538). It is a phase II randomized, double-blind, placebo-controlled, and multicenter trial with patients randomized to one of three groups of ($n = 21$ per group): (1) 1 mg/m^2/day rapamycin + riluzole, (2) 2 mg/m^2/day rapamycin + riluzole, or (3) placebo + riluzole, in addition to mainstay supportive care in accordance with established ALS guidelines (e.g., nutrition, respiratory, motor function, communication).[88] The primary endpoint is the number of T regulatory cells (Tregs), which are cells that are produced in the thymus and regulate the immune response against self-cells and foreign antigens, and, given that rapamycin activates Tregs, quantifying the change between baseline and 18 weeks of treatment will help monitor rapamycin activity. There are a number of secondary endpoints that will be quantified, such as safety and tolerability, minimum dose required for detection of rapamycin in the cerebrospinal fluid, modulation of inflammatory and mTOR markers, ALSFRS-R, and quality of life. The treatment of ALS with rapamycin, considering its importance in modulation of TDP-43 clearance and correction of cytoplasmic mislocalization of TDP-43, is a very interesting therapeutic approach to follow in the years to come.

Colchicine (Colcrys, Mitigare, Takeda Pharmaceuticals), plant alkaloid, inhibits inflammation, disrupts microtubule formation, induces autophagy, and is used for gout

A multicenter RCT of colchicine is in progress to determine its efficacy in ALS (*Colchicine for Amyotrophic Lateral Sclerosis: a Phase II, Randomized, Double Blind, Placebo Controlled, Multicenter Clinical Trial;* ClinicalTrials.gov *identifier:* NCT03693781). The 54-week study included 54 patients, 18 in the placebo + Riluzole 100 mg, and 18 in one of two different dosages of the drug (0.01 mg/kg/day + Riluzole 100 mg or 0.05 mg/kg/day + Riluzole 100 mg). As a primary outcome, the investigators will measure change in ALSFRS-R (at 8, 18, 30, and 54 weeks), and as secondary outcomes they will determine: overall survival, adverse effects, Forced Vital Capacity. In ALS patient fibroblasts and lymphoblasts, they will determine if autophagy is increased, stress granules properties, levels of insoluble molecules (including full length and truncated TDP-43, SQSTM1/p62, UBQLN, OPTN). In extracellular vesicles from plasma and CSF they will quantify TDP-43, hyperphosphorylated TDP-43, SQSTM1/p62, UBQLN, and OPTN and inflammatory biomarkers (IL18, its endogenous inhibitor IL-18BP, MCP1, and IL17). In plasma they will also study neurodegenerative biomarkers (creatinine, albumin, CK, and vitamin D in plasma as markers of disease severity; phosphorylated neurofilaments heavy chain). The study is expected to be completed in December of 2021.

Targeting motor neuron hyperexcitability in ALS

Mexiletine (Mexitil, Teva Pharmaceuticals), sodium channel blocker

Sodium channel blockers, including mexiletine, was able to prevent both motor neuron cell death and intracellular nitroxidative stress caused by exposure of wild-type primary spinal cord cells to conditioned medium obtained from astrocytes with different ALS motoneuron-linked mutations (including $SOD1^{G93A}$, $SOD1^{G86R}$, and or TDP-43^{A315T}).[91] Mexiletine is an FDA-approved antiarrhythmic drug and a lidocaine analogue. It is utilized as a local anesthetic, and it downmodulates persistent sodium ion channel conductance. Neuromuscular hyperexcitability and involuntary fasciculations or transient muscle twitches, overwhelmingly afflict ALS patients and the majority of patients suffer from debilitating muscle cramping. In a study of 60 patients with sALS, mexiletine (300 or 900 mg/day) treatment for 12 weeks assuaged muscle cramps intensity and frequency, but did not improve the rate of disease progression, ALSFRS-R, and, in terms of tolerability, the dose of 300 mg/day was well tolerated, while the higher dose was not.[92]

In another double-blind crossover trial (150 mg twice daily with a 1-week screening period, 2-week treatment interval, followed by a 1-week washout of drug and second treatment), the severity and frequency of muscle cramps, likewise significantly decreased, whereas there was no change in fasciculations.[93]

A few limitations of these important studies, which reported efficacy against muscle cramps, were the study duration, small number of enrolled patients, absence of quality-of-life questionnaires, and power to detect survival. In a follow-up study (*Effect of Mexiletine on Cortical Hyperexcitability in Sporadic Amyotrophic Lateral Sclerosis (SALS);* ClinicalTrials.gov *identifier:* NCT02781454), some of these questions will be further addressed.

Oxaloacetate in ALS (TOALS) (OAA, benaGene, Terra Biological LLC.), glutamate excitotoxicity and mitochondrial dysfunction

Oxaloacetate is part of the TCA cycle, derived from the ATP-dependent carboxylation of pyruvate during carbohydrate scarcity (instead of pyruvate undergoing oxidative decarboxylation to acetyl-CoA). OAA (1000 mg b.i.d., n = 30, with 15 in OAA group) has been shown in AD to improve PET FDG glucose uptake in various brain areas and glutathione in the (frontal)parietal regions, and was well tolerated, although, during the 1 month trial duration, cognitive scores did not improve.[94]

A small (n = 24) and short-term (28 days) open-label clinical trial of OAA in ALS is underway (*Trial of Oxaloacetate in ALS;* ClinicalTrials.gov *identifier:* NCT04204889). The testing parameter are dose-limiting toxicities (500 mg twice per day orally and to a maximum of 2500 mg twice per day), serum platelet TDP-43 and indicators of mitochondrial dysfunction (Annexin V/Mitotracker in lymphocytes to determine: apoptosis/mitochondrial number; Mitochondrial membrane potential and Mitochondrial superoxide as causes of cellular oxidative stress; Glutamate-plasma as an OAA-specific biomarker), MR spectroscopy of glutathione in the brain, adverse events, and pharmacokinetics of OAA concentration (at baseline, 1–4 h after administration). The short duration will not be suitable for determining disease progression. This study indicates a status of recruiting, while also projecting a completion date of February 2021 and it will be interesting to see how modulation of TDP-43 is associated with the study's clinical outcomes.

Perampanel (FYCOMPA, Eisai R&D Management Co., Ltd), a selective noncompetitive AMPA receptor antagonist, inhibits glutamate neurotransmission excitotoxicity

Downregulation of adenosine deaminase acting on RNA 2 (ADAR2) occurs in sporadic ALS (sALS).[95] ADAR2 is an RNA adenosine-to-inosine (A-to-I) mRNA editing enzyme. It catalyzes the mRNA deamination of adenosine into inosine. The inosine is subsequently decrypted by ribosomes as guanosine (most often), adenosine, or, on rare occasions, as uracil.[96] A-to-I editing occurs at millions of noncoding sites throughout the human transcriptome as well as in approximately several 1000's of coding regions. It creates new isoforms, alters complementary binding, regulates RNA, and adds diversity to the proteome by forming new codes not directly transmitted in the DNA sequence.

Defects in A-to-I editing have been associated with neurological diseases.[97] ADAR2 A-to-I editing regulates the influx of Ca^{2+} through α-amino-3-hydroxy-5-methyl-4-isoxazole propionic acid (AMPA) receptors. However, in sALS, ADAR2 is deficient and the majority of patients with sALS have a GluA2 subunit of the glutamate AMPA receptor that remains unedited.[95] A consequence of loss of glutamate AMPA receptor editing is that Ca^{2+} permeability is elevated, which has been linked with TDP-43 pathology and glutamate-induced excitotoxicity in motor neurons.[98,99]

Perampanel is a noncompetitive antagonist of the AMPA receptor, which has received FDA-approval for the treatment of seizures. When the AMPA receptor was knocked out in mouse motor neurons, a model for sALS, perampanel markedly attenuated neuropathology, motor neuron death, and TDP-43 nucleocytoplasmic mislocalization.[100]

This preclinical study has led to a phase 2 clinical trial of perampanel in sALS (*Perampanel for Sporadic Amyotrophic Lateral Sclerosis (ALS): A Multicenter, Randomized, Double-blind, Placebo-controlled, Parallel-group Phase 2 Trials;* ClinicalTrials.gov *Identifier:* NCT03019419). Patients (n = 66) were assigned to receive a placebo, 4 and 8 mg per day for 48 weeks perampanel and the primary endpoint will be the change in ALS functional rating scale, manual muscle tests, and percent-predicted forced vital capacity. No interim results have been reported, but are projected to be completed by the end of 2023.[101]

Therapeutics based on small peptides and small molecule inhibitors

Small molecule inhibitors that block accumulation of ALS-associated RNA-binding proteins (RBPs), e.g., TDP-43, in stress granules in the brain of ALS or FTD patients may be clinically beneficial in treating these and other incurable neuropathological conditions.[102] In a screen of small molecule libraries molecules to search for those that could prevent persistent stress granule formation in proliferating cells (HEK293xT, NPCs and in human iPSC-derived motor neurons), Fang et al. identified a novel class of small molecules which possess a planar aromatic moiety and which modulated the size as well as the number of stress granules. The authors used the PubChem database, from the National Center for Biotechnology Information, to annotate targets for these and other lead small molecules, and found, for example, that the molecules with planar structures were often DNA intercalating molecules (such as mitoxantrone, quinacrine, doxorubicin, and daunorubicin). Some of these small molecules were capable of inhibiting cytoplasmic TDP-43 aggregates, suggesting possible clinical importance in the treatment of various TDP-43 neuropathology, and the next steps will be to perform RCT to determine if these agents can be used therapeutically.

PU-AD (Samus Therapeutics), antiepichaperome small molecule inhibitor which targets heat shock protein 90

During periods of cellular stress, chaperone proteins assist in protein folding (e.g., heat shock protein 90, Hsp90) or direct aberrant proteins to be degraded (e.g., Hsp70).[103] However, during chronic stress, for example, age-related, inflammation, high temperature, or mutations that lead to oxidative stress, a number of chaperone, co-chaperones, adaptor, and folding proteins (collectively called "chaperomes") can increase interactions with one another. The chaperomes can form stable networks, the "epichaperome," by strong interacting networks at Hsp90 and Hsp70 nucleation sites.[104] The epichaperome impairs many proteins from interacting with their normal protein partners (called "protein connectivity−based dysfunction") and from folding into correct conformations, including protein implicated in neurodegenerative disorders like tau or TDP-43.[103] PU-AD is a small molecule epichaperome inhibitor targeting Hsp90 that crosses the blood−brain barrier. A phase 2a RCT will commence in January 2021 to investigate the pharmacokinetics, safety, and effects on disease progression and disease-related biomarkers of PU-AD in ALS (*A Randomized, Double-blind, Placebo-controlled Pilot Study to Evaluate the Biological Activity, Safety, and Pharmacokinetics of PU-AD in Subjects With Amyotrophic Lateral Sclerosis (ALS);* ClinicalTrials.gov *identifier:* NCT04505358). ALS patients (n = 30) will be randomized to receive either PU-AD 30 mg q.d. or placebo and the stable treatment they are already receiving at the beginning of the study. At 12−24 weeks, the primary outcomes to be measured will be ALSFRS-R, Slow Vital Capacity, Handgrip strength, 6 Minute Walk Test, and Rasch-Built Overall ALS Disability Scale, while the secondary endpoints will be in blood (NfL, pNfH, and SOD1), in CSF (NfL, pNfH, tau, p-tau231 GFAP, TDP 43, and C9orf72) and in urine (p75ECD), all compared to baseline values and the projected completion date is February 2022. PU-AD is also being investigated in AD (*A Phase 2A Study to Evaluate Safety, Tolerability and Pharmacological Effects of PU AD in Subjects With Mild Alzheimer's Disease (AD) Dementia;* ClinicalTrials.gov *Identifier:* NCT04311515). In this study patients (n = 75) will be treated with active PU-AD or placebo SyrSpend on a 1:1 ratio qd. The outcome measures will be obtained at baseline and over a 6-month period. The primary outcome measures are FDG-PET and Adverse Events. The secondary outcome measures are Tau PET, Ratio of p-Tau to t-tau in CSF, Alzheimer's disease composite score, Global Statistical Test, and Pharmacokinetics of PU-AD. Although TDP-43 is involved in AD etiology, it was not indicated to be a measured biomarker, yet this will be an interesting study to follow when results become available.

GM604 (GM6, Alirinetide, Genervon Biopharmaceuticals, LLC), multitarget neuropeptide restores TDP-43 to normal endogenous levels

It has been known for since at least the early 1900s that vertebrate skeletal muscles produce factors that sustain the viability of motoneurons. For example, Hollyday and Hamburger showed that, by transplanting a supernumerary leg on a developing chick embryo, the lumbar lateral motor column neurons in the spinal cord—which normally undergo 40% motoneuron death and "exuberant connection" pruning—resist cell death on the side where the extra limb is attached.[105] In further support, Shorey et al. demonstrated that unilateral limb bud extirpations/amputation resulted in nervous system cell death, and later it was shown that this was due to limb-derived neurotrophic factors with limb removal.[106–108] These "hypothanasic"/antiapoptotic factors may be useful in motoneuron disease to hinder neuronal cell death.

Neurotrophic factors that are produced in muscle have different properties as those produced in the brain (e.g., Nerve Growth Factor, or brain-derived neurotrophic factor).[109] Within one active site of the skeletal muscle–derived Motoneuronotrophic Factor 1 is a subunit from which a six amino acid peptide analogue (H-Phe-Ser-Arg-Tyr-Ala-Arg-OH), named GM604, was synthesized, which is capable of crossing the blood–brain barrier and which was shown in mice subjected to transient ischemia to prevent neurological damage, decrease apoptosis and inflammation, and induce neurogenesis.[110,111] The GM604 neuropeptide is called a master regulator because of its pleiotropic functions which are suggestive of benefiting patients with neurodegenerative diseases. In fact, it is shown to decrease TDP-43 plasma levels, cytoplasmic aggregates, and nuclear-to-cytoplasmic localization.[111,112] Swindell et al. discovered 77 ALS-associated genes altered by GM604.[111]

A small phase 2 RCT of GM604 was conducted in 12 ALS patients diagnosed within 2 years of the trial (n = 8 in GM604 group and n = 4 in the placebo group). One ALS, compassionate treatment patient (who was quadriplegic and on a ventilator) was evaluated before and after GM604 treatment, but was not part of the trial. The study was conducted to test safety and tolerability of GM604, effects on modulating ALS biomarkers (included TDP-43) and to look for improvement of clinical progression (*GM604 Phase 2A Randomized Double-blind Placebo Controlled Pilot Trial in Amyotrophic Lateral Disease (ALS);* ClinicalTrials.gov *Identifier:* NCT01854294).[112] Clinical progression was determined by ALSFRS-R, Forced Vital Capacity, Time Up and Go, Muscle Strength, and mortality. The study duration was 3 months and patients received bolus treatment 3 times per week for 2-week or a comparator placebo.

There were no serious adverse reactions, and when compared to historic controls to increase power, the GM604-treated patients showed significant improvements in ALSFRS-R functional decline ($P = .005$), respiratory function, and these ALS biomarkers: TDP-43, $P = .008$; Tau, $P = .037$; SOD1, $P = .009$. The compassionately treated patient also experienced clinical improvement in speech, fluid intake, mouth suctioning, and the ALS biomarkers.

It will be important to confirm these encouraging findings in a larger RCT, while following patients for a longer duration and it appears that several earlier concerns may have been addressed.[113]

Reldesemtiv (CK-2127107; Cytokinetics and Astellas Pharma), troponin activator

Reldesemtiv was identified in a high-throughput screen in search of troponin activators, using fast and slow skeletal myofibrils, and is a small molecule, fast skeletal muscle troponin activator.[114] It was developed in hopes of treating motor neuropathies like ALS and spinal muscular atrophy. It was reported to appreciably increase rotarod running performance in a rat model of left anterior descending coronary artery heart failure.[115] Contraction of sarcomere, the essential unit of skeletal muscle, mainly relies upon Ca^{2+} signaling. Reldesemtiv enables muscle contraction with less Ca^{2+} nerve stimulation required because it makes muscles more sensitive to Ca^{2+} by delaying the release of Ca^{2+} from troponin.[114] Ca^{2+} binding to troponin removes tropomyosin from the myosin binding site on actin (normally blocked by tropomyosin), so that actin–myosin motor complex can form and muscles can contract, enhancing skeletal muscle force and endurance.

Reldesemtiv has FDA and European Medicines Agency (EMA) approval as an orphan drug indicated for spinal muscular atrophy, and was awarded in December 2020, by the EMA, the same orphan status for the treatment of ALS after the FDA fast tracked in following the results of the RCT (*A Phase 2, Multi-Center, Double-Blind, Randomized, Dose-Ranging, Placebo-Controlled Study to Evaluate the Efficacy, Safety and Tolerability of CK-2127107 in Patients With Amyotrophic Lateral Sclerosis (ALS);* ClinicalTrials.gov *identifier:* NCT03160898). The RCT was designed to test the safety, dose response (at 150, 300, or 450 mg, two times daily of reldesemtiv vs. placebo) as well as the efficacy in ALS patients ($n = 458$) over a 12 week period.[116] The study endpoints were Slow Vital Capacity, change in baseline of ALSFRS-R and Slope of Muscle Strength, with P-values of 0.10, 0.01, and 0.20, respectively. Although statistical significance was not met, Cytokinetics reported that the drug was fast-tracked by the FDA for ALS because all three endpoints showed a trend in the direction of improvement and near statistical significance for ALSFRS-R, in addition to demonstrating good tolerability in terms of adverse reactions. It will be important to follow the long-term results of ALS patients treated to determine whether reldesemtiv increase overall survival.

Antisense oligonucleotide and viral therapies for ALS

Antisense oligonucleotides (ASOs) are single-stranded nucleic acid sequences which bind to complementary RNA sequences and may cause target RNA to be cleaved via the recruitment of RNase H to RNA-DNA duplexes. ASO can alter RNA splicing, miRNA processing, or enhance protein levels through binding of natural antisense transcripts, as well as inducing modifications to RNA.[117]

TDP-43 antisense oligonucleotides

TDP-43's sequestering in ALS and FTD is associated with coaggregation of TDP-43 with poly glycine-arginine protein poly(GR) and it is to the repeat expansion region of the poly(GR) protein that ASO have been targeted.[15] ALS-associated TDP-43 mutations often cluster within the protein's low complexity domains or RNA recognition motifs. RNA-binding proteins, such as TDP-43, often have low complexity domains and use liquid–liquid phase separation to bind to RNA, and these regions promote aggregation of the TDP-43 protein.[118] Stress granules are membraneless organelles in the cytoplasm which, during times of cellular stress, prevent unnecessary protein synthesis by sequestering ribosomal subunits, translation initiation factors, and, in fALS, mutated RNA-binding proteins like TDP-43. In an optogenetics study in which TDP-43 homo-oligomerization at the low complexity domain was inducible by exposure to blue light, it was found that the controllable induction of liquid–liquid phase separation was linked with the formation of pathological TDP-43 inclusions (neurotoxic optoTDP-43).[118] Mann et al. demonstrated, importantly, not only a mechanistic link between aberrant TDP-43 phase transitions, that could be averted by RNA binding, but also that exposing neurons to TPD-43-targeted oligonucleotides decreased inclusions and was neuroprotective, as well as demonstrating the utility of neurotoxic optoTDP-43 as a new model system for the study of ALS.[118] Their findings were also consistent with a report that ASO-mediated depletion of SG components is able to ameliorate neurotoxicity present in a TDP-43 rodent model.[119]

Therapeutics that target TDP-43 induced alterations in energy metabolism

Metformin (Glucophage, Bristol-Myers Squibb)

TDP-43 contributes to misregulation of energy metabolism.[120] In one pre-clinical trial conducted with the SOD1(G93A) murine model of ALS it appeared that metformin did not merit further study in human clinical trials for ALS as a result of a finding that metformin induced a dose-dependent decrease in survival in this model.[121] But in a subsequent study of C9orf72 ALS/FTD BAC mice, metformin was found to significantly improve disease phenotypes in these mice, such as brake, brake/stance, and brake/stride, increased center time by open field tests, decreased glial fibrillary acidic protein (GFAP, a marker of neuroinflammation), prevention of motor neuron degeneration in the lumbar spinal cord, reduction in Gly-Ala aggregates in the retrosplenial cortex, and decreased levels of soluble glycine-proline.[122] Also, Repeat-associated non-AUG (RAN) proteins accumulate in patient brains and contribute to a growing number of neurodegenerative diseases.[123] It is still not known why RAN proteins do not require AUG start codons to initiate protein synthesis from their mRNA (as is the case for the majority of eukaryotic mRNA). Recently Banez-Coronel and Ranum demonstrated that several RAN proteins are increased by activated (phosphorylated) PKR and that, in HEK293T cells, metformin can impede the production of several RAN neurotoxic proteins (including polyAla, polyLPAC, and polyGP RAN) in a mechanism that involves markedly reducing activated PKR.[123] It was also reported that exogenous reporter expression of the C-terminal prion-like domain of TDP-43 in HeLa cells increased RAN translation in a mechanism involving phosphorylation of the α subunit of eukaryotic initiation factor-2 and triggering of stress granules formation.[124]

There is an underway an open label phase 2 clinical trial of metformin in subjects with C9orf72-positive ALS (*A Single-Center, Open Label Study to Assess the Safety and Tolerability of Metformin in Subjects With C9orf72 Amyotrophic Lateral Sclerosis Over 24 Weeks of Treatment;* ClinicalTrials.gov *Identifier:* NCT04220021). The study investigators will monitor the treatment-induced adverse events, change in RAN protein levels in cerebrospinal fluid (CSF) and ALSFRS-R score at a starting dose of 500 mg, increased incrementally by 500 mg every week to a maximal dosage of 2000 mg in divided doses twice daily. This study will be quite interesting to follow, although it will not determine TDP-43 modulation, but would give an indication of whether the alterations in metabolism in neuropathology can be ameliorated through the repurposing of a well-known diabetic medication for neurodegenerative diseases.

AMX0035 (proprietary drug combination consisting of sodium phenylbutyrate/PB/Buphenyl and Taurursodiol/TUDCA, sodium Phenylbutyrate–Taurursodiol, PB-TURSO, Amylyx Pharmaceuticals, Inc)

In the CENTAUR RCT, ALS patients diagnosed within the past 18 months (n = 136) received either placebo or a 2:1 ratio of sodium phenylbutyrate (an aromatic fatty acid) to taurursodiol (a bile acid derivative) (3 g of sodium phenylbutyrate to 1 g of taurursodiol), once daily for 3 weeks and then twice daily for the duration of the 6-month study (*Evaluation of the Safety, Tolerability, Efficacy and Activity of AMX0035, a Fixed Combination of Phenylbutyrate (PB) and Tauroursodeoxycholic Acid (TUDCA), for the Treatment of ALS;* ClinicalTrials.gov *Identifier:* NCT03127514). Primary outcomes were the ALSFRS-R Slope, adverse events, and discontinuation rate, while secondary outcomes were Accurate Testing of Limb Isometric Strength, Slow Vital Capacity, Survival, tracheostomy and hospitalizations, and blood and imaging biomarkers.

At the completion of the 24-week trial, it was found that the ALSFRS-R decline was significantly slowed in treated patients versus placebo controls ($P = .03$), and that drug-related side effects were primarily gastrointestinal, but that secondary endpoints were not different between these groups.[125]

There are several additional trials registered for AMX0035, two more for ALS and one for AD. The CENTAUR participants (n = 40) who were treated with AMX0035 will be able to continue treatment for an additional year and any emergent adverse events will be accessed. The participants will be offered the option to take the ALSFRS-R (*AMX0035 for Amyotrophic Lateral Sclerosis—Compassionate Use Protocol;* ClinicalTrials.gov *identifier:* NCT04516096, *to be completed in 2023*). In an open label extension to the CENTAUR trial, the CENTAUR-OLE trial, study participants will continue to be monitored and survivability and long-term safety will be studied (*Open Label Extension Study of AMX0035 in Patients With ALS;* ClinicalTrials.gov *identifier:* NCT03488524, *to be completed in 2021*). The primary outcome is quantity of adverse events and secondary outcomes are number of hospitalizations, ALSFRS-R, ATLIS Strength Measurement, Slow Vital Capacity, number of gastric tubes placed, and number of patients of permanent invasive ventilation, and the study duration will be 30 months.

The AD (and mild cognitive impairment) study will evaluate safety, tolerability, biological and imaging markers, cognition and psychiatric symptoms in 100 participants over a 24-week period (*Phase II Study to Assess the Safety, Tolerability, and Target Engagement of AMX0035, a Fixed Combination of Sodium Phenylbutyrate and Tauroursodeoxycholic Acid for the Treatment of Alzheimer's Disease;* ClinicalTrials.gov *Identifier:* NCT03533257, *to be completed in December 2020*). At 6 months, comparing active participants and controls, the primary outcome measures will be the Quantity of Adverse Events, while the secondary outcome measures will be levels of hippocampal and whole brain atrophy (as assessed by volumetric MRI), Cognition (using ADAS-Cog), NPI, Functional MRI Imaging, CSF, and plasma biomarkers. The anticipated completion date is April 2021 and no interim results are available at this time.

Dietary molecular interventions: ketone bodies, medium-chain triglycerides, and high-caloric fatty diet

It has been reported that in ALS there is a decrease in energy uptake[126] while energy expenditure is increased.[127] This imbalance in energy supply and demand causes a net energy deficiency, which can culminate in metabolic failure and neuronal cell death.[128] In addition, there are several mitochondrial respiratory—chain complexes that have been shown to be decreased in AD,

HD, and PD brain tissue.[129] In the spinal cord of ALS patients, as well as in animal models of ALS, mitochondrial dysfunction is a characteristic hallmark and it has been shown that there is a metabolic shift from glucose to fatty acid oxidation.[130]

In TDP-43 mutant animal models and postmortem brain samples of ALS and FTD patients, mitochondrial impairment is evidenced by the increase in ROS, decrease in the mitochondrial membrane potential and number of cristae, and an abnormal ultrastructure of cristae as observed via electron microscopy.[131] It has also been proposed that underlying many incurable neurological diseases are metabolic deficiencies which lead to the degeneration of specific, susceptible neuronal subtypes.[132] In patients with AD, using FDG-PET imaging, it has been shown that the uptake of glucose is impaired in several brain regions of (including the hippocampus, posterior cingulate, precuneus, and prefrontal cortex).[133] In AD, PD, ALS, and HD, neuronal degeneration first begins in one or more subcortical regions (e.g., striatal, hippocampal or spinal motor neurons, varies depending on disease) and ultimately progresses to the loss of cortical neurons.[134] The neurons and glia that are most vulnerable to pathological changes are often those that are more morphologically complex, such as neurons with long range projections, or those with numerous synaptic connections.[132] The vulnerability of such neurons is, in part, related to the greater energetic demands that the maintenance of these complex cell-types requires. Some examples include CA1 neurons with high firing rates or Fast-Fatigable motor neurons with large axons, as well as glia and vascular cells.[132] In light of the metabolic aberrancies which occur in many patients with neurogenic disorders, metabolic therapy has been, in recent years, investigated as a possible way to preserve neurons and glia and to remediate some of the most prevalent neuropathological manifestations.

Ketogenic diet

In preclinical studies using a SOD1-G939 ALS mouse model, a ketogenic diet improved survival of motor neurons in the spinal cord and also decreased muscle deterioration.[135] It is also known that medium-chain triglyceride (MCT) are metabolized into ketone bodies (β-hydroxybutyrate, acetoacetate, and acetone), which thereby provides an energy substrate for the brain and other organs of the body as an alternative or supplemental fuel to glucose.[86,133] There are several reasons that support the possible benefit of ketone bodies in ALS. These include the antiinflammatory properties of the ketone bodies, including β-hydroxybutyrate, through the activation of HCA2 receptor and inhibition of IL-1β and IL-18 along with the inhibition of reactive oxygen species and decreasing the excitatory neurotransmitter glutamate.[136] To follow-up on these preclinical studies, RCT will be necessary to ascertain if the ketogenic diet is helpful addition in the treatment of ALS and neural disorders to alter the metabolic impairments that are widely observed.

MCT

In addition to the ketogenic diet, another way in which to obtain ketone bodies is through MCT. In the SOD1G939 ALS study, the authors showed that caprylic triglyceride, a type of MCT, was similarly able to protect spinal motor neurons and to ameliorate mitochondrial defects.[135]

A systematic review (registered on www.crd.york.ac.uk/prospero as CRD42019128311), consisting of 10 randomized clinical trials, ranging from 45 to 10 days, was undertaken to ascertain whether keto-neurotherapeutics as a metabolic intervention has efficacy in AD and mild cognitive impairment.[137] The types of interventions included the consumption of very low carbohydrate ketogenic meals or a variety of products containing high amounts of MCT (such as coconut and palm-kernel oils), which are metabolized into ketone bodies. These interventions resulted in increased levels of blood ketones and cerebral ketones utilization and, importantly, in improvements on cognitive performance tests. At both short- and long-term intervention durations, there was an improvement in the Alzheimer's Disease Assessment Scale-Cognitive test, while, after long-term ketogenic intervention, episodic and secondary memory also improved. Patients with APOE ε4(+) had delayed improvements compared to those who were APOE ε4(−).

A small single arm, open label clinical trial was completed by 16 (of $n = 20$) Japanese patients with mild to moderate AD to test the effects on cognitive functioning of chronic administration of chronic administration of 50 g of ketogenic formula with 20 g of MCTs (KetonFormula) for 12 weeks, along with the patients normal diet and stable antidementia medications (no modification in dose was permitted).[133] Blood ketone levels (acetoacetate and β-hydroxybutyrate) and cognitive functions were tested on weeks 4, 8, and 12. The cognitive function tests consisted of logical memory (immediate and delayed), digital and spatial span (parts of the Wechsler Memory Scale-Revised, WMS-R), block-design and digit-symbol coding tests (a part of the Wechsler Adult Intelligence Scale-3rd, WAIS-III), the Stroop test, and the Trail Making Test (TMT). The investigators determined that the experimental group, compared to the baseline, had a significant improvement in cognitive processing speed, and verbal memory verbal memory (immediate and delayed). Acetoacetate and β-hydroxybutyrate did not meet significance compared to baseline (possibly due to time interval between blood collection and ketogenic formula intake). An earlier study by the same authors found that in elderly participants without dementia, a ketogenic meal (which they showed elevated plasma β-hydroxybutyrate levels) improved a range of metrics from the Wechsler Memory Scale-Revised and Wechsler Adult Intelligence Scale, III, the letter-number sequencing and Trail-making test, including visual attention, memory, and digit-symbol coding.[138]

High-caloric fatty diet

Unexpectedly, a delay in ALS progression has been reported in several clinical studies of patients with concomitant type 2 diabetes, high body mass index, and dyslipidemia and it is suggested that altered metabolism in ALS and FTD might offset some pathological commonalities in these diseases.[136] Jawaid et al. offer several plausible explanations for how metabolic dysregulation might be protective in ALS and FTD. For example, stress granules, which are induced by exercise and decreased glucose, colocalize with TDP-43 inclusions in ALS and FTD, and play a role in the disease. Also, AMP-activated protein kinase (AMPK) is a nutrient sensor which is activated by low nutrient conditions, such as low glucose, and this protein has been implicated in the cytoplasmic sequestering of TDP-43. There is a double-blind RCT (*Safety and Tolerability Ultra-high-caloric Food Supplements in Amyotrophic Lateral Sclerosis (ALS) (TOLCAL);* ClinicalTrials.gov *Identifier:* NCT02306590, *also identified as LIPCAL-ALS study*) evaluating the effects of a high calorie fatty diet (HCFD), as drinkable beverage, which was a provision of 45 g of extra daily fat, as an add-on to the standard of care (100 mg riluzole) in patients with ALS (n = 201).[139] In a subset of fast progressing ALS and motor neuron disease patients, post-hoc analysis showed that the experimental HCFD group experienced a significant increase in survival, the primary outcome measure.[139]

In future studies it would be very interesting to test whether modulation of energy metabolism, such as via HCFD, is able to decrease stress granules, modulate AMPK, and ameliorate mitochondrial defects associated with altered TDP localization, aggregate formation, and functionality in subset of patients that achieved a survival benefit compared to those that did not.

Complementary and alternative medicine approaches for the treatment of neuropathological condition

Mucuna pruriens *(common name Velvet Bean) and* Withania somnifera *(common name Ashwagandha)*

Two Ayurvedic medicines in widespread use, *Mucuna pruriens* (Mpe) and *Withania somnifera* (Wse), have reported utility in the treatment of neurodegenerative diseases[140–142] In a partiahl TDP-43 protein depletion *Drosophila* model of ALS (achieved through the GAL4-Upstream Activation Sequence system in which knockdown of TAR-43 homolog was targeted to motoneurons), both Mpe and Wse (supplemented in food at 0.1% w/w) ameliorated ALS-associated disease phenotypes. Mpe and Wse improved the hyperactive anomalies in ambulation and climbing behavior, dysregulated sleep (in two age cohorts, 3–6 or 10–15 day-old male flies), and Mpe (but not Wse) corrected the evoked postsynaptic potential (ePSP) response to electrical stimulation of the dorsal longitudinal muscle (DLM) neuromuscular junction in TDP-43 mutant flies and the mutant flies showed a significant decrease in amplitude and an increase in latency.[140]

Studies on sleep disturbances in neurodegenerative diseases and links to TDP-43

Disruption of normal sleep oscillations, such as cortical slow oscillations, hippocampal ripples, and thalamocortical spindles impairs declarative memory and memory consolidation.[143] The elderly and patients diagnosed with neurodegenerative disorders, such as ALS, are particularly prone to sleep disruptions.

C9of72 is a ubiquitously expressed protein in the brain. It contains a hexanucleotide sequence (GGGGCC) which is repeated a variable number of times, depending on genotype and brain region somatic mosaicism. For example, in ~95% of people without neurodegenerative disease, the number of hexanucleotide repeats occurs fewer than 11 times, whereas this microsatellite is expanded to greater than 30 times in a large number of patients with ALS (~50% of fALS and ~10% of sALS) and familial FTD (~25%), such as TDP-43 pathology (FTD-TDP).[144] The expanded repeat occurs in Europe and North American but is rare in Asia and the Middle East.[145] The C9orf72 expansion is also involved in p-TDP-42 and dipeptide repeat protein (DPR) inclusions.[145] From the C9orf72 expansion region translation begins rather than at the normal ATG codon (a type of translation called repeat-associated non-ATG or RAN). RAN produces five known types of DPRs (poly-GA, poly-GR poly-PR poly-AP, and poly-GP) of which two are the particularly implicated in neurotoxicity (Poly-GR and poly-PR).[146]

In patients with rapid eye movement sleep disorders, including TDP-43opathies, such as FTD and ALS, the G4C2 C9orf72 repeat expansion may be involved. The suprachiasmatic nucleus (SCN) (in the hypothalamus) and the pineal gland, which synthesizes melatonin and serves as the main executor of the SCN, are two principle brain regions that govern circadian sleep/wake cycles. The SCN suppresses or stimulates the pineal synthesis of melatonin according to the light/dark cycle, leading to a decreased or increased tendency to sleep.[145] In pinealocytes and SCN neurons from C9orf72 cases, DPR inclusions were numerous, while pTDP-43 pathology was not observed.[145]

However, in *Drosophila*, in a comparison between flies expressing either wild type or mutant variants of TDP-43 in motor neurons or glia, total sleep is decreased in TDP-43 mutant variants, while ALS-linked mutant TDP-43 variants have more frequent sleep bouts,[147] which seems to implicate TDP-43 in sleep aberrations, although perhaps in brain regions other than the primary regions that regulate the sleep cycle (i.e., the SCN and pineal gland).

Interestingly, it was recently reported that targeting the poly-GR of the C9orf72 gene with antisense oligonucleotides suppressed aberrant poly-GR gain-of-function and inhibited cytoplasmic TDP-43 aggregate formation in stress granules as well as prevented neurodegeneration in a C9orf72 ALS transgenic mouse model, which has a number of behavior disturbances, including hyperactivity and anxiousness.[148]

In a behavioral disorder that occurs during sleep, patients with Rapid eye movement (REM) sleep behavior disorder (RBD) present with vocalizations and sometimes injurious or violent behaviors while sleeping, and, although rare, have nearly a 90% risk of developing a neurodegenerative disorder, predominantly PD, and RBD has been treated with low-dose clonazepam or high-dose melatonin taken at bedtime.[149] Interestingly, in ALS melatonin has been proposed as a possible treatment. In one preclinical study, high dose melatonin was studied for neuroprotective ability in ALS using motor neuron cells (NSC-34) induced to die with glutamate, ALS $SOD1^{G93A}$-transgenic mice, and sALS patients ($n = 31$) treated for 2 years with 300 mg/day melatonin.[150] That study demonstrated that cell death was diminished in the motoneuron cells, while survival was increased in the $SOD1^{G93A}$-transgenic mice and the treatment in patients was tolerated well and reduced oxidative stress surrogate markers (serum protein carbonyls). In another preclinical study to test standalone treatment of Riluzole compared to melatonin in an ALS mouse model from which ex vivo hypoglossal motoneurons from adult mouse brainstem slices (obtained after mice displayed disease onset), and also in a wild-type versus ALS-mutant $SOD1^{G93A}$-transfected neuroblastoma cells (SH-SY5Y), the side-by-side comparison indicated that Riluzole had standalone efficacy, in terms of inhibition of excitotoxicity and prevention of cell death, while melatonin did not, showing mixed results.[151]

There were no clinical trials of melatonin in ALS found on ClinicalTrials.gov at the time of this chapter (January 2021); however, there are current studies in process for PD and AD. As TDP-43 has been shown to play a role, as discussed above, in sleep disturbances in ALS and other neurodegenerative diseases, it will be of great interest to follow the ongoing preclinical and RCT of melatonin and other new sleep modulating compounds to determine how they alter TDP-43 proteinopathy in relation to the treatment outcomes of these exceedingly complex disorders.

Combinatorial and integrative approaches to neurodegenerative diseases

Polytherapy has been proposed as treatment plans that may be able to address the multiple contributing factors that give rise to types and subtypes of neurodegenerative diseases. Even in situations where monotherapies fail to perform well as a standalone drug, multiintervention combinations may work synergistically, such that several agents may afford clinical benefit not evident as a monotherapy.

For example, metabolic enhancement for neurodegeneration (MEND) is an individualized, combinatorial program that was anecdotally studied in a small pilot study of AD (n = 10), wherein 9 of the 10 patients were ApoE4+, and this trial was demonstrated to reverse some aspects of neurocognitive disease in AD, with 6 of the patients returning to work.[152] Specifically, the treatment protocol ("Therapeutic System 1.0") consisted of the following: a low glycemic, low grain and antiinflammatory diet, 12 h nightly fast (3 h prior to bedtime), stress reduction (such as yoga or meditation), exercise (30–60 min each day, between 4 and 6 days per week), 8 h nightly sleep (treatment for sleep apnea, if present, 0.5 mg po qhs melatonin and, if still awakening, 500 mg po 3 times per week Tryptophan), evaluation and/or treating heavy metal toxicity and pathogens, brain stimulation, and a cocktail of various herbal and nutritional supplements for cognition (Ashwagandha, Bacopa monnieri, magnesium-L-threonate, Vitamins D3, K2, H. erinaceus or Acetyl-L-carnitine, Citicoline, DHA), hormonal balance, GI health (prebiotics and probiotics), mitochondrial function, and to decrease inflammation (curcumin; DHA/EPA) and oxidative stress (Me–B12, MTHF, P5P). The authors reported that 9 of the 10 patients showed significant and durable improvements compared to the initial presentation (such as resuming employment and on Montreal Cognitive Assessment or MemTrax tests) and the follow-up time was up to 2.5 years.

A single arm, expanded case study of the MEND program enrolled 100 patients with mild or subjective cognitive impairment (MCI or SCI, respectively), or AD.[153] The patients in this study were accessed at the beginning and end of treatments using Mini-Mental Status Examination (MMSE) Montreal Cognitive Assessment (MoCA), or Saint Louis University Mental Status (SLUMS) exams and there was a mean improvement of 4.9 points (SD 2.6), which is notable as the trend toward deterioration rather than improvement is normally observed, and, as the authors note, it will be important to confirm these findings in an RCT.

One example of a multitarget strategy to treat dementia is a retrospective, observational study of dementia patients with probable AD (n = 344) which evaluated the addition of Chinese herbal medicine, the GRAPE formula, to traditional AD therapy, consisting of donepezil, a cholinesterase inhibitor (5–10 mg once a day), or memantine, an NMDA antagonist (10–20 mg once a day), or both medications (*A Retrospective Study to Investigate the Additive Effectiveness of Chinese Herbal Medicine in Alzheimer's Disease;* ClinicalTrials.gov *identifier:* NCT03221894).[154] The herbal formula was devised to increase energy and circulation, while inhibiting inflammation and the GRAPE formula (Beijing Tcmages Pharmaceutical Co., LTD) consisted of (Ren shen/Panax ginseng, 10 g/day), (Di huang/Rehmannia glutinosa,

30 g/day), (Cang pu/Acorus tatarinowii, 10 g/day), (Yuan zhi/Polygala tenuifolia, 10 g/day), (Yin yanghuo/Epimedium brevicornu, 10 g/day), (Shan zhuyu/Cornus officinalis, 10 g/day), (Rou congrong/Cistanche deserticola, 10 g/day), (Yu jin/Curcuma aromatica, 10 g/day), (Dan shen/Salvia miltiorrhiza, 10 g/day), (Dang guiAngelica sinensis, 10 g/day), (Tian ma/Gastrodia elata, 10 g/day), and (Huang lian/Coptis chinensis/berberine 10 g/day) and was administered twice daily in dissolved in 150 mL hot water. The primary outcome was the MMSE and secondary outcomes, the Activities of Daily Living, Clinical Dementia Rating and NPI and mild (MMSE 21–26) and moderate (MMSE 10–20) AD patients were followed up to 24 months and assessed at approximately 3 month intervals. The herbal granules added to standard AD medications was shown to outperform standard medication alone, and the improvement was seen both in mild AD (2.60 point higher MMSE, $P = .012$) and moderate AD (2.12 points higher MMSE, $P = .025$).[154] Interestingly, one of the components of the GRAPE formula, berberine, has been reported to inhibit mTOR/p70S6K pathway and activate autophagy, which reduced insoluble TDP-43 and inhibited TDP-43 aggregate formation in a cell culture model of FTD and ALS (Neuro 2a murine neuroblastoma transfected with truncated C-terminal human TDP-43 fragments), suggesting possible clinical usefulness for these conditions.[155]

International interest has been growing in the area of integrating conventional and traditional therapeutics, and in using multiinterventions for improving clinical outcomes in severe neurodegenerative disorders.[156–159] Although it is difficult to ascribe activity to any particular component of complex plant extracts, the long traditional history of use suggests importance and effects and safety profiles of candidates can be further investigated. One interesting study described a computation strategy to identify drug combinations which could point toward ability to prevent or delay neurodegenerative disorders based on use of electronic medical records (n = approx. 50,000), 1900 FDA-approved drugs and OTC supplements, which were grouped into 53 categories by mechanism, and related these to 300 clinical trials.[160] The groundbreaking work in polytherapies for neuropathology, ways to predict possible combinations will be important to keep abreast of (Fig. 9.1).

Figure 9.1 provides a summary of the various neurodegenerative disease interventions described throughout this book chapter and the associated mechanisms of actions discussed. The immunostaining photomicrograph of TDP-43 is from the hippocampal dentate gyrus of an autopsy specimen from a patient with FTD, showing, in some neurons, the loss of TDP-43 in the nucleus and cytoplasmic inclusions (adapted from Ref.: Jensflorian, CC BY-SA 3.0 <https://creativecommons.org/licenses/by-sa/3.0>, via Wikimedia Commons). ASO schematic is derived from Ref: Biosyn89, CC BY-SA 3.0 <https://creativecommons.org/licenses/by-sa/3.0>, via Wikimedia Commons.

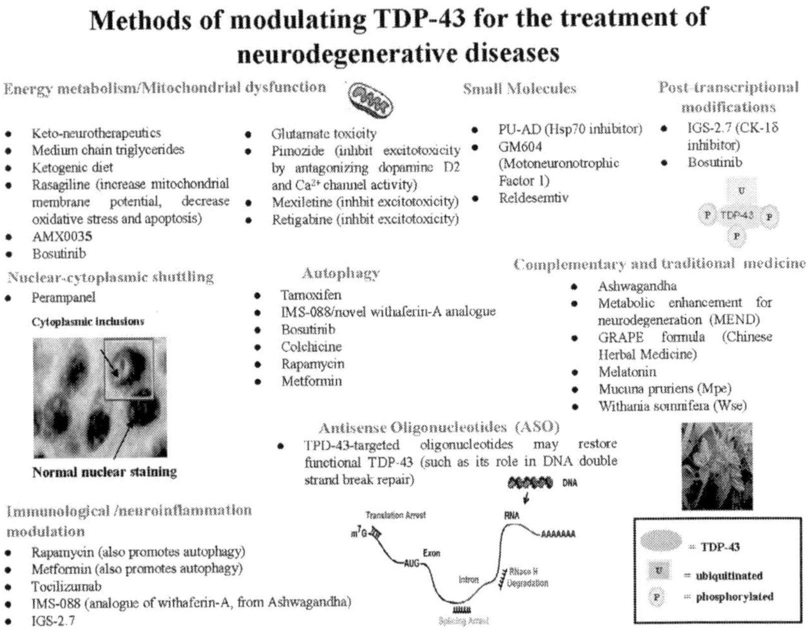

FIGURE 9.1 Methods of modulating TDP-43 for the treatment of neurodegenerative diseases.

Discussion and future directions

We have reviewed some of the multidisciplinary avenues that point toward the discovery of new treatment options for neuropathological diseases for which present-day therapies are inadequate. TDP-43 is regarded as a central mechanistic target that is implicated in many neurodegenerative conditions. It has many pleiotropic cellular functions and appears justly viewed as a nodal protein that may be modulated for clinical purposes. Many treatment strategies now on the horizon for the treatment of neurodegenerative disease show evidence for modulating TDP-43's loss- or gain-of-function properties. The future vision and direction of neurodegenerative disease research must necessarily include studies of the efficacy of the upcoming and ongoing preclinical and clinical trials, drug repurposing, prevention, use of plant neurobioactive compounds, and individualized and multiphasic approaches. Some of these studies will aim to directly explore associations with hallmark biomarkers such as TDP-43, in addition to metabolic and imaging markers, along with clinical metrics specific for the particular pathology. Of paramount importance is to devise integrative approaches that may give patients with severe neurological disorders an improved quality of life and a realistic hope for long-term survival.

List of abbreviations

AD Alzheimer's disease
ADAS-cog AD Assessment Scale-Cognitive Subscale
ALS amyotrophic lateral sclerosis
AMPK AMP-activated protein kinase
APP amyloid precursor protein
ASO antisense oligonucleotides
bvFTD behavioral variant FTD
C9ORF72 chromosome 9 open reading frame 72
CAFS function and survival
CTD C-terminal domain
CTD IDR, intrinsically disordered region
CTE chronic traumatic encephalopathy
CTFs C-terminal fragments
DLM dorsal longitudinal muscle
DPR dipeptide repeat protein
DSB double-strand break
EMA European Medicines Agency
ePSP evoked postsynaptic potential
fALS familial amyotrophic lateral sclerosis
FTD frontotemporal dementia
FTD-TDP TDP-43 pathology
FUS fused in sarcoma
HCFD high calorie fatty diet
HD Huntington's disease
HPA hypothalamic−pituitary−adrenal
Hsp90 heat shock protein 90
kDa kilo Daltons
LATE limbic predominant age-related TDP-43 encephalopathy
LB Lewy bodies
LBD Lewy bodies dementia
LMN lower motor neurons
MCI mild or cognitive impairment
MCT medium-chain triglyceride
MEND metabolic enhancement for neurodegeneration
MMSE Mini-Mental Status Examination
MoCA Montreal Cognitive Assessment
Mpe *Mucuna pruriens*
MSA multiple system atrophy
MSQ mental status questionnaire
NES a nuclear export signal
NfL neurofilament light chain
NLS nuclear localization signal
NPI neuropsychiatric inventory
NTD N-terminal domain
ORAC Oxygen Radical Antioxidant Capacity assay
PD Parkinson's disease

PLS primary lateral sclerosis
PMA progressive muscular atrophy
PS Perry syndrome
PVN paraventricular
RAN repeat-associated non-AUG
RCT randomized clinical trial
RGNEF Rho guanine nucleotide exchange factor
ROS reactive oxygen species
RRMI and RRM2 RNA recognition motifs 1 and 2, respectively
RWE real-world evidence
sALS sporadic amyotrophic lateral sclerosis
SCI subjective cognitive impairment
SLUMS Saint Louis University Mental Status
SOD1 superoxide dismutase
SON supraoptic nuclei
SVC slow vital capacity
TARDBP transactive response DNA-binding protein gene
TDP-43 transactive response DNA-binding protein 43
TMT trail making test
UMN upper motor neurons
WAIS-III Wechsler Adult Intelligence Scale-3rd
WMS-R Wechsler Memory Scale-Revised
Wse *Withania somnifera*

Conflicts of interests

The author declares that no competing interests exist and warrants that there are no financial or personal interests, motivations, or affiliations that would potentially cause a conflict in the judgments and opinions expressed in this chapter review. The views and opinions expressed herein do not necessarily reflect the views and opinions of the Icahn School of Medicine at Mount Sinai. This work is original, nonredundant, and has not been duplicated or published elsewhere.

Author contributions

This manuscript was conceived, drafted, and written exclusively by Pasha Apontes.

Acknowledgments

The author wishes to express her appreciation for the valuable edits and discussion of the manuscript with Dr. Manoj Kumar Jaiswal. I would also like to express my sincere thanks to Drs. Trupti Vardam and Vijay Kumar for their valuable comments on my draft chapter.

References

1. Hergesheimer RC, Chami AA, Assis DR de, et al. The debated toxic role of aggregated TDP-43 in amyotrophic lateral sclerosis: a resolution in sight? *Brain* 2019;**142**(5):1176–94. https://doi.org/10.1093/brain/awz078.
2. Espay A, Vizcarra J, Marsili L, et al. Revisiting protein aggregation as pathogenic in sporadic Parkinson and Alzheimer diseases. *Neurology* 2019;**92**:329–37.
3. Neumann M, Sampathu DM, Kwong LK, et al. Ubiquitinated TDP-43 in frontotemporal lobar degeneration and amyotrophic lateral sclerosis. *Science* 2006;**314**(5796):130–3. https://doi.org/10.1126/science.1134108.
4. Kawahara Y, Mieda-Sato A. TDP-43 promotes microRNA biogenesis as a component of the Drosha and Dicer complexes. *Proc Natl Acad Sci USA* 2012;**109**(9):3347–52. https://doi.org/10.1073/pnas.1112427109.
5. Ou SH, Wu F, Harrich D, García-Martínez LF, Gaynor RB. Cloning and characterization of a novel cellular protein TDP-43, that binds to human immunodeficiency virus type 1 TAR DNA sequence motifs. *J Virol* 1995;**69**(6):3584–96. https://doi.org/10.1128/jvi.69.6.3584-3596.1995.
6. Garcia JA, Wu FK, Mitsuyasu R, Gaynor RB. Interactions of cellular proteins involved in the transcriptional regulation of the human immunodeficiency virus. *EMBO J* 1987;**6**(12):3761–70. https://doi.org/10.1002/j.1460-2075.1987.tb02711.x.
7. de Boer EMJ, Orie VK, Williams T, et al. TDP-43 proteinopathies: a new wave of neurodegenerative diseases. *J Neurol Neurosurg Psychiatry* November 11, 2020. https://doi.org/10.1136/jnnp-2020-322983.
8. Jo M, Lee S, Jeon Y-M, Kim S, Kwon Y, Kim H-J. The role of TDP-43 propagation in neurodegenerative diseases: integrating insights from clinical and experimental studies. *Exp Mol Med* 2020;**52**(10):1652–62. https://doi.org/10.1038/s12276-020-00513-7.
9. Suk TR, Rousseaux MWC. The role of TDP-43 mislocalization in amyotrophic lateral sclerosis. *Mol Neurodegener* 2020;**15**(1). https://doi.org/10.1186/s13024-020-00397-1.
10. Bolognesi B, Faure AJ, Seuma M, Schmiedel JM, Tartaglia GG, Lehner B. The mutational landscape of a prion-like domain. *Nat Commun* 2019;**10**(1):4162. https://doi.org/10.1038/s41467-019-12101-z.
11. Arnold ES, Ling S-C, Huelga SC, et al. ALS-linked TDP-43 mutations produce aberrant RNA splicing and adult-onset motor neuron disease without aggregation or loss of nuclear TDP-43. *Proc Natl Acad Sci USA* 2013;**110**(8):E736–45. https://doi.org/10.1073/pnas.1222809110.
12. Mitra J, Guerrero EN, Hegde PM, et al. Motor neuron disease-associated loss of nuclear TDP-43 is linked to DNA double-strand break repair defects. *SSRN Electron J* 2018. https://doi.org/10.2139/ssrn.3155763.
13. Jiang L-L, Xue W, Hu H-Y. *Solution structure of the N-terminal domain of TDP-43.* Worldwide Protein Data Bank; 2017. https://doi.org/10.2210/pdb5x4f/pdb.
14. Mejzini R, Flynn LL, Pitout IL, Fletcher S, Wilton SD, Akkari PA. ALS genetics mechanisms, and therapeutics: where are we now? *Front Neurosci* 2019;**13**. https://doi.org/10.3389/fnins.2019.01310.

15. Nagano S, Jinno J, Abdelhamid RF, et al. TDP-43 transports ribosomal protein mRNA to regulate axonal local translation in neuronal axons. *Acta Neuropathol* 2020;**140**(5):695−713. https://doi.org/10.1007/s00401-020-02205-y.
16. Mitra J, Hegde ML. A commentary on TDP-43 and DNA damage response in amyotrophic lateral sclerosis. *J Exp Neurosci* 2019;**13**. https://doi.org/10.1177/1179069519880166.
17. Wolozin B. The evolution of phase-separated TDP-43 in stress. *Neuron* 2019;**102**(2):265−7. https://doi.org/10.1016/j.neuron.2019.03.041.
18. Yang C, Tan W, Whittle C, et al. The C-terminal TDP-43 fragments have a high aggregation propensity and harm neurons by a dominant-negative mechanism. *PLoS One* 2010;**5**(12):e15878. https://doi.org/10.1371/journal.pone.0015878. Linden R, ed.
19. Pesiridis G, Lee V, Trojanowski J. Mutations in TDP-43 link glycine-rich domain functions to amyotrophic lateral sclerosis. *Hum Mol Genet* 2009;**18**:R156−62.
20. Wright GSA, Watanabe TF, Amporndanai K, et al. Purification and structural characterization of aggregation-prone human TDP-43 involved in neurodegenerative diseases. *iScience* 2020;**23**(6):101159. https://doi.org/10.1016/j.isci.2020.101159.
21. Afroz T, Hock E-M, Ernst P, et al. Functional and dynamic polymerization of the ALS-linked protein TDP-43 antagonizes its pathologic aggregation. *Nat Commun* 2017;**8**(1). https://doi.org/10.1038/s41467-017-00062-0.
22. Arai T. Significance and limitation of the pathological classification of TDP-43 proteinopathy: classification of TDP-43 proteinopathy. *Neuropathology* 2014;**34**(6):578−88. https://doi.org/10.1111/neup.12138.
23. Kumar V, Wahiduzzaman, Prakash A, et al. Exploring the aggregation-prone regions from structural domains of human TDP-43. *Biochim Biophys Acta Proteins Proteom* 2019;**1867**:286−96.
24. Conicella AE, Dignon GL, Zerze GH, et al. TDP-43 α-helical structure tunes liquid−liquid phase separation and function. *Proc Natl Acad Sci USA* 2020;**117**(11):5883−94. https://doi.org/10.1073/pnas.1912055117.
25. Jiang L-L, Xue W, Hong J-Y, et al. The N-terminal dimerization is required for TDP-43 splicing activity. *Sci Rep* 2017;**7**(1). https://doi.org/10.1038/s41598-017-06263-3.
26. Jaiswal MK. Riluzole and edaravone: a tale of two amyotrophic lateral sclerosis drugs. *Med Res Rev* 2018;**39**(2):733−48. https://doi.org/10.1002/med.21528.
27. Fernandopulle M, Wang G, Nixon-Abell J, et al. Inherited and sporadic amyotrophic lateral sclerosis and fronto-temporal lobar degenerations arising from pathological condensates of phase separating proteins. *Hum Mol Genet* 2019;**28**(R2):R187−96. https://doi.org/10.1093/hmg/ddz162.
28. Wang Y-T, Kuo P-H, Chiang C-H, et al. The truncated C-terminal RNA recognition motif of TDP-43 protein plays a key role in forming proteinaceous aggregates. *J Biol Chem* 2013;**288**(13):9049−57. https://doi.org/10.1074/jbc.M112.438564.
29. Li Q, Yokoshi M, Okada H, Kawahara Y. The cleavage pattern of TDP-43 determines its rate of clearance and cytotoxicity. *Nat Commun* 2015;**6**(1):6183. https://doi.org/10.1038/ncomms7183.
30. Berning BA, Walker AK. The pathobiology of TDP-43 C-terminal fragments in ALS and FTLD. *Front Neurosci* 2019;**13**:335. https://doi.org/10.3389/fnins.2019.00335.
31. Igaz LM, Kwong LK, Xu Y, et al. Enrichment of C-terminal fragments in TAR DNA-binding protein-43 cytoplasmic inclusions in brain but not in spinal cord of frontotemporal lobar degeneration and amyotrophic lateral sclerosis. *Am J Pathol* 2008;**173**(1):182−94. https://doi.org/10.2353/ajpath.2008.080003.

32. Prasad A, Bharathi V, Sivalingam V, Girdhar A, Patel BK. Molecular mechanisms of TDP-43 misfolding and pathology in amyotrophic lateral sclerosis. *Front Mol Neurosci* 2019;**12**. https://doi.org/10.3389/fnmol.2019.00025.
33. Winton MJ, Igaz LM, Wong MM, Kwong LK, Trojanowski JQ, Lee VM-Y. Disturbance of nuclear and cytoplasmic TAR DNA-binding protein (TDP-43) induces disease-like redistribution sequestration, and aggregate formation. *J Biol Chem* 2008;**283**(19):13302–9. https://doi.org/10.1074/jbc.m800342200.
34. French RL, Grese ZR, Aligireddy H, et al. Detection of TAR DNA-binding protein 43 (TDP-43) oligomers as initial intermediate species during aggregate formation. *J Biol Chem* 2019;**294**(17):6696–709. https://doi.org/10.1074/jbc.RA118.005889.
35. Ayala YM, Conti LD, Avendaño-Vázquez SE, et al. TDP-43 regulates its mRNA levels through a negative feedback loop. *EMBO J* 2010;**30**(2):277–88. https://doi.org/10.1038/emboj.2010.310.
36. Koyama A, Sugai A, Kato T, et al. Increased cytoplasmic TARDBP mRNA in affected spinal motor neurons in ALS caused by abnormal autoregulation of TDP-43. *Nucleic Acids Res* 2016;**44**(12):5820–36. https://doi.org/10.1093/nar/gkw499.
37. Jovičić A, Paul JW, Gitler AD. Nuclear transport dysfunction: a common theme in amyotrophic lateral sclerosis and frontotemporal dementia. *J Neurochem* 2016;**138**:134–44. https://doi.org/10.1111/jnc.13642.
38. Barbieri EM, Shorter J. TDP-43 shape shifts to encipher FTD severity. *Nat Neurosci* 2019;**22**(1):3–5. https://doi.org/10.1038/s41593-018-0299-6.
39. Laferrière F, Maniecka Z, Pérez-Berlanga M, et al. TDP-43 extracted from frontotemporal lobar degeneration subject brains displays distinct aggregate assemblies and neurotoxic effects reflecting disease progression rates. *Nat Neurosci* 2019;**22**(1):65–77. https://doi.org/10.1038/s41593-018-0294-y.
40. Wang P, Wander CM, Yuan C-X, Bereman MS, Cohen TJ. Acetylation-induced TDP-43 pathology is suppressed by an HSF1-dependent chaperone program. *Nat Commun* 2017;**8**(1):82. https://doi.org/10.1038/s41467-017-00088-4.
41. Vanden BL, Callaerts P, Dermaut B. TDP-43-mediated neurodegeneration: towards a loss-of-function hypothesis? *Trends Mol Med* 2014;**20**:66–71.
42. Polymenidou M, Lagier-Tourenne C, Hutt K, et al. Long pre-mRNA depletion and RNA missplicing contribute to neuronal vulnerability from loss of TDP-43. *Nat Neurosci* 2011;**14**:459–68.
43. Lagier-Tourenne C, Polymenidou M, Hutt K, et al. Divergent roles of ALS-linked proteins FUS/TLS and TDP-43 intersect in processing long pre-mRNAs. *Nat Neurosci* 2012;**15**:1488–97.
44. Wenqiang C, Lonskaya I, Hebron M, et al. Parkin-mediated reduction of nuclear and soluble TDP-43 reverses behavioral decline in symptomatic mice. *Hum Mol Genet* 2014;**23**:4960–9.
45. McCoy MJ, Fire AZ. Intron and gene size expansion during nervous system evolution. *BMC Genom* 2020;**21**(1):360. https://doi.org/10.1186/s12864-020-6760-4.
46. Wang J, Ho W, Lim K, et al. Cell-autonomous requirement of TDP-43, an ALS/FTD signature protein, for oligodendrocyte survival and myelination. *Proc Natl Acad Sci USA* 2018;**115**:E10941–50.
47. Arai T, Hasegawa M, Akiyama H, et al. TDP-43 is a component of ubiquitin-positive tau-negative inclusions in frontotemporal lobar degeneration and amyotrophic lateral sclerosis. *Biochem Biophys Res Commun* 2006;**351**(3):602–11. https://doi.org/10.1016/j.bbrc.2006.10.093.

48. Takei K, Takahashi F, Liu S, Tsuda K, Palumbo J. Post-hoc analysis of randomised placebo-controlled, double-blind study (MCI186-19) of edaravone (MCI-186) in amyotrophic lateral sclerosis. *Amyotroph Lateral Scler Front Degener* 2017;**18**(Suppl. 1):49–54. https://doi.org/10.1080/21678421.2017.1361443.
49. Yoshino H, Kimura A. Investigation of the therapeutic effects of edaravone a free radical scavenger, on amyotrophic lateral sclerosis (Phase II study). *Amyotroph Lateral Scler* 2006;**7**(4):247–51. https://doi.org/10.1080/17482960600881870.
50. Abe K, Itoyama Y, Sobue G, et al. Confirmatory double-blind parallel-group, placebo-controlled study of efficacy and safety of edaravone (MCI-186) in amyotrophic lateral sclerosis patients. *Amyotroph Lateral Scler Front Degener* 2014;**15**(7–8):610–7. https://doi.org/10.3109/21678421.2014.959024.
51. Abe K, Aoki M, Tsuji S, et al. Safety and efficacy of edaravone in well defined patients with amyotrophic lateral sclerosis: a randomised, double-blind, placebo-controlled trial. *Lancet Neurol* 2017;**16**(7):505–12. https://doi.org/10.1016/S1474-4422(17)30115-1.
52. Paganoni S, Hendrix S, Dickson SP, et al. Long-term survival of participants in the CENTAUR trial of sodium phenylbutyrate-taurursodiol in amyotrophic lateral sclerosis. *Muscle Nerve* October 30, 2020. https://doi.org/10.1002/mus.27091.
53. Lacomblez L, Bensimon and G, Meininger V, Leigh PN, Guillet P. Dose-ranging study of riluzole in amyotrophic lateral sclerosis. *Lancet* 1996;**347**(9013):1425–31. https://doi.org/10.1016/s0140-6736(96)91680-3.
54. Fang T, Khleifat AA, Meurgey J-H, et al. Stage at which riluzole treatment prolongs survival in patients with amyotrophic lateral sclerosis: a retrospective analysis of data from a dose-ranging study. *Lancet Neurol* 2018;**17**(5):416–22. https://doi.org/10.1016/s1474-4422(18)30054-1.
55. Hinchcliffe M, Smith A. Riluzole: real-world evidence supports significant extension of median survival times in patients with amyotrophic lateral sclerosis. *Degener Neurol Neuromuscul Dis* 2017;**7**:61–70. https://doi.org/10.2147/DNND.S135748.
56. Wang H, Guo W, Mitra J, et al. Mutant FUS causes DNA ligation defects to inhibit oxidative damage repair in Amyotrophic Lateral Sclerosis. *Nat Commun* 2018;**9**(1):3683. https://doi.org/10.1038/s41467-018-06111-6.
57. Baratchi S, Kanwar RK, Kanwar JR. Survivin: a target from brain cancer to neurodegenerative disease. *Crit Rev Biochem Mol Biol* 2010;**45**(6):535–54. https://doi.org/10.3109/10409238.2010.516740.
58. Charriaut-Marlangue C. Apoptosis: a target for neuroprotection. *Therapies* 2004;**59**(2):185–90. https://doi.org/10.2515/therapie:2004035.
59. Mitra J LHM. The role of TDP-43 in genome repair and beyond in amyotrophic lateral sclerosis. In: Hegde M L, editor. *Amyotrophic lateral sclerosis - recent advances and therapeutic challenges*. IntechOpen; 2020. https://doi.org/10.5772/intechopen.92696.
60. Wang I-F, Tsai K-J, Shen C-KJ. Autophagy activation ameliorates neuronal pathogenesis of FTLD-U mice. *Autophagy* 2013;**9**(2):239–40. https://doi.org/10.4161/auto.22526.
61. Chen P-C, Hsieh Y-C, Huang C-C, Hu C-J. Tamoxifen for amyotrophic lateral sclerosis. *Medicine* 2020;**99**(22):e20423. https://doi.org/10.1097/md.0000000000020423.
62. Pozzi S, Thammisetty SS, Codron P, et al. Virus-mediated delivery of antibody targeting TAR DNA-binding protein-43 mitigates associated neuropathology. *J Clin Invest* 2019;**129**(4):1581–95. https://doi.org/10.1172/JCI123931.

63. Kumar S, Phaneuf D, Cordeau P, Boutej H, Kriz J, Julien J-P. Induction of autophagy mitigates TDP-43 pathology and translational repression of neurofilament mRNAs in mouse models of ALS/FTD. *Mol Neurodegener* 2021;**16**(1):1. https://doi.org/10.1186/s13024-020-00420-5.

64. Okano H, Yamanaka S. iPS cell technologies: significance and applications to CNS regeneration and disease. *Mol Brain* 2014;**7**(1):22. https://doi.org/10.1186/1756-6606-7-22.

65. Fujimori K, Ishikawa M, Otomo A, et al. Modeling sporadic ALS in iPSC-derived motor neurons identifies a potential therapeutic agent. *Nat Med* 2018;**24**(10):1579–89. https://doi.org/10.1038/s41591-018-0140-5.

66. Morimoto S, Takahashi S, Fukushima K, et al. Ropinirole hydrochloride remedy for amyotrophic lateral sclerosis - protocol for a randomized, double-blind, placebo-controlled, single-center, and open-label continuation phase I/IIa clinical trial (ROPALS trial). *Regen Ther* 2019;**11**:143–66.

67. Wainger BJ, Kiskinis E, Mellin C, et al. Intrinsic membrane hyperexcitability of amyotrophic lateral sclerosis patient-derived motor neurons. *Cell Rep* 2014;**7**(1):1–11. https://doi.org/10.1016/j.celrep.2014.03.019.

68. Imamura K, Izumi Y, Watanabe A, et al. The Src/c-Abl pathway is a potential therapeutic target in amyotrophic lateral sclerosis. *Sci Transl Med* 2017;**9**.

69. Okano H, Yasuda D, Fujimori K, Morimoto S, Takahashi S. Ropinirole, a new ALS drug candidate developed using iPSCs. *Trends Pharmacol Sci* 2020;**41**:99–109.

70. Statland J, Moore D, Wang Y, et al. Rasagiline for amyotrophic lateral sclerosis: a randomized, controlled trial. *Muscle Nerve* 2019;**59**:201–7.

71. Baker S, Lindeman K. Rethinking historical controls. *Biostatistics* 2001;**2**:383–96.

72. Ludolph A, Schuster J, Dorst J, et al. Safety and efficacy of rasagiline as an add-on therapy to riluzole in patients with amyotrophic lateral sclerosis: a randomised, double-blind, parallel-group, placebo-controlled, phase 2 trial. *Lancet Neurol* 2018;**17**:681–8.

73. Patten SA, Aggad D, Martinez J, et al. Neuroleptics as therapeutic compounds stabilizing neuromuscular transmission in amyotrophic lateral sclerosis. *JCI Insight* 2017;**2**(22). https://doi.org/10.1172/jci.insight.97152.

74. Watermeyer T, Robb C, Gregory S, Udeh-Momoh C. Therapeutic implications of hypothalamic-pituitaryadrenal-axis modulation in Alzheimer's disease: a narrative review of pharmacological and lifestyle interventions. *Front Neuroendocrinol* 2021;**60**:100877. https://doi.org/10.1016/j.yfrne.2020.100877.

75. Grinevich V, Neumann ID. Brain oxytocin: how puzzle stones from animal studies translate into psychiatry. *Mol Psychiatry* 2021;**26**:265–79. https://doi.org/10.1038/s41380-020-0802-9.

76. Diodati D, Cyn-Ang L, Kertesz A, Finger E. Pathologic evaluation of the supraoptic and paraventricular nuclei in dementia. *Can J Neurol Sci* 2012;**39**:213–9.

77. Makkinejad N, Schneider J, Yu J, et al. Associations of amygdala volume and shape with transactive response DNA-binding protein 43 (TDP-43) pathology in a community cohort of older adults. *Neurobiol Aging* 2019;**77**:104–11.

78. Nag S, Yu L, Wilson RS, Chen E-Y, Bennett DA, Schneider JA. TDP-43 pathology and memory impairment in elders without pathologic diagnoses of AD or FTLD. *Neurology* 2017;**88**(7):653–60. https://doi.org/10.1212/wnl.0000000000003610.

79. Finger E, MacKinley J, Blair M, et al. Oxytocin for frontotemporal dementia: a randomized dose-finding study of safety and tolerability. *Neurology* 2015;**84**:174–81.

80. Wobst HJ, Delsing L, Brandon NJ, Moss SJ. Truncation of the TAR DNA-binding protein 43 is not a prerequisite for cytoplasmic relocalization and is suppressed by caspase inhibition and by introduction of the A90V sequence variant. *PLoS One* 2017;**12**(5):e0177181. https://doi.org/10.1371/journal.pone.0177181. Buratti E, ed.
81. Liachko NF, McMillan PJ, Guthrie CR, Bird TD, Leverenz JB, Kraemer BC. CDC7 inhibition blocks pathological TDP-43 phosphorylation and neurodegeneration. *Ann Neurol* 2013;**74**(1):39–52. https://doi.org/10.1002/ana.23870.
82. Martínez-González L, Rodríguez-Cueto C, Cabezudo D, et al. Motor neuron preservation and decrease of in vivo TDP-43 phosphorylation by protein CK-1δ kinase inhibitor treatment. *Sci Rep* 2020;**10**:4449.
83. Gu J, Hu W, Tan X, et al. Elevation of casein kinase 1ε associated with TDP-43 and tau pathologies in Alzheimer's disease. *Brain Pathol* 2020;**30**:283–97.
84. Neumann M, Kwong LK, Lee EB, et al. Phosphorylation of S409/410 of TDP-43 is a consistent feature in all sporadic and familial forms of TDP-43 proteinopathies. *Acta Neuropathol* 2009;**117**(2):137–49. https://doi.org/10.1007/s00401-008-0477-9.
85. Khalid S, Ampie L, Kelly R, Ladha S, Dardis C. Immune modulation in the treatment of amyotrophic lateral sclerosis: a review of clinical trials. *Front Neurol* 2017;**8**:486.
86. Apontes P, Leontieva O, Demidenko Z, Li F, Blagosklonny M. Exploring long-term protection of normal human fibroblasts and epithelial cells from chemotherapy in cell culture. *Oncotarget* 2011;**2**:222–33.
87. Blagosklonny MV. Rapamycin for longevity: opinion article. *Aging* 2019;**11**(19):8048–67. https://doi.org/10.18632/aging.102355.
88. Mandrioli J, D'Amico R, Zucchi E, et al. Rapamycin treatment for amyotrophic lateral sclerosis: protocol for a phase II randomized, double-blind, placebo-controlled, multicenter, clinical trial (RAP-ALS trial). *Med Baltim* 2018;**97**:e11119.
89. Caccamo A, Majumder S, Deng JJ, Bai Y, Thornton FB, Oddo S. Rapamycin rescues TDP-43 mislocalization and the associated low molecular mass neurofilament instability. *J Biol Chem* 2009;**284**(40):27416–24. https://doi.org/10.1074/jbc.m109.031278.
90. Barmada SJ, Serio A, Arjun A, et al. Autophagy induction enhances TDP43 turnover and survival in neuronal ALS models. *Nat Chem Biol* 2014;**10**(8):677–85. https://doi.org/10.1038/nchembio.1563.
91. Rojas F, Cortes N, Abarzua S, Dyrda A, van ZB. Astrocytes expressing mutant SOD1 and TDP43 trigger motoneuron death that is mediated via sodium channels and nitroxidative stress. *Front Cell Neurosci* 2014;**8**:24.
92. Weiss M, Macklin E, Simmons Z, et al. A randomized trial of mexiletine in ALS: safety and effects on muscle cramps and progression. *Neurology* 2016;**86**:1474–81.
93. Oskarsson B, Moore D, Mozaffar T, et al. Mexiletine for muscle cramps in amyotrophic lateral sclerosis: a randomized, double-blind crossover trial. *Muscle Nerve* 2018;**58**(1):42–8. https://doi.org/10.1002/mus.26117 [published online ahead of print, 2018 Mar 6].
94. Vidoni ED, Choi I, Lee P, et al. Safety and target engagement profile of two oxaloacetate doses in Alzheimer's patients. *Alzheimers Dement* 2021;**17**(1):7–17. https://doi.org/10.1002/alz.12156.
95. Kawahara Y, Ito K, Sun H, Aizawa H, Kanazawa I, Kwak S. RNA editing and death of motor neurons. *Nature* 2004;**427**(6977). https://doi.org/10.1038/427801a. 801-801.
96. Licht K, Hartl M, Amman F, Anrather D, Janisiw MP, Jantsch MF. Inosine induces context-dependent recoding and translational stalling. *Nucleic Acids Res* 2019;**47**(1):3–14. https://doi.org/10.1093/nar/gky1163.

97. Rosenthal J, Seeburg P. A-to-I RNA editing: effects on proteins key to neural excitability. *Neuron* 2012;**74**:432−9.
98. Selvaraj BT, Livesey MR, Zhao C, et al. C9ORF72 repeat expansion causes vulnerability of motor neurons to Ca2+-permeable AMPA receptor-mediated excitotoxicity. *Nat Commun* 2018;**9**(1):347. https://doi.org/10.1038/s41467-017-02729-0.
99. Yamashita T, Kwak S. The molecular link between inefficient GluA2 Q/R site-RNA editing and TDP-43 pathology in motor neurons of sporadic amyotrophic lateral sclerosis patients. *Brain Res* 2014;**1584**:28−38. https://doi.org/10.1016/j.brainres.2013.12.011.
100. Akamatsu M, Yamashita T, Hirose N, Teramoto S, Kwak S. The AMPA receptor antagonist perampanel robustly rescues amyotrophic lateral sclerosis (ALS) pathology in sporadic ALS model mice. *Sci Rep* 2016;**6**:28649.
101. Aizawa H, Kwak S. [Perampanel for sporadic amyotrophic lateral sclerosis]. *Brain Nerve* 2019;**71**:1270−8.
102. Fang M, Markmiller S, Vu A, et al. Small-molecule modulation of TDP-43 recruitment to stress granules prevents persistent TDP-43 accumulation in ALS/FTD. *Neuron* 2019;**103**:802−819.e11.
103. Inda MC, Joshi S, Wang T, et al. The epichaperome is a mediator of toxic hippocampal stress and leads to protein connectivity-based dysfunction. *Nat Commun* 2020;**11**(1):319. https://doi.org/10.1038/s41467-019-14082-5.
104. Rizzolo K, Houry WA. Multiple functionalities of molecular chaperones revealed through systematic mapping of their interaction networks. *J Biol Chem* 2019;**294**(6):2142−50. https://doi.org/10.1074/jbc.TM118.002805.
105. Hollyday M, Hamburger V. Reduction of the naturally occurring motor neuron loss by enlargement of the periphery. *J Comp Neurol* 1976;**170**(3):311−20. https://doi.org/10.1002/cne.901700304.
106. Levi-Montalcini R. The nerve growth factor 35 years later. *Science* 1987;**237**(4819):1154−62. https://doi.org/10.1126/science.3306916.
107. Low LK, Cheng H-J. Axon pruning: an essential step underlying the developmental plasticity of neuronal connections. *Philos Trans R Soc B Biol Sci* 2006;**361**(1473):1531−44. https://doi.org/10.1098/rstb.2006.1883.
108. Calderó J, Prevette D, Mei X, et al. Peripheral target regulation of the development and survival of spinal sensory and motor neurons in the chick embryo. *J Neurosci* 1998;**18**(1):356−70. https://doi.org/10.1523/JNEUROSCI.18-01-00356.1998.
109. Oppenheim R, Haverkamp L, Prevette D, McManaman J, Appel S. Reduction of naturally occurring motoneuron death in vivo by a target-derived neurotrophic factor. *Science* 1988;**240**(4854):919−22. https://doi.org/10.1126/science.3363373.
110. Yu J, Zhu H, Ko D, Kindy MS. Motoneuronotrophic factor analog GM6 reduces infarct volume and behavioral deficits following transient ischemia in the mouse. *Brain Res* 2008;**1238**:143−53. https://doi.org/10.1016/j.brainres.2008.08.053.
111. Swindell WR, Bojanowski K, Kindy MS, Chau RMW, Ko D. GM604 regulates developmental neurogenesis pathways and the expression of genes associated with amyotrophic lateral sclerosis. *Transl Neurodegener* 2018;**7**(1):30. https://doi.org/10.1186/s40035-018-0135-7.
112. Kindy M, Lupinacci P, Chau R, Shum T, Ko D. A Phase 2A randomized, double-blind, placebo-controlled pilot trial of GM604 in patients with Amyotrophic Lateral Sclerosis (ALS Protocol GALS-001) and a single compassionate patient treatment (Protocol GALS-C). *F1000Research* 2017;**6**:230. https://doi.org/10.12688/f1000research.10519.1.

113. The ALSUntangled Group. ALSUntangled No. 34: GM604. *Amyotroph Lateral Scler Front Degener* 2016;**17**(7–8):617–21. https://doi.org/10.3109/21678421.2016.1164940.
114. Russell AJ, Hartman JJ, Hinken AC, et al. Activation of fast skeletal muscle troponin as a potential therapeutic approach for treating neuromuscular diseases. *Nat Med* 2012;**18**(3):452–5. https://doi.org/10.1038/nm.2618.
115. Hwee DT, Kennedy AR, Hartman JJ, et al. The small-molecule fast skeletal troponin activator, CK-2127107, improves exercise tolerance in a rat model of heart failure. *J Pharmacol Exp Ther* 2015;**353**(1):159–68. https://doi.org/10.1124/jpet.114.222224.
116. Shefner JM, Andrews JA, Genge A, et al. A phase 2, double-blind, randomized, dose-ranging trial of reldesemtiv in patients with ALS. *Amyotroph Lateral Scler Front Degener* September 24, 2020:1–13. https://doi.org/10.1080/21678421.2020.1822410.
117. Wurster CD, Ludolph AC. Antisense oligonucleotides in neurological disorders. *Ther Adv Neurol Disord* 2018;**11**. https://doi.org/10.1177/1756286418776932.
118. Mann J, Gleixner A, Mauna J, et al. RNA binding antagonizes neurotoxic phase transitions of TDP-43. *Neuron* 2019;**102**:321–338.e8.
119. Becker LA, Huang B, Bieri G, et al. Therapeutic reduction of ataxin-2 extends lifespan and reduces pathology in TDP-43 mice. *Nature* 2017;**544**(7650):367–71. https://doi.org/10.1038/nature22038.
120. Floare M, Allen S. Why TDP-43? Why not? Mechanisms of metabolic dysfunction in amyotrophic lateral sclerosis. *Neurosci Insights* 2020;**15**. 2633105520957302.
121. Kaneb HM, Sharp PS, Rahmani-Kondori N, Wells DJ. Metformin treatment has No beneficial effect in a dose-response survival study in the SOD1G93A mouse model of ALS and is harmful in female mice. *PLoS One* 2011;**6**(9):e24189. https://doi.org/10.1371/journal.pone.0024189. Cai H, ed.
122. Zu T, Guo S, Bardhi O, et al. Metformin inhibits RAN translation through PKR pathway and mitigates disease inC9orf72ALS/FTD mice. *Proc Natl Acad Sci USA* 2020;**117**(31):18591–9. https://doi.org/10.1073/pnas.2005748117.
123. Banez-Coronel M, Ranum LPW. Repeat-associated non-AUG (RAN) translation: insights from pathology. *Lab Invest* 2019;**99**(7):929–42. https://doi.org/10.1038/s41374-019-0241-x.
124. Cheng W, Wang S, Mestre AA, et al. C9ORF72 GGGGCC repeat-associated non-AUG translation is upregulated by stress through eIF2α phosphorylation. *Nat Commun* 2018;**9**(1):51. https://doi.org/10.1038/s41467-017-02495-z.
125. Paganoni S, Macklin EA, Hendrix S, et al. Trial of sodium phenylbutyrate–taurursodiol for amyotrophic lateral sclerosis. *N Engl J Med* 2020;**383**(10):919–30. https://doi.org/10.1056/NEJMoa1916945.
126. Ahmed R, Irish M, Piguet O, et al. Amyotrophic lateral sclerosis and frontotemporal dementia: distinct and overlapping changes in eating behaviour and metabolism. *Lancet Neurol* 2016;**15**:332–42.
127. Bouteloup C, Desport J-C, Clavelou P, et al. Hypermetabolism in ALS patients: an early and persistent phenomenon. *J Neurol* 2009;**256**(8):1236–42. https://doi.org/10.1007/s00415-009-5100-z.
128. Vandoorne T, De BK, Van DBL. Energy metabolism in ALS: an underappreciated opportunity? *Acta Neuropathol* 2018;**135**:489–509.
129. Pathak D, Berthet A, Nakamura K. Energy failure: does it contribute to neurodegeneration? *Ann Neurol* 2013;**74**:506–16.
130. Szelechowski M, Amoedo N, Obre E, et al. Metabolic reprogramming in amyotrophic lateral sclerosis. *Sci Rep* 2018;**8**(1):3953. https://doi.org/10.1038/s41598-018-22318-5.

131. Wang P, Deng J, Dong J, et al. TDP-43 induces mitochondrial damage and activates the mitochondrial unfolded protein response. *PLoS Genet* 2019;**15**(5):e1007947. https://doi.org/10.1371/journal.pgen.1007947. Hamilton BA, ed.

132. Muddapu V, Dharshini S, Chakravarthy V, Gromiha M. Neurodegenerative diseases - is metabolic deficiency the root cause? *Front Neurosci* 2020;**14**:213.

133. Ota M, Matsuo J, Ishida I, et al. Effects of a medium-chain triglyceride-based ketogenic formula on cognitive function in patients with mild-to-moderate Alzheimer's disease. *Neurosci Lett* 2019;**690**:232–6. https://doi.org/10.1016/j.neulet.2018.10.048.

134. Zhou J, Gennatas E, Kramer J, Miller B, Seeley W. Predicting regional neurodegeneration from the healthy brain functional connectome. *Neuron* 2012;**73**:1216–27.

135. Zhao W, Varghese M, Vempati P, et al. Caprylic triglyceride as a novel therapeutic approach to effectively improve the performance and attenuate the symptoms due to the motor neuron loss in ALS disease. *PLoS One* 2012;**7**(11):e49191. https://doi.org/10.1371/journal.-pone.0049191. Pandey U, ed.

136. Jawaid A, Khan R, Polymenidou M, Schulz PE. Disease-modifying effects of metabolic perturbations in ALS/FTLD. *Mol Neurodegener* 2018;**13**(1):63. https://doi.org/10.1186/s13024-018-0294-0.

137. Grammatikopoulou MG, Goulis DG, Gkiouras K, et al. To keto or not to keto? A systematic review of randomized controlled trials assessing the effects of ketogenic therapy on Alzheimer disease. *Adv Nutr* June 29, 2020. https://doi.org/10.1093/advances/nmaa073.

138. Ota M, Matsuo J, Ishida I, et al. Effect of a ketogenic meal on cognitive function in elderly adults: potential for cognitive enhancement. *Psychopharmacology* 2016;**233**(21–22):3797–802. https://doi.org/10.1007/s00213-016-4414-7.

139. Ludolph AC, Dorst J, Dreyhaupt J, et al. Effect of high-caloric nutrition on survival in amyotrophic lateral sclerosis. *Ann Neurol* 2020;**87**(2):206–16. https://doi.org/10.1002/ana.25661.

140. Maccioni R, Setzu M, Talani G, et al. Standardized phytotherapic extracts rescue anomalous locomotion and electrophysiological responses of TDP-43 *Drosophila melanogaster* model of ALS. *Sci Rep* 2018;**8**:16002.

141. Katzenschlager R, Evans A, Manson A, et al. Mucuna pruriens in Parkinson's disease: a double blind clinical and pharmacological study. *J Neurol Neurosurg Psychiatry* 2004;**75**:1672–7.

142. Dar N, Hamid A, Ahmad M. Pharmacologic overview of Withania somnifera, the Indian ginseng. *Cell Mol Life Sci* 2015;**72**:4445–60.

143. Salfi F, D'Atri A, Tempesta D, De GL, Ferrara M. Boosting slow oscillations during sleep to improve memory function in elderly people: a review of the literature. *Brain Sci* 2020;**10**.

144. Balendra R, Isaacs AM. C9orf72-mediated ALS and FTD: multiple pathways to disease. *Nat Rev Neurol* 2018;**14**(9):544–58. https://doi.org/10.1038/s41582-018-0047-2.

145. Dedeene L, Van SE, Vandenberghe R, Van DP, Poesen K, Thal D. Circadian sleep/wake-associated cells show dipeptide repeat protein aggregates in C9orf72-related ALS and FTLD cases. *Acta Neuropathol Commun* 2019;**7**:189.

146. Gittings LM, Boeynaems S, Lightwood D, et al. Symmetric dimethylation of poly-GR correlates with disease duration in C9orf72 FTLD and ALS and reduces poly-GR phase separation and toxicity. *Acta Neuropathol* 2019;**139**(2):407–10. https://doi.org/10.1007/s00401-019-02104-x.

147. Estes PS, Daniel SG, Mccallum AP, et al. Motor neurons and glia exhibit specific individualized responses to TDP-43 expression in a Drosophila model of amyotrophic lateral sclerosis. *Dis Model Mech* 2013;**6**(3):721–33. https://doi.org/10.1242/dmm.010710.

148. Chew J, Gendron T, Prudencio M, et al. Neurodegeneration. C9ORF72 repeat expansions in mice cause TDP-43 pathology, neuronal loss, and behavioral deficits. *Science* 2015;**348**:1151–4.

149. Howell MJ, Schenck CH. Rapid eye movement sleep behavior disorder and neurodegenerative disease. *JAMA Neurol* 2015;**72**(6):707. https://doi.org/10.1001/jamaneurol.2014.4563.

150. Weishaupt JH, Bartels C, Pölking E, et al. Reduced oxidative damage in ALS by high-dose enteral melatonin treatment. *J Pineal Res* 2006;**41**(4):313–23. https://doi.org/10.1111/j.1600-079X.2006.00377.x.

151. Jaiswal MK. Riluzole but not melatonin ameliorates acute motor neuron degeneration and moderately inhibits SOD1-mediated excitotoxicity induced disrupted mitochondrial Ca^{2+} signaling in amyotrophic lateral sclerosis. *Front Cell Neurosci* 2017;**10**. https://doi.org/10.3389/fncel.2016.00295.

152. Bredesen DE. Reversal of cognitive decline: a novel therapeutic program. *Aging* 2014;**6**(9):707–17. https://doi.org/10.18632/aging.100690.

153. Bredesen DE, Sharlin K, Jenkins D, et al. Reversal of cognitive decline: 100 patients. *J Alzheimer's Dis Park* 2018;**08**(05). https://doi.org/10.4172/2161-0460.1000450.

154. Shi J, Ni J, Lu T, et al. Adding Chinese herbal medicine to conventional therapy brings cognitive benefits to patients with Alzheimer's disease: a retrospective analysis. *BMC Complement Altern Med* 2017;**17**(1):533. https://doi.org/10.1186/s12906-017-2040-5.

155. Chang C-F, Lee Y-C, Lee K-H, et al. Therapeutic effect of berberine on TDP-43-related pathogenesis in FTLD and ALS. *J Biomed Sci* 2016;**23**(1):72. https://doi.org/10.1186/s12929-016-0290-z.

156. Catanesi M, Caioni G, Castelli V, Benedetti E, d'Angelo M, Cimini A. Benefits under the sea: the role of marine compounds in neurodegenerative disorders. *Mar Drugs* 2021;**19**(1):24. https://doi.org/10.3390/md19010024.

157. Khan A, Jahan S, Imtiyaz Z, et al. Neuroprotection: targeting multiple pathways by naturally occurring phytochemicals. *Biomedicines* 2020;**8**(8). https://doi.org/10.3390/biomedicines8080284.

158. Feng S-T, Wang Z-Z, Yuan Y-H, Sun H-M, Chen N-H, Zhang Y. Mangiferin: a multipotent natural product preventing neurodegeneration in Alzheimer's and Parkinson's disease models. *Pharmacol Res* 2019;**146**:104336. https://doi.org/10.1016/j.phrs.2019.104336.

159. Briffa M, Ghio S, Neuner J, et al. Extracts from two ubiquitous Mediterranean plants ameliorate cellular and animal models of neurodegenerative proteinopathies. *Neurosci Lett* 2017;**638**:12–20. https://doi.org/10.1016/j.neulet.2016.11.058.

160. Mayburd AL, Koivogui M, Baranova A. Pharmacological signatures of the reduced incidence and the progression of cognitive decline in ageing populations suggest the protective role of beneficial polypharmacy. *PLoS One* 2019;**14**(11):e0224315. https://doi.org/10.1371/journal.pone.0224315. Khanin R, ed.

Further reading

1. Barmada SJ, Skibinski G, Korb E, Rao EJ, Wu JY, Finkbeiner S. Cytoplasmic mislocalization of TDP-43 is toxic to neurons and enhanced by a mutation associated with familial amyotrophic lateral sclerosis. *J Neurosci* 2010;**30**(2):639–49. https://doi.org/10.1523/jneurosci.4988-09.2010.

Index

Note: 'Page numbers followed by "f" indicate figures and "t" indicate tables.'

R

S

T

U

Y

Z

Made in the USA
Monee, IL
13 September 2026

63138370R00149